ANSYS

Structural Analysis Using the ANSYS Mechanical APDL Release 2020 R1 Environment

ANSYS

Structural Analysis Using the ANSYS Mechanical APDL Release 2020 R1 Environment

R.B. Choudary

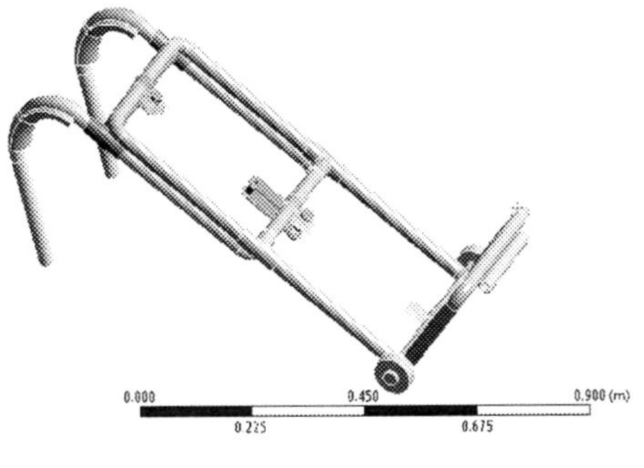

TECHSAR™
NEW DELHI

Published and Printed by
Techsar LLP
4435–36/7, Ansari Road, Daryaganj
New Delhi-110 002 (India)
E-mail: tech_info@techsarworld.com
Website: www.techsarworld.com

ISBN 978-93-90620-46-3

Reprint 2026

Dedication

Dedicated to all who have stood by me in the tests of times particularly Smt. Jyothi Latha, my wife, Ms. Bindu Madhavi, my daughter.

PREFACE

ANSYS is a large and complex program with many different capabilities. Learning the software and documentation of the various features can be a big task for a beginner. This book is intended to help the new ANSYS user in getting started. It teaches the basics of ANSYS in the following areas:

- ANSYS capabilities, basic terminology and the GUI
- Building or importing solid models and meshing
- Applying loads, solving and reviewing results
- Carrying out a complete ANSYS analysis
- Productivity enhancement tools.

The four chapters in this book introduce the reader to effective finite element problem solving by demonstrating the use of the comprehensive ANSYS FEM software in a series of step-by-step tutorials.

The steps given do not provide every single mouse click, but, hopefully all steps needed will be apparent. Also, the commands can be entered directly at the command line instead of using the menu picks. In this book, however, the commands are not provided. They can be obtained, however, by selecting "HELP" on a related dialog box.

Topics covered include ANSYS Basics, ANSYS Operations, Modeling and Meshing and Structural Analysis.

R.B. Choudary

ACKNOWLEDGEMENTS

I thank Mr. B. Ajayaram, Project Manager, L&T Technologies Services, Hyderabad, Mr. Thodeti Vijaya Bhanu, CADD Solutions, Vijayawada, Dr. N. Gopi, Nagarjuna University, Sri B. Krishna Murthy, Asst. Prof., SITE, Sri V.V.N. Satish, Hardware Engineer, SITE for the technical help they rendered during the preparation of the book. I thank Prof. M. Venkateswara Rao, for creating an opportunity to prepare the updated version.

This concise book is an abridged version of ANSYS help manuals. It is expected to be very much helpful for beginners. I take this opportunity to acknowledge my indebtedness to M/s Swanson Analysis Systems, Inc., USA and ANSYS Inc., India for giving permission to use the menus, dialog boxes, and plots of ANSYS academic teaching software in the book. Numerous tutorials available in the web on the subject were consulted during the preparation of the book. I am highly indebted to the eminent authors and their publishers whose works have been consulted.

I am very much obliged to my wife Smt. Jyothi Latha and daughter Ms. Bindu Madhavi for their patience and perseverance in shouldering the domestic responsibilities and providing the moral support during the writing of this book.

The lack of a simple book prompted the author to prepare this present work. The author does not pretend to claim any originality for the material. He does claim some degree of it in the presentation, and in his efforts to combine good features of previous works in this field. The sources are appropriately acknowledged.

Any suggestion, criticism, comment towards improvement of the book will be thankfully acknowledged and appreciated and those may be freely addressed to the publisher.

25 September 2020

Vijayawada **R.B. Choudary**

SYMBOLS AND ABBREVIATIONS

SYMBOLS

EX, EY, EZ	Young's modulus
GXY	Shear modulus
KXX, KYY, KZZ	Thermal conductivity
PRXY	Poisson's ratio

ABBREVIATIONS

APDL	ANSYS Parametric Design Language
CAD	Computer Aided Design
CAE	Computer Aided Engineering
CAM	Computer Aided Manufacturing
CPU	Central Processing Unit
CS	Coordinate System
DOF	Degrees of Freedom
DDAM	Dynamic Design Analysis Method
FEM	Finite Element Method
FSI	Fluid-Structure Interaction
GUI	Graphical User Interface
ICCG	Incomplete Cholesky Conjugate Gradient
IGES	Initial Graphics Exchange Specification
JCG	Jacobi Conjugate Gradient
MDOF	Million Degrees of Freedom
PCG	Preconditioned Conjugate Gradient
PSD	Power Spectral Density
QMR	Quasi-Minimal Residual
UIDL	User Interface Design Language
2-D	2-Dimensional
3-D	3-Dimensional

CONVENTIONS

This book uses several conventions to help in identifying various types of information:

Type style or text	Indicates
BOLD	Upper case, bold text indicates command names (such as K, DDELE, etc.)
Bold>Bold	Bold text with mixed case indicates GUI picks. An angle bracket (>) separating GUI picks indicates a GUI path, or series of picks (for example, **Main Menu> Solution)**
ITALICS	Upper case italic letters indicate command arguments for numeric values (such as *VALUE, INC, TIME*). On some commands, non-numeric convenience labels (for example, *ALL* and *F*) can also be entered for these arguments
Italics	Initial upper case italic letters indicate command arguments for alphanumeric values (for example, *Lab* or *Fname*). The book also uses italic text for emphasis
`TYPEWRITER`	Typewriter font indicates command input listings, ANSYS output listings, and text that an ANSYS user enters
Underlined	Mouse button to be clicked in the ANSYS GRAPHICS window

CONTENTS

CHAPTER 1 ANSYS BASICS

ANSYS is a sophisticated and comprehensive finite element program. It has capabilities in many different physics fields such as static structural, nonlinear, thermal, implicit and explicit dynamics, fluid flow, electromagnetics, and electric field analysis (Fig. 1.1). Obviously because ANSYS is a huge program with so many capabilities it is impossible to cover everything in this small book. This book will give an introduction as to how the ANSYS program works and how these basic skills will be applicable to any type of analysis within ANSYS. The most important concepts in using ANSYS will be addressed here in a compressed format. The key to becoming productive in any computer aided engineering program is to start to think like the program thinks, to get the big picture of how it works in general.

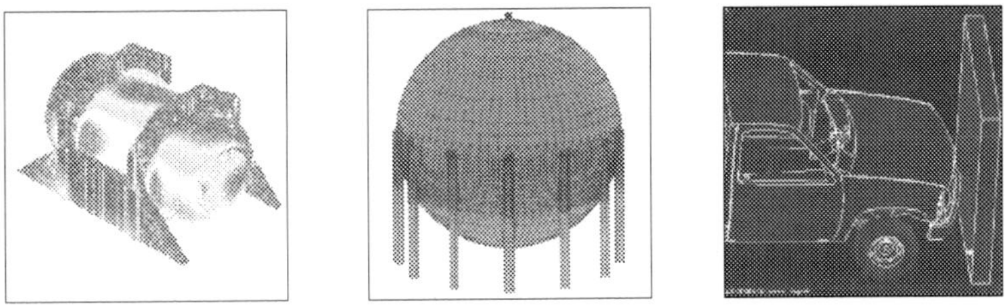

Fig. 1.1 Some applications of ANSYS.

ANSYS is an integrated program with all operations performed under one Graphic User Interface, GUI. Creating the model, running it, and postprocessing the results are all done without leaving the ANSYS environment.

There are many different ways of working within ANSYS. Like every software, ANSYS is driven by commands. It gives easy access to these commands. These commands are simple to use; just a keyword followed by several arguments.

New ANSYS users generally don't use scripting to start with and just want to find out how to do what they want within the GUI environment. That's where this book starts as well. Each key concept is explained as briefly as possible.

1.1 FINITE ELEMENT METHOD

It is very helpful to an engineer using ANSYS to have at least some idea of how the finite element method (FEM) works. What happens internally when ANSYS is told to Solve? Many terms associated with finite element modeling in general heard over and over again are based on the common numerical algorithms. After all, the finite element method is purely a mathematical technique.

The FEM was developed more by engineers during the 1950s and 1960s using physical insight than by mathematicians using abstract methods. The underlying theory is over 100 years old. It was the basis for pen-and-paper calculations in the evaluation of suspension bridges and steam boilers. It was first applied to problems of stress analysis. In all applications the analyst seeks to calculate a field quantity: in stress analysis it is the displacement field or the stress field; in thermal analysis it is the temperature field or the heat flux; in fluid flow it is the stream function or the velocity potential function; and so on. Results of greatest interest are usually peak values of either the field quantity or its gradients. The FE method is a way of getting a numerical solution to a specific problem. An FE analysis does not produce a formula as a solution, nor does it solve a class of problems. Also, the solution is approximate unless the problem is so simple that a convenient exact formula is already available.

When there is a structure like a simply supported beam, it's easy enough to grab an engineering handbook or textbook and look up the equations needed to calculate the bending moments and displacements. However, if there is a structure that is even just a little more complicated, like a stepladder (Fig. 1.2) for example, one will not be able to go to the engineering handbook and look up the stepladder equation. Sure, it's possible to derive some governing equation(s) for the step-ladder but this might be a lot of trouble. For even more complex structures it quickly becomes impossible to do so.

Fig. 1.2 A stepladder and its FE model.

This is where the finite element technique comes in. In this method an arbitrarily complex structure that cannot be described with an equation is broken down into small pieces, each of which is simple enough on its own that it can be described with an equation. While the actual structure theoretically contains an infinite number of small pieces it could be broken down into, this method breaks it down into a finite number of pieces of a specific size that analyst can choose. The smaller the pieces, the closer the approximation is to reality. Then the response of the entire structure is calculated by essentially summing up the response of each little piece. These small pieces are called *elements*, and each element is connected to its neighbor by *nodes* at each of its vertices.

More and more these days solid part is meshed to create nodes and elements. It's important to realize that the solid CAD part is not the finite element model; it is just a vehicle used to tell ANSYS where analyst wants the nodes and elements to go. The finite element method works only with the nodes and elements.

The FE method involves cutting a structure into several elements (pieces of the structure), describing the behavior of each element in a simple way, and then connecting elements together again. This process results in a set of simultaneous algebraic equations. In stress analysis these equations are equilibrium equations of the connection points. There may be several hundred or several thousand such equations, which means that computer implementation is mandatory. A more sophisticated description of the FE method regards it as piecewise polynomial interpolation. That is, over an element, a field quantity such as displacement is interpolated from values of the field quantity at the connection points. By connecting elements together, the field quantity becomes interpolated over the entire structure in piecewise fashion, by as many polynomial expressions as there are elements. The best values of the field quantity at the connection points are those that minimize some function such as total energy. The minimization process generates a set of simultaneous algebraic equations for values of the field quantity at the connection points.

The power of the FE method is its versatility. The object analyzed may have arbitrary shapes, arbitrary supports and arbitrary loads. Such generality does not exist in classical analytical methods. For example, temperature-induced stresses are difficult to analyze with classical methods, even when the structure geometry and the temperature field are both simple. The FE method treats thermal stresses as easily as stresses induced by mechanical load, and the temperature distribution itself can be calculated by FE.

The theory of FE includes matrix manipulations, numerical integration, equation solving, and other procedures carried out automatically by commercial software. The analyst may see only hints of these procedures as the software processes data. The analyst deals mainly with preprocessing (describing loads, supports, materials, and generating the FE mesh) and postprocessing (sorting output, listing and plotting of results).

There are various levels of engineering abstraction built into different kinds of elements. A solid element is fully defined by its geometry in 3D space (think of it as a small block cut out of the part), while shells (also called plates) are idealized by a surface and so need a thickness to be assigned for them to fully understand what they are. A beam element is a further abstraction. A simple line between two points defines it geometrically, and so analyst has to supply its area, moment of inertia, torsional stiffness, etc.

All of the basic calculations in the finite element method are performed at the nodes. What happens inside an element between the nodes is controlled by the element *shape function.* This is an assumed distribution of the primary variable and can be linear, quadratic, cubic, etc. For structural analysis the primary variable is the displacement and is also called the *degree of freedom* (DOF). A general three-dimensional structural model has six degrees of freedom at each node, translations in the X, Y and Z directions, and rotations about those three axes. For a thermal analysis the degrees of freedom are the temperatures at each node.

The fundamental equation of the finite element method is $[K]\{u\} = \{F\}$, where $[K]$ is the stiffness matrix, $\{u\}$ is the displacement vector, and $\{F\}$ is the applied load vector. This is analogous to a simple spring (Fig. 1.3) that obeys Hooke's law (stress = Young's modulus × strain), and is the essence of "small deflection theory", or linear analysis. Small deflection theory means that the displacement calculated for the applied load can be extrapolated to any value of load, and that the stiffness of the structure does not change for any value of applied load. Strictly speaking, this assumption is never exactly true, but for a very large class of analyses it has traditionally been considered good enough.

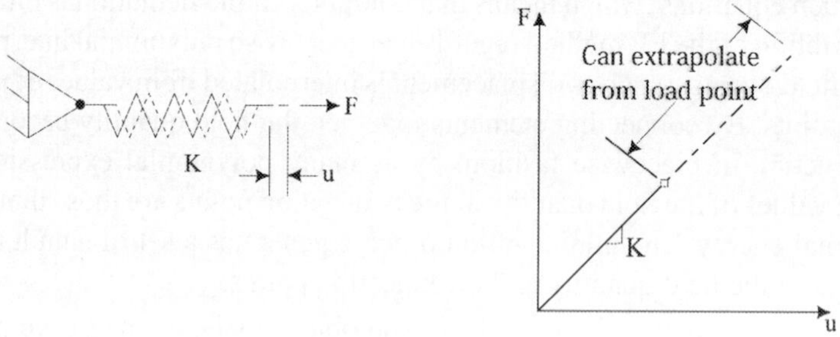

Fig. 1.3

1.2 INTRODUCTION TO ANSYS

Engineers can perform the following tasks using finite element analysis software ANSYS:

- Transferring CAD models of structures, products, components, or systems or building computer models
- Applying design performance conditions or working loads
- Studying responses, such as stress levels or temperature distributions
- Optimizing a design
- Do prototype testing in situations where it otherwise would be impossible.

The ANSYS offers a Graphic User Interface (GUI) that permits easy, interactive access to functions, commands and documentation. An intuitive menu system helps the analyst steer through the ANSYS program. S/he can input data using a mouse, a keyboard, or both.

This chapter provides basic guidelines for using the ANSYS program: starting and stopping the analysis, using the online help system, using and customizing the GUI, etc.

1.3 MECHANICAL APDL ENVIRONMENT

There are two basic levels in ANSYS – Begin level and Processor level. The Begin level acts as a doorway into and out of the program. It is also used for performing operations such as changing the jobname, clearing the database, and copying binary files. When the analyst first enters the program, s/he is at the Begin level.

At the Processor level, many processors are available. Each processor is a set of functions that performs a specific analysis task. The model is built in the general preprocessor (PREP7). Loads are applied and the solution is obtained in the solution processor (SOLUTION). The results of a solution are assessed in the general postprocessor (POST1). An additional postprocessor, POST26, permits to evaluate solution results at specific points in the model as a function of time.

1.3.1 Entering a Processor

Select a processor to enter it from the Main Menu of the GUI. Choose **Main Menu>Preprocessor** to enter into PREP7.

1.3.2 Exiting from a Processor or from the Program

Pick **Main Menu> Finish** to return to the Begin level from a processor. Without returning to the Begin level, the user can move from one processor to another. Simply pick the processor necessary to do so.

Pick **Utility Menu> File> Exit** to leave ANSYS. By default, ANSYS saves the model and loads automatically and writes them to the database file, `Jobname. DB`. If a backup of the current database file already exists, ANSYS writes it to Jobname.DBB. Other portions of the database can be saved using options in the dialog box. The user can as well quit without saving.

In order to input a new file, select **Utility Menu> File> Clear & Start New** to clear the current file from memory, then select a file to input. In order to return to processing the original file, select **Utility Menu> File> Read Input** from... , select the name of the file to resume from and select the **OK** button.

The processing of an input file can be stopped as it is being input. Most files will display the Process Status window at the top of the screen. Use the **STOP** button on the Process Status window to terminate the input of a file. Stopping file input is useful when a binary file is input unintentionally.

1.3.3 Database

ANSYS stores all input data and results data in one large database in an structured manner. The main advantage of the database is any specific data item can quickly and easily be listed, displayed, modified, or deleted.

No matter which processor the analyst is in, s/he works with the same database. This gives basic access to the model and loads portions of the database from anywhere in the program."Basic access" refers the ability to select, list, or display an item.

1.3.3.1 Defining or Deleting Database Items

The analyst must be in the right processor to define or delete items from the database. Nodes, elements, and other geometry can be defined only in PREP7, the general preprocessor. Loads can be specified and applied in either the PREP7 or the SOLUTION processor. However, geometry items can be selected, listed, or displayed from anywhere in ANSYS.

1.3.3.2 Saving the Database

The analyst should frequently save copies of database to a file as it contains all the input data. Pick **Utility Menu> File> Save as Jobname.DB** to do this.This choice writes the database to the file `Jobname.DB`. **SAVE** command has the following options:

- the model data only
- the model and solution data
- the model, solution and preprocessing data.

Pick **Utility Menu> File> Save as** to specify a different file name. Any save operation first writes a backup of the present database file to `Jobname.DBB`. If a `Jobname.DBB` file already exists, the new backup file overwrites it.

1.3.3.3 Restoring Database Contents

Pick **Utility Menu> File> Resume Jobname.DB** to restore data from the database file. This reads the file `Jobname.DB`. Pick **Utility Menu> File> Resume from** to specify a different file name.

The database can be saved or resumed from anywhere in ANSYS, including the Begin level. A resume operation replaces the data currently in memory with the data in the named database file. Using the save and resume operations together is helpful when "testing" a function or command.

1.3.3.4 Session Editor

During an analysis, it may be necessary to modify or delete commands entered since the last **SAVE** or **RESUME**. The session editor is used for displaying and editing the sequence of operations performed since the last **SAVE** or **RESUME** command. The analyst can modify command parameters, delete whole sections of text or save a portion of the command string to a separate file. The session editor (Fig. 1.4) can be accessed by choosing **Main Menu> Session Editor**.

Fig. 1.4 The Session Editor.

The following file operations are available in the session editor:

- **OK:** Enters the series of operations displayed in the window below. Use this option to input the command sequence after modifying it
- **Save:** Saves the command sequence displayed in the window below to a separate file. The program names the file `Jobnam000.cmds`, with each subsequent save operation incrementing the filename by one digit. The **INPUT** command is used to reenter the saved file
- **Cancel:** Dismisses this window and returns to the program
- **Help:** Displays the command reference for the **UNDO** command.

The Session Editor is available in GUI mode only.It is a powerful ANSYS Undo. ANSYS keeps a running copy of the log file from the last save command. By modifying the values in the Session Editor window and clicking OK, the modified commands are read back into ANSYS. If no **SAVE** or **RESUME** command has been issued during the analysis, all commands from the current session will be executed, including the start160.ans file, if present.

1.3.3.5 Clearing the Database

In building a model, sometimes it may be necessary to clear out the database contents and start afresh. Choose **Utility Menu> File> Clear & Start New** to do so. This method clears the database stored in memory. Clearing the database has the same effect as leaving and re-entering the program, but does not require to exit.

1.3.4 Program Files

The program writes and reads many files for data storage and retrieval. File names follow this pattern: `Name. Ext`. Name defaults to the jobname, which the analyst specifies while entering the program or by choosing **Utility Menu> File> Change Jobname**. The default jobname is `FILE (or file)`. Ext is a unique, two- to four-character identifier that identifies the contents of the file. For example, `Jobname.DB` is the database file, `Jobname.EMAT` is the element matrix file, and `Jobname.GRPH` is the neutral graphics file.

1.3.4.1 File Types

The filename can contain up to 248 characters, including the directory name, and the extension can contain up to eight characters. ANSYS can process blanks in file or directory names, so blank spaces are permitted in object names. When an object has a blank space in its name, always enclose the name in a pair of single quotes.

Use back slashes (\) instead of (/) slashes in directory names. For example, a typical directory path looks like this:

```
C:\Program Files\ANSYS Inc\ANSYS
Student\v201\ANSYS\apdl\start.ans
```

1.3.4.2 Jobname.LOG File

The `Jobname.LOG` file (the session log) is mainly important, as it provides a complete log of the session. This file opens immediately when the analyst enters ANSYS, and it records all commands executed. S/he can read the `Jobname.LOG` file, view it while in ANSYS, edit it, and input it later.

The program always appends log data to the log file instead of overwriting it. If the analyst changes the jobname while in a session, the log file name does not change to the new jobname.

1.3.5 Communicating with ANSYS

The easiest way to communicate with the ANSYS is by using the intuitive menu system, called the GUI.

1.3.5.1 Communicating via the GUI

The GUI consists of windows, menus, dialog boxes and other components that permit to feed input data and execute functions simply by picking buttons with a mouse or typing in responses to prompts. All users, both beginner and advanced, find it convenient to use the GUI for interactive work.

1.3.5.2 Communicating via Commands

Commands are the instructions that instruct the program. The program offers numerous commands, each meant for a specific function. Most commands are allied with specific processors, and work in respective processor(s).The analyst can either type in the appropriate command or access that function from the GUI to use it.

Program commands have a specific format. A typical command consists of a command name in the first field, usually followed by a comma and several more fields (containing arguments). A comma separates each field. For example, the **F** command, which applies a force at a node, looks like this: **F** commmand

```
F,NODE,Lab,VALUE
```

To apply an X-direction force of 2000 at node number 376, the **F** command would read as follows:

```
F,376,FX,2000
```

Command arguments may take a number or an alphanumeric labels, depending on their function. In the **F** command example described previously, `NODE` and `VALUE` are numeric arguments, but `Lab` is an alphanumeric argument. In this book, numeric arguments appear in all uppercase italic letters (as in `NODE` and `VALUE`), and alphanumeric arguments appear in initial uppercase italic format (as in `Lab`).

1.3.5.3 Abbreviations

If a GUI function is used frequently, it can be renamed or abbreviated to a string of up to eight alphanumeric characters using: **Utility Menu> Macro> Edit Abbreviations** or **Utility Menu>MenuCtrls> Edit Toolbar**. In the GUI, abbreviations appear as push buttons on the Toolbar. They can be executed with a quick click of the mouse.

1.3.5.4 Command Macro Files

A frequently used sequence of commands can be recorded in a macro file, thus creating a custom command. If an unrecognized command is entered, the program searches for a macro file by that name (with an extension of **.MAC** or **.mac**).

1.4 RUNNING THE ANSYS PROGRAM

The ANSYS program can be run in interactive mode or in batch mode. In interactive mode, the analyst exchanges information with the program continuously. A command can be executed by selecting its menu path in the GUI or by typing it in directly. The ANSYS program processes the command in real time. Interactive mode permits to use the GUI, online help, and various tools to create the model in the graphics window and modify it.

In batch mode, a file of commands is submitted to the ANSYS. A batch job can be run in the background while doing other work on the system. Batch mode is useful when there is no need to interact with the program, such as during the solution phase. The ANSYS Launcher can be used to set options for an ANSYS run.

1.4.1 Mechanical APDL Product Launcher

1.4.1.1 Starting an ANSYS Session from the Start Menu/Launcher

Use the Mechanical APDL Product Launcher (Fig. 1.5) to run an ANSYS program, one of the auxiliary programs, or to access a modifiable ANSYS file. In order to access ANSYS functionality, including the launcher, choose **Start > All Programs > ACP 2020 R1** and select the appropriate option. This causes the launcher to appear automatically.

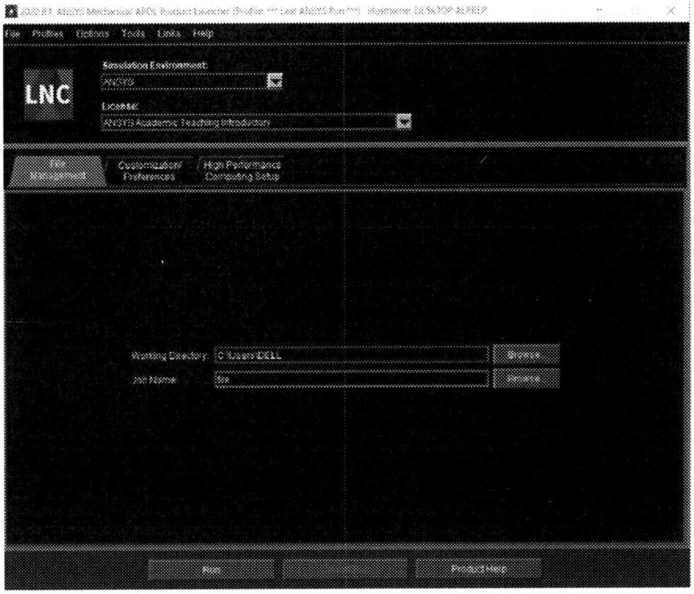

Fig. 1.5 Mechanical APDL product launcher.

Start Menu Options on Windows

Choose **Start > All Programs > ANSYS 2020 R1** New > **Mechanical APDL Product Launcher 2020 R1** to activate the launcher on Windows.

The File Management Tab

This tab contains the information necessary to manage the files, such as location of the working directory and job name. The available options will differ depending on the simulation environment selected. In the **ANSYS** simulation environment, the analyst can specify:

Working directory

Sets the directory in which the ANSYS run will be executed. The program writes files it generates to this directory. Type the new directory name in the **Working Directory** text box to change the working directory or press the Browse button to display a file selection dialog box. The working directory can be specified by defining the **ANSYS201_WORKING_DIRECTORY** environment variable. If **ANSYS201_WORKING_DIRECTORY** is set, the launcher will use that setting as the default. However, any working directory that is specified via the launcher (such as a profile) overrides the **ANSYS201_WORKING_DIRECTORY** environment variable setting.

Jobname

Defines the base filename used for all files generated by the ANSYS run. The initial jobname defaults to `file`. It can be changed to any alphanumeric string up to 32 characters long.

The Customization/Preferences Tab

The settings under this tab permit to specify detailed settings about the working environment, such as memory settings, parallel/distributed processing settings, custom executables, and additional parameters. The available options will differ depending on the simulation environment selected on the first tab.

On selecting the **ANSYS** or **ANSYS Batch** simulation environment, the analyst can specify:

Custom Memory Settings

Specifies the use of custom memory settings rather than using the default memory model.Select custom memory settings in order to set the `-m` or `-db` options.

Total Workspace (MB)

Specifies the initial allocation of memory requested for the ANSYS run. It defaults to 2 GB (2048 MB). This value is adequate for most modest-sized models.

Database (MB)

Specifies the portion of total memory that the database will use for the initial allocation. It defaults to 1GB (1024 MB). This value is adequate for most modest-sized models.

Graphics Device Name

Sets the graphics device (Fig. 1.6). The GUI requires a terminal that supports graphics. The graphics device can be WIN32 (default), WIN32c, or 3D. The 3D graphics driver facilitates to rotate a shaded view of the model. Most newer graphics cards can handle this well.

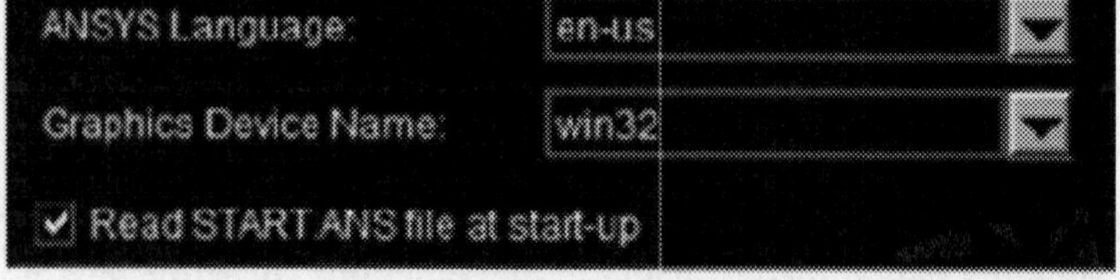

Fig. 1.6

Read start.ans at start-up

Indicates whether the `start201.ans` file is read at start-up. Include commands to be executed when the program starts up in the start201.ans file. Uncheck the box to not read the `start201.ans` at start-up.

1.4.1.2 Launcher Menu Options

Launcher-specific controls can be found under different launcher menu options. The menu options are:

File: Use the **File** menu to exit.

Profiles: Use the **Profiles** menu to save a specific launcher configuration. The profile the analyst is currently running is displayed in the launcher title bar. An asterisk (*) follows the profile name if any settings are changed from what is saved.

- Click **Save Profile** to save the current launcher configurations. A name can be entered for this profile. The last run profile will automatically be set as the default unless the analyst specifies otherwise. Check the **Set as Default** box to set a subsequent profile as the default.
- Click **Load Profile** to select the last run profile or one of the saved profiles.
- Click **Manage Profiles** to rename a profile, specify a different default profile, unset a default profile, or delete a profile.

If the analyst chooses to use the detailed working directory browse dialog box, it shows what files are contained in each directory, allows creating a new directory, and allows filtering on file type. The standard browse dialog displays only the directories from which to choose.

By default, the launcher saves existing output files from batch runs. When a new file is created, the old one is renamed to `jobname.out.<timestamp>` when the batch job is started. The timestamp is in the format YYYY-MM-DD@hh_mm_ss and reflects the date that the file was renamed, not the date the original file was created. Uncheck the **Save Existing Output Files From Batch Runs** item under the **Options** menu on the launcher to overwrite the output file rather than renaming it.

Use the **Reset ANSYS GUI Configuration** option to return the ANSYS GUI to its default configuration if the menus or toolbars have been rearranged. This option is especially useful if an ANSYS GUI component has been moved out of the screen's viewable area.

The **Delete All Settings/Profiles** option deletes all profiles saved for this release and sets all Options menu settings back to the defaults. The launcher will also close automatically.

1.4.2 Interactive Mode

In interactive mode, the analyst works with menus and dialog boxes to drive the ANSYS program. ANSYS graphics capabilities, online help, and other tools, such as wizards can be accessed easily.

The standard ANSYS GUI is the default. The Utility Menu, Standard Toolbar, Input Window, ANSYS Toolbar, Main Menu, Graphics Window, Status Area, and Output Window are shown in this layout. The toolbars can be resized. The overall size of the GUI, the font, and the color can be modified.

Double-click on the following types of files from the Windows Explorer to execute the ANSYS or DISPLAY programs:

- Double-click on a `.db` or `.dbb` file to execute the ANSYS program. When executed in this way, ANSYS will set the filename to the filename of the `.db` file and the working directory to the directory where the .db file resides.
- Double-click on a `.grph` or `.f33` file to execute the DISPLAY program. The first plot that appears in the file will be loaded into DISPLAY automatically.

1.5 GUI

The GUI is an easy and intuitive way to interact with the program. The GUI mode is often referred to as *interactive mode*. The GUI allows to perform an analysis interactively with little or no knowledge of individual commands. Each GUI option activates one or more commands to perform the requested operation. Most operations can be performed interactively, but some operations are available only by entering commands in the command input area or via a batch file. The program records all commands issued in the input history file (`Jobname.LOG`).

ANSYS can be run through the GUI and commands input window, or by creating an ASCII text file for input, or both. It is often more desirable to use a combination of both graphical methods and input files. An input file is often useful for defining the initial model geometry, characteristics, and constraints. This is because the input file can be easily changed to correct the errors that are inevitable when setting up a new ANSYS model. Once the input is correct and the model geometry and material characteristics are read into ANSYS, it is then desirable to examine the results using either direct command input with the command input window, or the graphical menu selections.

1.5.1 GUI Components

Dialog boxes are used to provide input for a particular function. The type of dialog box provided depends on the type of input required. The input might require any of the following: text entry box, check button, radio button, option button, single-selection list, multiple-selection list, two-column selection list, a tabbed box, or a tree structure. Other components of a dialog box include action buttons such as **OK**, **Apply**, and **Cancel**.

1.5.1.1 Text Entry Boxes

In a text entry box (Fig. 1.7) numbers or character strings are entered. If text already exists in the text entry box, overwrite it by highlighting the existing text. Press and drag the left mouse button to highlight existing text. Double-clicking the left mouse button on a word highlights the word, and triple-clicking it highlights the entire string.

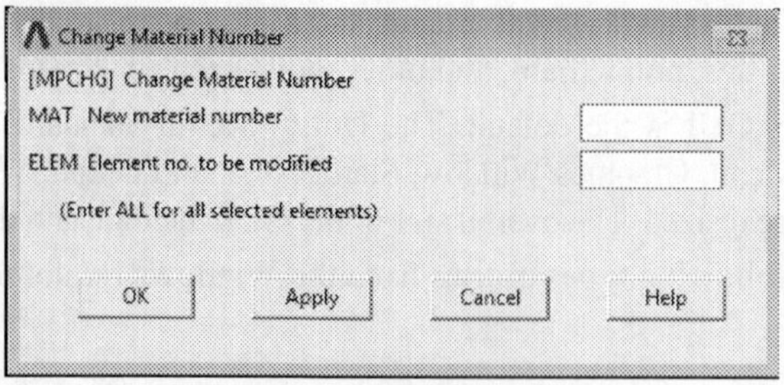

Fig. 1.7 Text entry box.

1.5.1.2 *Check Buttons*

Check buttons are square buttons that enable or disable features. Click on it with the left mouse button to change a check button from ON to OFF or vice versa. The buttons under "Preferences for GUI Filtering" (Fig. 1.8) are check buttons.

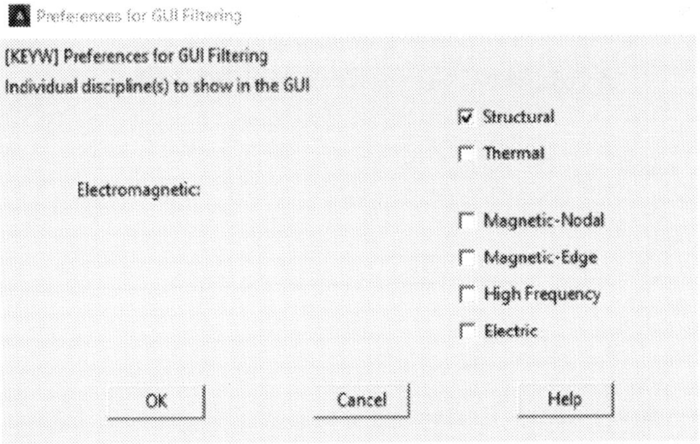

Fig. 1.8 Check and radio buttons.

1.5.1.3 *Radio Buttons*

Radio buttons are diamond shaped buttons used to select one of several options. One button is always "On" in a set of radio buttons. Clicking on the desired item makes that button active and turns off other buttons. The buttons shown under "Structural discipline options" (Fig. 1.8) are radio buttons.

1.5.1.4 *Option Buttons*

With option buttons (Fig. 1.9) the analyst selects an item from a pop-up menu that collapses into one button showing the active choice. Adopt the following procedure to select a different option:

1. Place the cursor on the button
2. Press the left mouse button and hold it
3. Drag the mouse through the pop-up menu until the desired item is highlighted

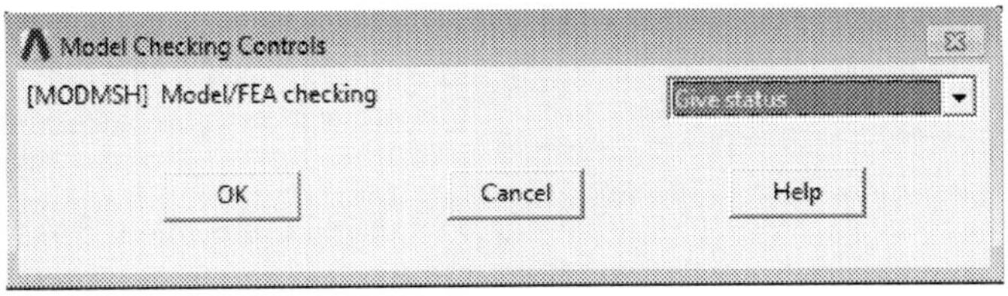

Fig. 1.9 Option buttons.

1.5.1.5 *Single-Selection Lists*

A single-selection list (Fig. 1.10) is used to select one option from a scrollable list. Clicking on the desired item highlights it and copies it to the Selection box, where it can be edited.

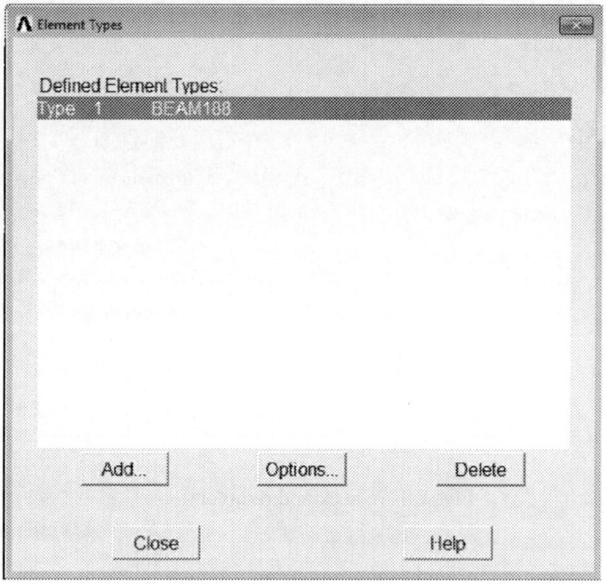

Fig. 1.10 Example of a single-selection list.

1.5.1.6 *Multiple-Selection Lists*

A multiple-selection list (Fig. 1.11) is used to select items from a list. It resembles the single-selection list except that there is no Selection box, and more than one item can be selected. Clicking on an item highlights and makes it active, and clicking the same item again deactivates it.

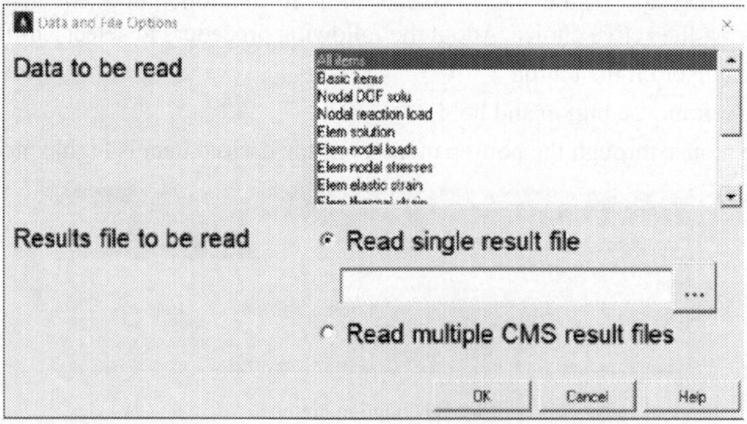

Fig. 1.11 Multiple-selection list.

1.5.1.7 *Two-Column Selection Lists*

One of several choices can be selected from a two-column selection list (Fig. 1.12). It resembles the single-selection list except the choices are grouped into categories. Select a category from the left column, then pick the desired item from the right column. Choices available in the right column vary according to the choice in the left column.

The GUI displays this type of list primarily when it is necessary to select from a large number of items.

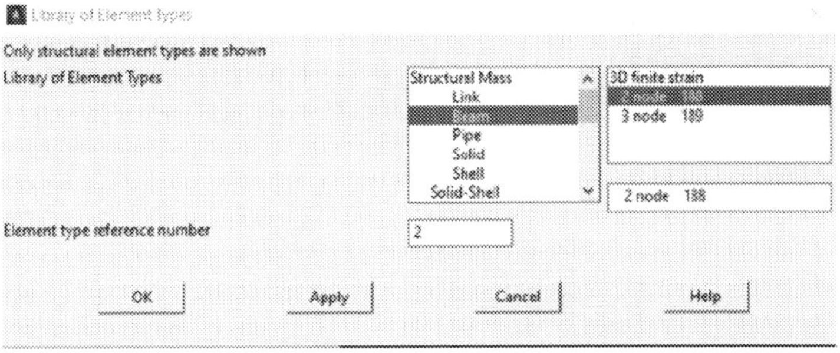

Fig. 1.12 Two-column selection list.

1.5.1.8 *Tabbed Dialog Boxes*

A tabbed dialog box (Fig. 1.13) presents groups of related commands in one location. Select an individual tab by clicking on it, or by using **Ctrl+Tab** to move between the tabs. Some tabbed dialog boxes require that the analyst navigates through each tab sequentially, with later tabs "grayed out" until s/he completes required tasks in an earlier tab.

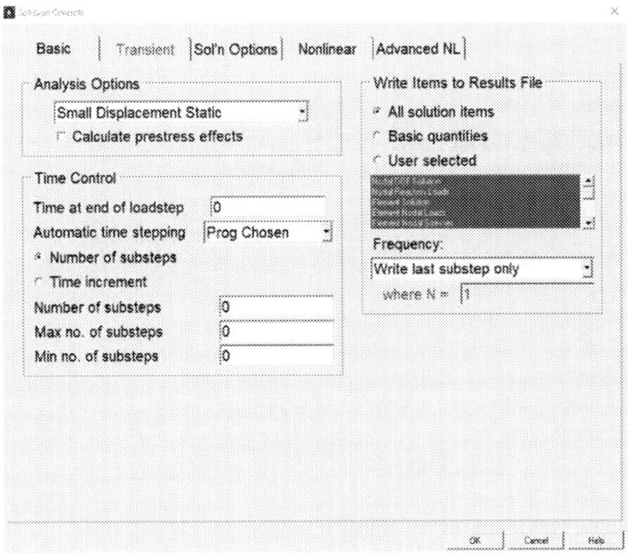

Fig. 1.13 Tabbed dialog box.

1.5.1.9 *Drop-Down List Boxes*

A drop-down list box (Fig. 1.14) provides a list of choices for an option. Make a selection by clicking on the small arrow on the right side of the list box and then scrolling to and clicking on the desired choice from the list.

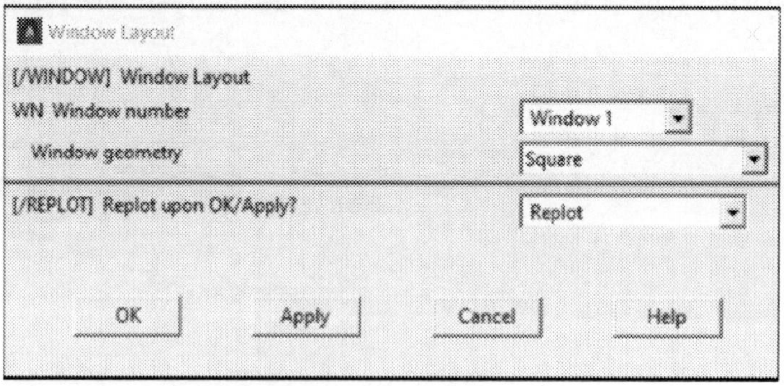

Fig. 1.14 Drop-down list box.

1.5.1.10 *Tree Structures*

A tree structure (Fig. 1.15) presents a hierarchical flow of logical choices from major categories to specific subcategories. Select a major category by double-clicking on the category. Subcategories appear listed vertically and indented beneath that major category. If a subcategory is selected in the same way, further subcategories are listed, and become more specific. Continually selecting subcategories yields the final item in the "branch" of the tree. At this point, instead of a categorical choice, enter or edit data to provide specific input.

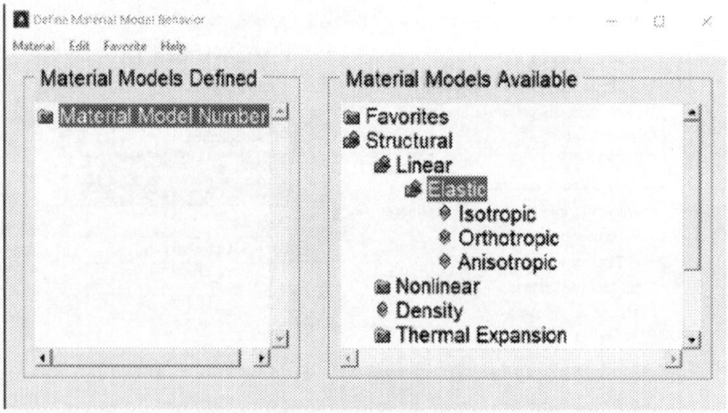

Fig. 1.15 Tree structures.

Double-click on the name of the final item to display a data input dialog box (Fig. 1.16) where the data is entered or edited.

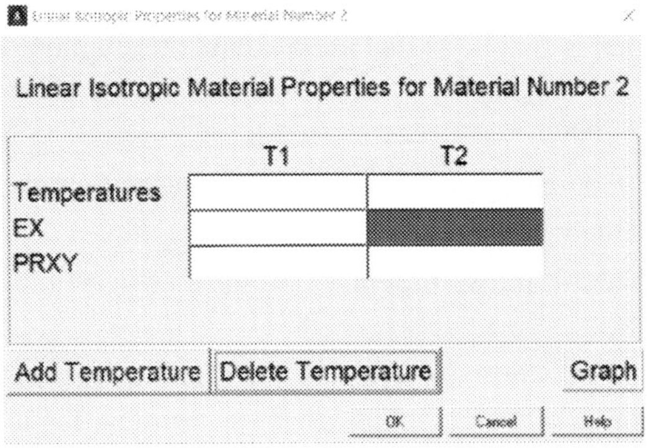

Fig. 1.16 Sample data input dialog box.

1.5.1.11 Using Action Buttons

A dialog box typically contains combinations of the following action buttons:

- **OK** – Applies the changes and closes the dialog box.
- **Apply** – Applies changes but does not close the dialog box. Use this button when it is necessary to make more than one change within a dialog box, or when it is necessry to execute a function more than once.
- **Reset** – The settings in the dialog box revert to the previous or default settings.
- **Cancel** – Closes the dialog box without applying any changes made. The difference between **Cancel** and **Reset** is that **Reset** does not close the dialog box.
- **Help** – Displays help information for the function being performed.

1.5.2 Layout of the GUI

If the analyst enters the program through the launcher, the GUI is activated automatically for interactive running. The GUI (Fig. 1.17) has seven areas:

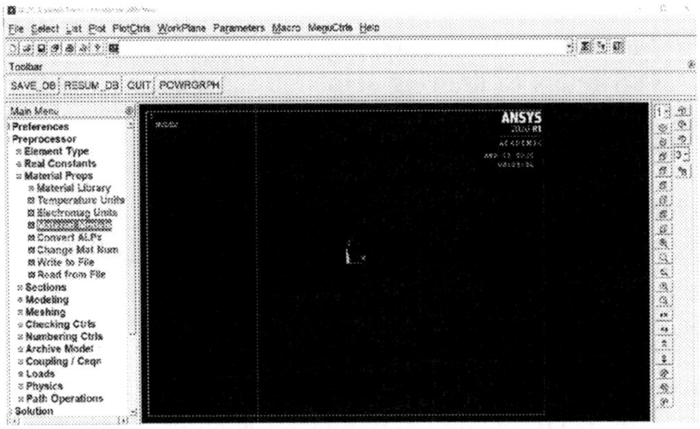

Fig. 1.17 The program GUI replace.

1. **Utility Menu** – Contains functions that are available throughout the ANSYS session, such as file controls, selections, graphic controls and parameters. The analyst can also exit the program through this menu.
2. **Standard Toolbar** – Contains push buttons that execute commonly used ANSYS commands. More push buttons can be added if desired.
3. **Command Input Area** – Shows program prompt messages and allows analyst to type in commands directly.
4. **Main Menu** – Contains the primary ANSYS functions, organized by preprocessor, solution, general postprocessor, and design optimizer. It is from this menu that the vast majority of modeling commands are issued.
5. **Graphics Window** – The Graphics Window is where graphics are shown and graphical picking can be made. It is here where analyst will graphically view the model in its various stages of construction and the ensuing results from the analysis.
6. **Status and Prompt Area** – Located at the bottom of the GUI, shows prompts and the status of the analysis. Typically, prompts for functions that involve graphical picking will be seen. Be sure to read the prompt so that proper entities can be picked in the proper order.This area also contains status information.
7. **Output Window** – The Output Window (Fig. 1.18) shows text output from the program, such as listing of data, etc. It is usually positioned behind the main window and can be put to the front if necessary.

The GUI is fairly easy to use. There is some terminology related to basic operation:

Resume: Resume opens a previously saved database. Resuming a database doesn't change the *jobname*. For example: start ANSYS with a *jobname* of *file*. Then resume *example.db*, do some work, then save. That save is done to *file.db*! This can be avoided by always resuming using the  icon on the toolbar. If example.db is opened using this method, it resumes the model and automatically changes the jobname to example.

Plotting: Contrary to the name, this has nothing to do with sending an image to a plotter or printer. Plotting in ANSYS refers to drawing something in the graphics window. Plot one type of entity (lines, elements, etc.) to the screen at a time. To plot more than one kind of entity use, "**Plot >Multiplot**", which by default will plot everything in the model at once.

Plot Controls: This refers to how the "plot" has to look on the screen (shaded, wireframe, entity numbers on or off, etc). Other plot control functions include sending an image to a graphics file or printer.

Raise hidden: If a dialog box disappears behind the main window, bring it back with this.

The analyst can resize the toolbar, Main Menu and Graphics Window, as well as the GUI. In order to resize the areas in the GUI, drag the borders around the areas of the GUI while holding down the left mouse button.

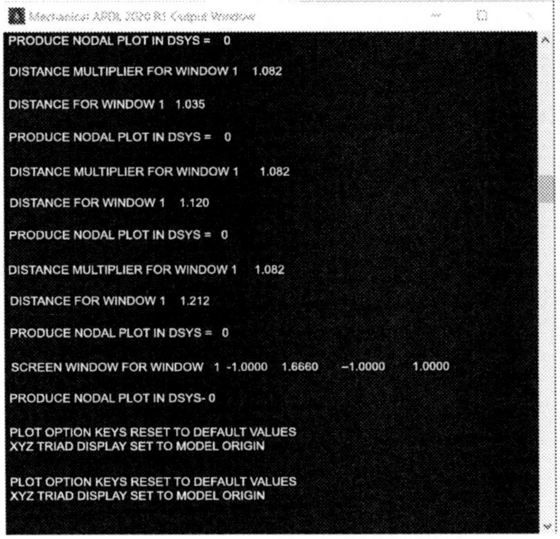

Fig. 1.18 Output window.

In order to change the overall size of the GUI, position the mouse on the corners of the GUI and drag it diagonally towards the center of the GUI while holding down the left mouse button. Save the GUI size settings by selecting **Utility Menu>MenuCtrls> Save Menu Layout**.

The GUI is iconified. Later to completely remove the program from the screen without terminating the session, the session can be brought back by restoring the icon.The remainder of this section describes each of the areas of the GUI.

1.5.2.1 The Utility Menu

The Utility Menu contains utility functions such as file controls, selecting, graphics controls, and parameters. Most of these functions can be executed anytime during the session.For example, while picking locations on the working plane to create keypoints, the analyst can select **Utility Menu>PlotCtrls> Pan-Zoom-Rotate>Iso** to change the view to an isometric view.

Each menu topic on the Utility Menu brings up a pull-down menu of subtopics, which in turn either cascade to a submenu (indicated by a >) or perform an action. The action may do any of the following:

- Immediately execute a function
- Bring up a dialog box (indicated by a ...)
- Bring up a picking menu (indicated by a +).

Figure 1.19 shows the pull-down list of subtopics seen when the **List** topic is selected.

Left mouse button is used to "pull down" a menu topic on the Utility Menu. Pressing and dragging the mouse button permits to move rapidly to the desired subtopic. Releasing the mouse button while it is on an "action" subtopic causes the program to perform that action.Clicking the left mouse button leaves the pull-down and cascading menus in place. The menus disappear when the analyst clicks on an action subtopic or elsewhere in the GUI.

Use the keyboard to pull down a menu topic and move to the desired function. Do so via the menu's *mnemonic* character, indicated by an underscore. Simply place the mouse cursor anywhere in the Utility Menu and press the ALT key and the mnemonic character simultaneously. For example, pressing ALT+F with the mouse cursor anywhere in the Utility Menu pulls down the File menu, since F is the mnemonic (indicated by the underscore in **File**).

Once the pull-down menu appears, the analyst can use the other mnemonic characters (without the ALT) or the arrow keys on the keyboard to navigate to the desired menu topic and the ENTER or RETURN key to "pick" it.

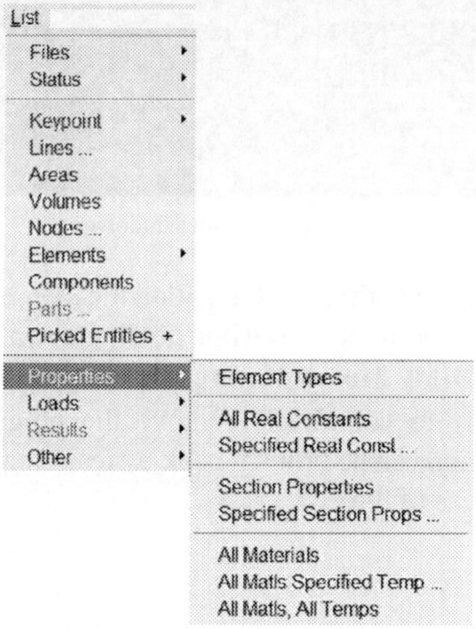

Fig. 1.19 Example of a pull-down, cascading menu.

The Utility Menu lists 10 topics. A brief description of each topic follows.

- **File** – Contains file and database related functions, such as clearing the database, saving it to a file, and resuming it from a file. Some of the functions under the **File** menu are valid at Begin level only. If the analyst selects such a function when not at Begin level, s/he will see a dialog box giving a choice of moving to Begin level and executing the function or cancelling the function.
- **S**elect – Includes functions that allow to select subsets of entities and to create components.
- **L**ist – Enables to list virtually any data item stored in the database. The analyst can also obtain status information about different areas of the program and list the contents of files residing on the system.
- **P**lot – Plots keypoints, lines, areas, volumes, nodes, elements, and other data that can be graphically displayed.
- Plot**C**trls – Includes functions which control the view, style, and other characteristics of graphics displays.

- The **Hard Copy** function helps in obtaining hard copies of the entire screen or just the Graphics Window.
- **W**orkPlane – Enables to toggle the working plane on or off and to move, rotate, and otherwise maneuver the working plane. The analyst can also create, delete, and switch coordinate systems by using this menu.
- P**a**rameters – Includes functions to define, edit, and delete scalar and array parameters.
- **M**acro – Allows to execute macros and data blocks. The analyst can also create, edit, and delete abbreviations, which appear as push buttons on the Toolbar.
- Me**n**uCtrls – Permits to create, edit, and delete abbreviations on the Toolbar Once the analyst adjusts the GUI to her/his liking, s/he can use the **Save Menu Layout** function to save the current GUI configuration.

1.5.2.2 Standard Toolbar

The Standard Toolbar (Fig. 1.20) contains a set of icon buttons that execute commonly used functions.

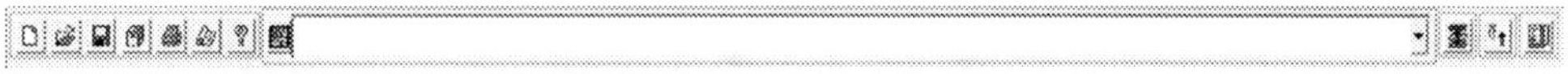

Fig. 1.20 Standard Toolbar.

By default, the Standard Toolbar is loaded and positioned when the analyst starts ANSYS. The Standard Toolbar is defined at start up, and cannot be modified during the session. The standard buttons and their functions include:

New Analysis: Saves and clears information for the existing analysis and starts a new analysis.

Open File: Reads database or input files to be read into ANSYS. The file type determines the operation.

Save Analysis: Saves the current analysis to a database file.

Pan-Zoom-Rotate: Opens the Pan-Zoom-Rotate dialog box.

Image Capture: Opens the image capture GUI.

Report Generator: Opens the report generator GUI.

Help: Displays the table of contents for the HTML-based help.

Raise Hidden: Raises hidden windows to the top of the application.

Reset Picking: Resets the picking menu if it becomes unresponsive or does not appear when invoked via a menu pick or a command in the Input Window.

Contact Manager: Opens the Contact Manager GUI.

1.5.2.3 Command Input Options

Although the GUI provides intuitive graphical access, the analyst can provide input to the program by typing in commands, even while the GUI is active. There are two modes available for directly entering commands during the analysis. The Single Line Input Window displays only one line and uses a drop-down window to display the command history. The Command Window can be used for more intensive command operations, such as pasting in multiple command strings or copying longer strings and selected commands from the history window. Both modes are described below.

The Single Line Input Window

The Input Window (Fig. 1.21) is the default display for all GUI operations. The Input Window is used to conveniently enter single commands. It is also used to access the history buffer without changing the overall configuration of the GUI. As commands are entered into the Input Window, dynamic command help appears in a box above the window. As the analyst types the letters, the command help displays the possible commands, and guides through the proper spelling and syntax of the command.

Fig. 1.21 Single line input window.

The history buffer contains all previously entered commands. Individual lines from the history buffer can be displayed in the Input Window and indexed with the up or down arrow keyboard keys. Once displayed in the window, the analyst can edit the entry and execute it by using the return key.

View and access the history buffer by clicking the down arrow on the right of the text entry box. A drop down list containing the entry history appears. Clicking the left mouse button on any line in the history buffer moves that line to the text entry box where the analyst can edit and execute it. A double click on any line in the history buffer automatically executes that line.

The vertical scroll bar at the right corner of the (unhighlighted) history buffer box allows to scroll through the history buffer. The up and down arrow keys can also be used to navigate the history buffer.

The Floating Command Window

Bring the Floating Command Window (Fig. 1.22) up by clicking on the keyboard icon at the left side of the Single Line Input Window. The Command Window is a floating window that can be resized and positioned easily for more complex command entry operations. When the command window is closed, all the operations will appear in the history buffer.

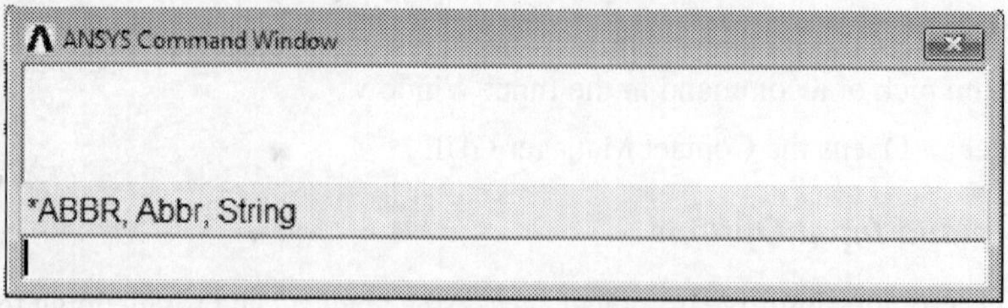

Fig. 1.22 The Floating Command Window.

When the command history is accessed with the command window, select commands from the buffer, and selectively enter them into input line, in any order of selection. Do this by holding down the **Ctrl** key and clicking on each command. As this is done, the commands appear in the input line below. Then edit the commands, changing parameters or syntax, before hitting enter to execute them.

The size and location of the Floating Command Window can be set to the desired orientation. Then use the "Save Menu Layout" feature to make it default. The Single Line Window will be available when the Command Window is closed, but will no longer be the default.

The Command Window provides a scrollable, editable text area where individual commands can be entered or pasted in long command strings. Multiple commands can also be selected from the history buffer area, either using the shift key to select large contiguous portions of the history, or the control key to select multiple, individual commands. Like the Input Window, dynamic command help is available; as command is typed, the probable command is displayed. This display shows the complete command syntax, and stays visible so that the help system can be checked.

1.5.2.4 The Toolbar

The Toolbar (Fig. 1.23) is a convenient area where push-buttons can be added for command, function and macro shortcuts. The Toolbar is a set of push buttons that execute commonly used functions. The toolbar can be set up to provide one-button access to the favorite and frequently used functions. It is loaded and positioned by default, and can be modified during the session. Some push buttons are predefined, but all others are to be defined by the analyst. That is, the analyst selects how many push buttons the Toolbar contains (to a maximum of 100) and what functions they execute.

Fig. 1.23 The toolbar.

Adding Buttons to the Toolbar

Additional push-buttons are added to the Toolbar by creating abbreviations. An abbreviation is simply an alias (up to 32 characters long) for a complete command or GUI function name. For example, **SAVE_DB** is an alias for the **SAVE** command and **RESUM_DB** is an alias for the **RESUME** command. Macros can also be added to the Toolbar by defining an abbreviation that executes the macro.

Creating Abbreviations

Select **Utility Menu>MenuCtrls> Edit Toolbar or Macro> Edit Abbreviations** to create an abbreviation. Both menu choices bring up the "Edit Toolbar/Abbreviations" dialog box shown in Fig. 1.24. The Toolbar immediately reflects any changes the analyst makes to the abbreviations using this dialog box.

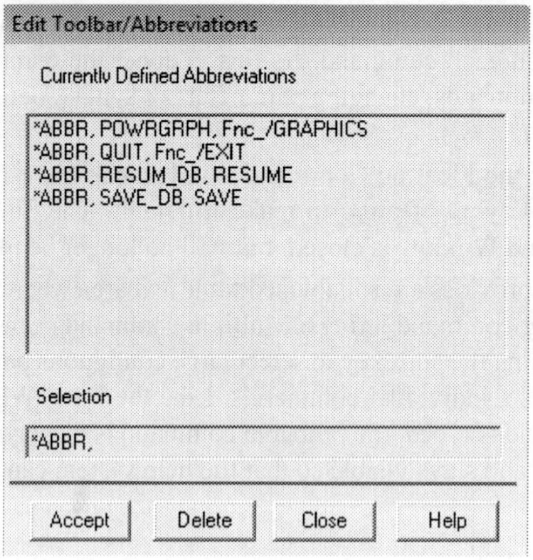

Fig. 1.24 Edit Toolbar / Abbreviations dialog box.

An abbreviation can also be created by issuing the ***ABBR** command from the Input Window. In this method, it is necessary to update the Toolbar manually by picking **Utility Menu>MenuCtrls> Update Toolbar**. The Toolbar can be sized to make it smaller or larger to fit the set of abbreviations.

The order in which abbreviations are defined determines the placement of the buttons on the Toolbar. After defining the buttons, they cannot be rearranged graphically within the GUI.

Buttons that execute commands or functions from a processor other than the current one will not work. For example, if the analyst is in PREP7 and picks a button for a POST1 command, s/he will receive an "unrecognized PREP7 command..." warning.

After creating own set of abbreviations in the Toolbar, the analyst can save them in a file by choosing either **Utility Menu>MenuCtrls> Save Toolbar** or **Utility Menu> Macro> Save Abbr**. Pick either **Utility Menu>MenuCtrls> Restore Toolbar or Utility Menu> Macro> Restore Abbr** to restore a set of abbreviations from a file.

1.5.2.5 Main Menu

The Main Menu (Fig.1.25) is where the analyst begins the analysis. It contains the analysis functions used to create the model. The Main Menu is arranged in a tree structure. This structure has progressive submenus accessible as the analyst proceeds through the steps of the analysis. Each menu topic in the main menu either expands to show more menu options (indicated by a boxed +) or performs an action (indicated by an icon preceding the menu topic). Click on the boxed + or the topic name to expand a menu topic to reveal the subtopics (the boxed + will turn into a boxed -). Continue expanding subtopics (also indicated by a boxed +) until the desired action is reached. The action name is preceded by an icon used to indicate the action:

- A square for a dialog box
- An arrow for a picking menu.

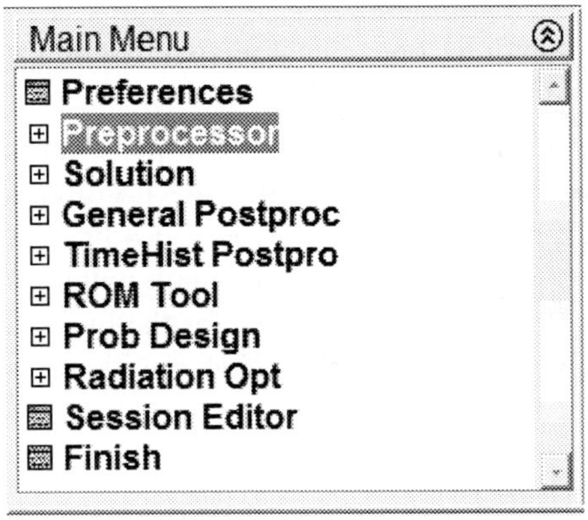

Fig. 1.25 Main Menu.

Using Preferences to Set Menu Content

One of the most useful customizations that can be performed from the GUI is to apply filtering to the menu choices. Filtering lets gray out, or completely hides many of the functions the analyst determines s/he won't need during the analysis. Use the preferences dialog box to adjust the filtering.

At the start of the program, no filtering is applied. Menu choices for all analysis disciplines are displayed. Menu items that are not applicable are grayed out, based on the context of the analysis. The grayed out items can be removed by choosing the appropriate discipline from within the "Preferences for GUI Filtering" dialog box. Not all menu options may be visible during the analysis. If the desired option is not available, check the preferences and element types.

For example, selecting **Thermal** suppresses structural, magnetic, and fluid element types in the "Element Types" dialog box, displacements, and potentials, etc. In the "Apply Loads" and "Delete Loads" menus, and so on. Another item is the methodology used for structural analysis: h-method or LS-DYNA Explicit. By default, the program uses h-method. The analyst can also change the menu hierarchy and customize it to her/his needs by using the User Interface Design Language (UIDL), an ANSYS-developed GUI language.

Additional Usability Features

On opening a Main Menu topic, the submenus stay in place until the analyst selects a different Main Menu topic. If a menu topic is not seen, scroll up or down in the Main Menu area until it is found or click on toggle button – to collapse subtopics that are no longer in use.

Individual topics in the Main Menu can be completely expanded to reveal each and every subtopic. Right-mouse click context-sensitive control is available within the Main Menu to expand and collapse the individual topics. Once an individual subtopic is completely expanded, the analyst can selectively close topics to have areas at different locations under each main topic expanded. As s/he does so, the amount of expansion accessed is recorded and replicated in the next time.

Menus can be set to automatically collapse and expand the subtopics. Use the "Collapse Siblings" feature (found in the right-mouse click menu) to set the menu expansion preferences.

When collapse is selected, the subtopics opened automatically collapse when another main topic is selected.

When many submenus are open, navigating the main menu can become confusing, especially when individual submenus are large enough to push the heading out of the viewing area. The Preprocessor subtopic alone has more than 800 nested subheadings beneath it. Use the same right-mouse click context-sensitive control to configure the main menu for selectable contrasting color display within each nested level. Designate any color for the menu text at each level, making the transition between levels easily detectable. This makes navigating within a subtopic, and up to higher levels much easier.

Main Menu Analysis Functions

Most of the functions in the main menu are "modal" with respect to each other; that is, the analyst must complete one function before starting the next. For example, for creating keypoints on the working plane, s/he cannot simultaneously create lines or mesh volumes. However, s/he can define or edit scalar parameters in the middle of creating keypoints since parameter functions are"modeless" Utility Menu functions.

The Main Menu functions will vary according to the product and type of analysis. The following functions are included in the default, non-filtered Main Menu:

- **Preprocessor** – Enters the 43 odelling 43 sor and contains PREP7 functions such as 43 odelling, meshing, and loads.
- **Solution** – Enters SOLUTION and contains SOLUTION functions such as analysis type and options, loads, load step options, and solution execution.
- **General Postproc** – Enters the general postprocessor and contains POST1 functions such as plotting and listing of results.
- **TimeHist Postpro** – Enters the time-history postprocessor and loads the Time History Variable Viewer. Contains POST26 functions such as defining, listing, and plotting of variables.

Additional Main Menu Utilities

The default main menu also contains the following dialog boxes that are available at any time during the analysis:

- **Session Editor** – Opens the session editor
- **Finish** – Exits the current processor and moves to the Begin level by executing the **FINISH** command.

1.5.2.6 The Graphics Window

The Graphics Window is where all graphic displays are drawn and all graphical picking is done. It is usually the largest of the GUI windows. Maintain the 4:3 width-to-height proportion while changing the size of the Graphics Window. Graphic displays are drawn in the Graphics Window when a plot is requested. In addition, graphics displays that are generated by immediate mode and XOR mode will be seen.

Immediate Mode

An *immediate mode* plot is one that is drawn automatically when the analyst creates, moves, reflects, or otherwise manipulates the model. It is only a temporary graphics display meant to give immediate feedback on the function just executed. As a result, an immediate mode plot has two drawbacks:

1. It will be destroyed if a menu or a dialog is brought up on top of it or if the GUI is iconified and restored.
2. Its scaling is based on the scaling for the last plot request, so if the new entity lies "outside"the boundaries of that scaled image, it will not appear in the Graphics Window. Issue a plot request to see the new entity.

Numbers and symbols drawn in immediate mode have a similar drawback: they will disappear when a plot is requested unless they are explicitly set to "on" via the appropriate functions under the **PlotCtrls** menu. Immediate mode can be turned off using the **Immediate Display** function under **Utility Menu>PlotCtrls> Erase Options**.

If a plot is requested "manually", the program calculates the graphics scaling such that the display fills the window. This is more of a"permanent" display in that it stays in place even if it is obscured by a dialog box or a menu or if the Graphics Window is iconified and then restored.

XOR Mode

The program uses this mode when something needs to be drawn or erased quickly without destroying whatever is currently being displayed in the Graphics Window. For example, XOR mode takes effect during graphical picking to highlight or unhighlight the item being picked. It is also used to display the working plane and for rubber-banding.

The advantage of using XOR mode is that it produces an instantaneous display without affecting the existing plot on the screen. The only drawback is that drawing in the same location a second time erases the display. For example, picking the same node or keypoint a second time erases the highlight. Similarly, with the working plane display on, requesting another plot without erasing the screen erases the working plane.

Capture Image Feature

A useful feature that permits to create "snapshots" of the Graphics Window is the Capture Image function (using the **Capture Image** button in the Standard Toolbar or **Utility Menu>PlotCtrls> Capture Image**). After an image is captured, it can be saved to a file and then restored in any session. Captured images are useful for comparing different views, sets of results, or other significant images simultaneously on the screen.

Mouse Functionality

Pressing the scroll wheel button (Fig. 1.26) is the same as a middle mouse button.

Picking Entities

Left Button: Picks an entity. Picking is cumulative, so it is not necessary to press control or shift to pick more than one entity. Click and hold the button, then move the cursor around until the entity under consideration is highlighted. When the button is released the highlighted entity will be selected.

Middle Button: Completes a selection. This is like clicking "Apply" in the picking dialog (also called "the picker").

Right Button: Toggles back and forth between "pick" and "unpick" mode. Cursor changes indicating what mode ANSYS is in.

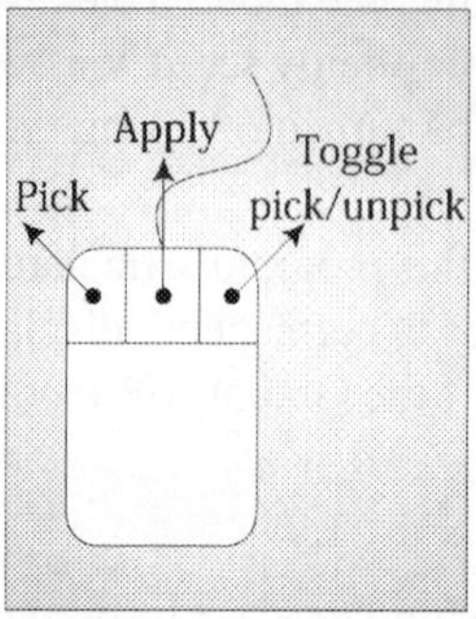

Fig. 1.26

Manipulating the Model View: (these defaults can be changed to different buttons if desired)

CTRL + Left Button: Pans the model side to side and up and down.

CTRL + Middle Button: Moves the mouse left and right to rotate about screen Z. Up and down zooms in and out.

CTRL + Right Button: Rotates the model.

Right Button: Click and drag the right button to zoom in using a window.

Rolling the scroll wheel also zooms in and out.

When the right button is clicked in the graphics area a pop-up menu with some very common graphics functions appears (Fig. 1.27) . ANSYS does not always refresh the graphics screen so *Replot* is *very* handy. *Fit* makes the whole model visible. *Zoom Back* will go back to the view the way it was just before zooming in. Along with some of the standard Pan-Zoom-Rotate functions, the analyst can also access many of the window control functions found in the PlotCrtrls section of the Utility Menu. Placing the cursor over the legend areas of the Graphics Window channels the context to access the legend control menus.

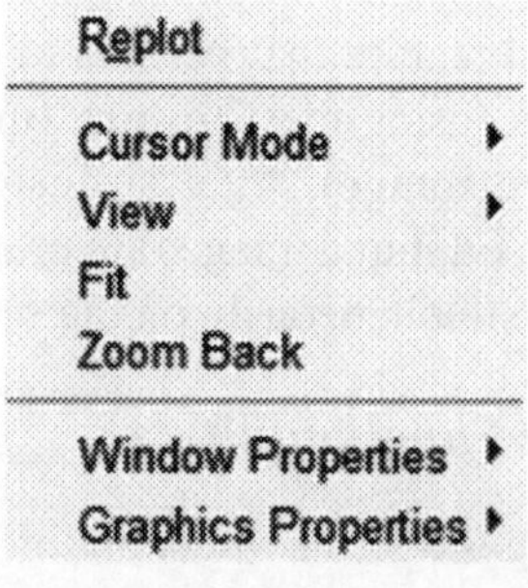

Fig. 1.27 Right Button Pop-up Menu.

1.5.2.7 *The Output Window*

The Output Window (Fig. 1.18) receives all text output from the program – command responses, notes, warnings, errors, and any other messages. It is usually positioned behind the GUI, but it can be raised to the front when necessary.

Sizing and Positioning the Output Window on Windows Systems

The size and position of the Output Window can be saved using:

1. Move the window to the desired location
2. Size the window to the desired dimensions
3. Select **Utility Menu** > **MenuCtrls** > **Save Menu Layout**

1.5.2.8 *Creating, Modifying and Positioning Toolbars*

When a session is begun, the start up routine reads a number of text files and scripts that set parameters and conditions for the session. Many of these files can be modified to provide a more customized level of operation. The `start160.ans` file is one such file. Call up toolbars, set their position and define their content in a similar fashion.

1.5.2.9 *Help System*

Much of the ANSYS documentation is available on-line. Type **help**, *command* to get help on any command. Another option to get help is to select HELP on the menu bar at the top of the screen. Information about the commands, theory, and element types used by ANSYS can be viewed in Fig. 1.28.

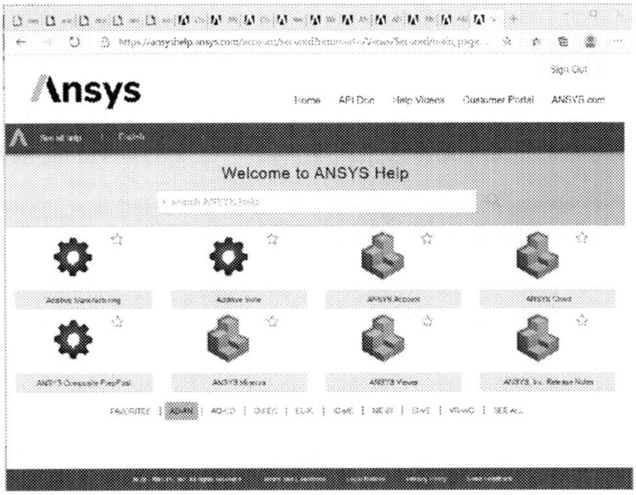

Fig. 1.28 Help system.

1.6 GRAPHICAL PICKING

Many functions in the ANSYS program involve graphical picking – using the mouse to identify model entities and coordinate locations. There are three types of graphical picking operations:

- *Locational* picking, where the coordinates of a new point are located.
- *Retrieval* picking, where existing entities are identified for subsequent operations.
- *Query* picking, where points are identified on the model for querying data.

Mouse Button Assignments for Picking

Mouse button assignments during any picking operation are as follows:

- The left button picks or unpicks the entity or location closest to the mouse pointer. When this button is pressed, the analyst can cross-examine the entity numbers before they are picked, and thus query data. Press and drag the left mouse button to preview the item being picked or unpicked.
- The middle button (shift-right button on a two-button mouse) is the same as the Apply button on the picker. A number of items can be picked, and no operations will take effect until **Apply** or **OK** is pressed. If there are no items currently picked, the middle button closes the picker.
- The right button toggles between pick and unpick modes. This button is the same as the Pick and Unpick toggle buttons on the picking menu.

Hot Spots

Hot spots are the locations on an entity that identify it for retrieval picking. For example, if two elements are adjacent, the element that is picked is the one whose hot spot is closest to the mouse pointer. For areas, volumes, and elements, the hot spot is at a location near the centroid. Lines have three hot spots: one in the middle and one near each end.

If the hot spots of two or more entities are coincident, picking the coincident entities brings up a "Multiple Entities" dialog box (Fig. 1.29). Press the **Next** or **Prev** button to "cycle through" the coincident entities until the desired entity is highlighted. Then click **OK** to pick that entity.

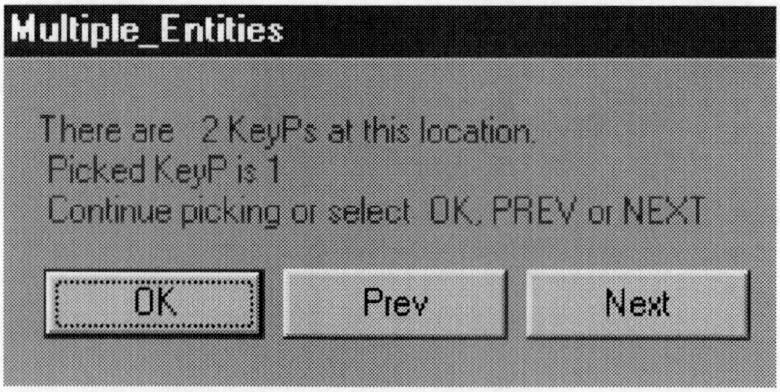

Fig. 1.29 Multiple Entities dialog box.

1.6.1 Locational and Retrieval Picking

Whenever graphical picking (clicking on a menu topic ending with the +symbol) is used, the GUI brings up a picking menu, sometimes known as the *picker*. Figure 1.30 shows the picking menus for locational and retrieval picking. In this example, creating keypoints by picking their locations on the working plane is a locational picking operation, and picking those keypoints to apply a load on them is a retrieval picking operation.

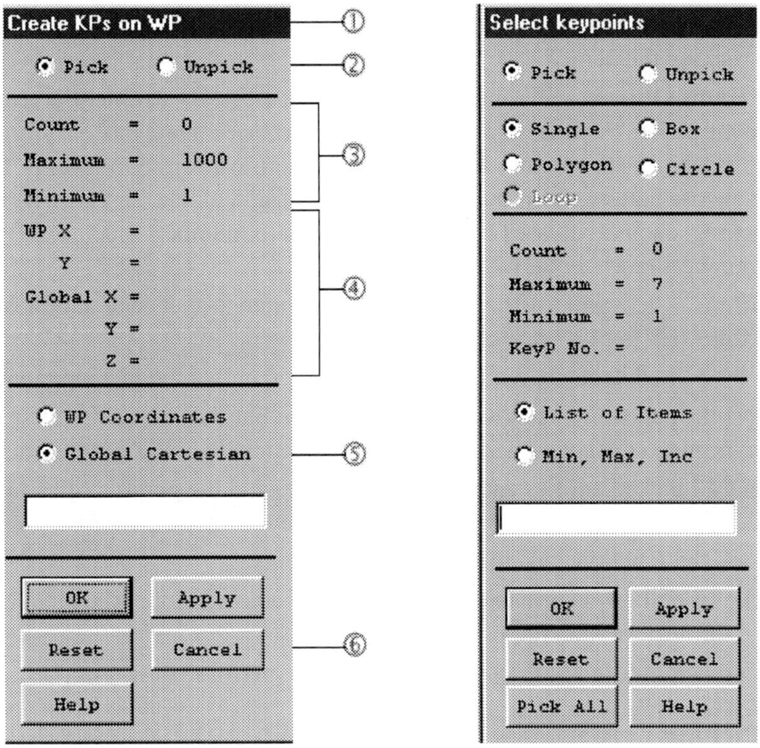

Fig. 1.30 Picking menus for locational and retrieval picking.

- **Function Title** – Identifies the function being performed.
- **Pick Mode** – Allows to *pick* or *unpick* a location or entity. Either these toggle buttons or the right mouse button can be used to switch between pick and unpick modes. The mouse pointer is shown as an up arrow for picking and a down arrow for unpicking.

For retrieval picking, choose among single, box, polygon, circle, and loop mode. In single pick mode, each click on the mouse picks one entity. With box, polygon, and circle modes, press and drag the mouse to enclose a set of entities in a box, polygon, or circle.

Loop mode is available for picking lines and areas only. With loop mode, when the analyst picks on a line, then the complete set of lines (or areas) defining a continuous loop including that line is also picked. This feature is useful when it is necessary to identify continuous lines in order to make an area. For all modes of retrieval picking, ANSYS shows the picked entities highlighted for visual clarity.

- **Pick Status** – Shows the number of items picked ("Count") and the minimum and maximum number of picks required for the function.
- **Picked Data** – Shows information about the item being picked. For locational picking, the working plane and global Cartesian coordinates of the point are shown. For retrieval picking, this area shows the entity number. This data can be seen by pressing and dragging the mouse in the graphics area. This allows to preview the information before releasing the mouse button and picking the item.
- **Keyboard Entry Options** – In some cases, it may be necessary to enter the required data by keyboard in the picker. For example, to specify a known coordinate location during locational picking, it may be easier to enter the coordinates than to use the mouse. In that case, the analyst can choose between working plane coordinates and global Cartesian coordinates. For retrieval picking, choose between entering a list of entity numbers (such as 1,21,343,...) and a range of numbers (such as 1,21,2).
- **Action Buttons** – This area of the menu contains buttons that take action on the picked entities, as follows:

OK – Applies the picked items to execute the function and closes the picking menu.

Apply – Applies the picked items to execute the function but does not close the picking menu. Use this button on the menu or stay in the graphics area and click the middle mouse button to apply.

Reset – Unpicks all picked entities and restores the menu and the graphics area to their state at the last Apply.

Cancel – Cancels the function and closes the picking menu.

Pick All – Picks all entities, executes the selected function, and closes the picking menu. This feature is available for retrieval picking only.

Help – Brings up help information for the function being performed.

1.6.2 Query Picking

The picker can be used to "Query" data on the model at any time during the analysis. Querying means that, for any picked point on the model, ANSYS will retrieve or calculate specified items from the database and display them. This information can be displayed in an output window, or it can be applied to the model as 3-D annotation. Query picking is a retrieval picking function, with the selection box items varying according to the requested data. The model and results query pickers are discussed in the following sections.

1.6.2.1 The Model Query Picker

The model query picker allows to access model information by picking displayed entities. It also performs basic computations to provide simple geometric/loading information. The model query picker can be used to obtain information about the model while building it. This is helpful when building on to an existing model, or for applying forces or loads that are dependent on the model data. The model query picker can be accessed by choosing **Utility Menu> List> Picked Entities**, at any time during the analysis.

Annotation

When the model query picker is used to obtain model information, ANSYS displays the information in a text window. That information can also be applied to the model as 3-D annotation by checking the "Generate 3D Anno" box on the picker. The annotation is applied with the appropriate units displayed, and arrows and leaders are attached where necessary. 3-D annotation, unlike 2-D, will attempt to retain the proper positioning on the appropriate entities, even when the model is rotated or resized.

The overall 3-D dimensions of the model are defined by a bounding box. If portions of the model's bounding box lie outside of the visible area of the graphics window, it can affect the placement of the 3-D annotations. Zooming out will usually overcome this problem.

Modify query annotation using **Utility Menu>PlotCtrls> Annotation> Create 3D Annotation**. Select "Options" to modify the typeface and move, copy, or resize the annotation.

Action Buttons

- **OK** – Enters all query data and either displays it in a window or applies it to the model. The model query picker is then closed.
- **Apply** – Enters all query data and displays it in a window or applies it to the model. The model query picker is left open for additional operations.
- **Reset** – Clears all query data.
- **Cancel** – Clears all query data and closes the model query picker.
- **Pick All** – Accesses all of the selected query items on the model.
- **Help** – Brings up this screen of help information.

Tips on Using the Model Query Picker

The model query picker is accessible at any time during the analysis. It is useful for determining model database information that would require hand calculations to determine. The computations performed by the picker are not displayed until **OK** or **APPLY** is pressed.

The items selected by the picker are highlighted before they are actually selected. Hold the left mouse button down and move the pointer around the model to ensure that the proper entity is selected. The analyst can select multiple items, with no computations being performed or annotation applied until OK is clicked. The commands and operations performed by the model query picker are not written to the log file.

The annotation that is generated during the preprocessing stage can be displayed at any time during the analysis. This information is not modified or updated during the course of the analysis, and can become invalid. For instance, when a shape is deformed by the applied forces, the area and volume annotations created before applying the force will not change, and the information displayed may be incorrect.

1.6.2.2 *Results Query Picker*

The most convenient way to review results for specific points on the model is by using the results query picker. Querying means that, for any picked point on the model, the ANSYS program will retrieve a specified results data value from the database and display it. Access the results query picker by choosing **Main Menu> General Postproc> Query Results**. The analyst can query nodal or element solution data.

When this GUI path is chosen, the results query picker for element solution data appears. This menu has most of the same items as the locational and retrieval picking menus:

- **Function Title** – Identifies the information being queried; either nodal, element, or subgrid solution data.
- **Pick Mode** – Allows to *pick* or *unpick* a location on the model to be queried. These toggle buttons or the right mouse button can be used to switch between pick and unpick modes. The mouse pointer is shown as an up arrow for picking and a down arrow for unpicking.
- **Picked Location** – Shows the number and location of the node, element, or subgrid point being picked for query. This information can be previewed by pressing and dragging the mouse in the graphics area. The picked location is highlighted on the model as a square on the displaced shape.
- **Queried Data** – Shows the queried solution data for the node, element, or subgrid point at the picked location. Use the Max or Min buttons to automatically obtain the maximum or minimum value of the query data over the entire model.

Annotation

The results query picker retrieves solution data from the database and displays it in a text window. That information can be applied to the model as 3-D annotation by selecting the "Generate 3D Anno" box on the picker. The annotation is applied with the appropriate units displayed, and arrows and leaders are attached where necessary. The overall 3-D dimensions of the model are defined by a bounding box. If portions of the model's bounding box lie outside of the visible area of the graphics window, it can affect the placement of the 3-D annotations. Zooming out will usually overcome this problem. Unlike the model query picker, the results query picker displays the actual data on the model as the mouse pointer is moved over the different entities. This function is available only from the General Postprocessor.

Action Buttons

- **OK** – Clears all query data and closes the results query picker.
- **Apply** – Clears all query data and brings up the Query Solution Data dialog box again. This button can be used on the picking menu or stay in the graphics area and click the middle mouse button to achieve the same effect.
- **Reset** – Clears all query data.
- **Cancel** – Clears all query data and closes the query picking menu.
- **Min** – Shows the minimum value of the query data over the entire model.

- **Max** – Shows the maximum value of the query data over the entire model.
- **Help** – Brings up this screen of help information.

1.7 CUSTOMIZING THE GUI

The GUI allows several levels of customization, ranging from a simple change in the sizes of the GUI and the areas in it to a more complex change in the menu hierarchy and design of the dialog boxes. GUI attributes which can be changed include the following items:

- The size of the GUI and the areas in it
- Colors and fonts
- The menus shown at GUI start-up
- The mouse and keyboard focus
- The menu hierarchy and dialog boxes.

1.7.1 Changing the GUI Layout

The ANSYS toolbar, Main Menu, and Graphics Window, as well as the overall size of the GUI can be resized. To resize the areas in the GUI, drag one of the borders around the areas of the GUI while holding down the left mouse button.

In order to change the overall size of the GUI, position the mouse on one of the corners of the GUI and drag it diagonally toward the center of the GUI while holding down the left mouse button. GUI size settings can be saved by selecting **Utility Menu>MenuCtrls> Save Menu Layout**.

1.7.2 Changing Colors and Fonts

The system-level fonts determine the default ANSYS GUI font component. ANSYS uses the fonts that are set at the Display Properties system level. The fonts in the ANSYS GUI are based on the system font settings. Use the **Utility Menu>MenuCtrls> Font** dialog to modify the GUI fonts. Once these settings are changed, use the **Utility Menu>MenuCtrls> Save Menu Layout** to save the settings. The saved settings will override the system fonts and become the ANSYS default.

The Main Menu fonts can also be modified by using the "right mouse button" – preferences function. At this level, specific fonts and colors are set in the Main Menu at a functional level. The analyst can change the overall fonts, and change the colors according to the progressive level as s/he opens the nested menu items.

1.7.3 Changing the GUI Components Shown at Start-Up

By default, the main areas of the GUI (Utility Menu, Main Menu, Standard Toolbar, ANSYS Toolbar, Input Window, Graphics Window, and Output Window) display when the GUI is activated.

1.7.4 Changing the Mouse and Keyboard Focus

The ANSYS program uses the concept of input focus to determine which window on the screen is active; that is, which window receives input from the mouse and the keyboard. A window may have *implicit focus* (also known as *pointer focus*), in which case moving the mouse pointer into that window makes it active. With *explicit focus*, it is necessary to move the pointer into the window and *click the left button* for that window to be active.

The ANSYS GUI also gives automatic keyboard focus to the Input Window. This means a command can be entered without first moving the mouse into the Input Window. If a dialog box is brought up, then keyboard focus is automatically redirected to it so the analyst can enter data without moving the mouse. This feature is sometimes called the *hot keyboard* feature.

1.8 USING THE ANSYS SESSION AND COMMAND LOGS

The ANSYS program records every command it executes, at two places: the session log file and the internal database command log.The session log file is a text file which is saved in the working directory.The database command log is saved in the ANSYS database. This log can be copied to a file at any time by choosing **Utility Menu> File> Write DB Log File**. Both files are command logs that can be used as input to the ANSYS program.

1.8.1 Session Log File

Every ANSYS session produces a session log named `Jobname.LOG`. (Jobname is determined by the jobname defined at ANSYS entry. The default jobname is `FILE` or `file`, depending on the operating system.The program opens the log file when the analyst first enters the program, and closes it when s/he exits the program.

The session log file provides a complete record of the ANSYS session in terms of commands and is quite valuable as a means of recovering from a system crash or catastrophic user mistake. By reading in a renamed copy of the log file or by submitting it as a batch file, the s/he can re-execute every command in the log file, recreating the database exactly as it existed previously. The log file is also useful as a debugging tool that can help to reveal any mistakes that might have been made in an ANSYS session.

Each new ANSYS session appends commands to the existing file `Jobname.LOG`. That is, the log file is not overwritten during a new ANSYS session, but added to. A "time stamp," consisting of the current date and time, is included so that the analyst can identify the start of each session. The entire log file can be listed during an interactive run by picking **Utility Menu> List> Files> Log File**. Because this file is in ASCII format, it can be viewed and edited readily using an external text editor.

1.8.2 Using the Database Command Log

ANSYS captures commands generated during an ANSYS session not only in the log file but also in memory. This in-memory version of the command history is called the internal database log. When the database is saved, the program saves this command log in the database file (`Jobname.DB`) along with the other database information.

Use **Utility Menu> File> Write DB Log File** to write the database command log to a named ASCII file. Then edit this file, make desired changes, and use the file as command input to the program. This capability is especially useful to use the command history that was created during an interactive session, but have somehow lost or corrupted the session log file (`Jobname.LOG`) that was associated with the database.

If the database is created in multiple sessions by saving and resuming the database file, the ANSYS program keeps the database log continuous by appending each new command that is processed. Therefore, the internal database log is not fragmented; it will represent the complete database.

1.8.3 Command Log File as Input

The procedure for re-executing the commands contained in a `Jobname.LOG` file or in the database log consists of three main steps:

1. Establish the command log file
2. Edit the command log file as desired
3. Read the edited file into the ANSYS session.

Step 1: Establish the Command Log File

The method to do this depends on whether the analyst uses the session log file or the database log.

Session Log File

Perform the following steps to establish a command log file from the session log file, (`Jobname.LOG`):

1. Rename or copy the session log file to a different name. This can be done at the system level or from within the program.
2. List the log file by choosing **Utility Menu** > **List** > **Files** > **Log File**.
3. Choose **File** > **Save** as from the Log File window.

Database Command Log

Pick **Utility Menu> File> Write DB Log File** to establish a command log file from the database log. A file name can be specified or the default name, `Jobname.LGW` can be executed.

Step 2: Edit the Command Log File

Sometimes, it is necessary to edit the command log file before using it as a program input. As the analyst edits the log file, s/he may want to add comments or indentation to improve its readability. Comments can be added to the log file by using comment commands or by using the comment character (!).

Step 3: Read in the Edited Log File

Pick **Utility Menu** > **File** > **Read Input from** to read in the edited command log file.

CHAPTER 2 ANSYS OPERATIONS

2.1 GETTING STARTED

There are three basic steps to start ANSYS from the Windows Start Menu.

ANSYS Workbench: This is a new GUI with an emphasis on CAD connectivity, ease of use, and easy management of assembly contact.

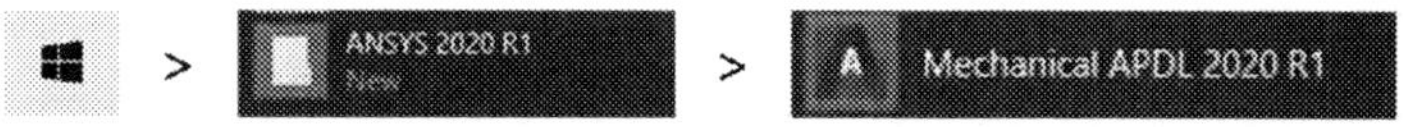

Fig. 2.1 Three steps to start ANSYS

ANSYS: This starts ANSYS in the traditional GUI. The program starts immediately using the settings last changed under the next item, 'Configure ANSYS Products'.

Configure ANSYS Products: This choice brings up the ANSYS Launcher.

ANSYS has many finite-element analysis capabilities, ranging from a simple, linear, static analysis to a complex, nonlinear, transient dynamic analysis. The process for a typical analysis involves three general tasks:

- Building the model
- Applying loads and obtaining the solution
- Reviewing the results.

2.1.1 Building the Model

Building a finite element model requires more time than any other part of the analysis. First, specify a jobname and analysis title. Then, use the preprocessor to define the element type(s), element real constant(s), material properties and the model geometry.

2.1.1.1 Specifying a Jobname and Analysis Title

This task is not required for an analysis, but is recommended.

Defining the Jobname

The *jobname* is a name that identifies an analysis job. When the analyst defines a jobname for an analysis, the jobname becomes the first part of the name of all files the program creates. The extension or suffix for these files' names is a file identifier such as `.DB`. Using a jobname for each analysis, ensures that no files are overwritten.

If a jobname is not specified, all files receive the name `FILE` or `file`, depending on the operating system. The default jobnamecan be changed using **Utility Menu> File> Change Jobname**.

Defining an Analysis Title

Utility Menu> File> Change Title defines a title for the analysis. The program includes the title on all graphics displays and on the solution output.

Defining Units

The program does not assume a system of units for the analysis. Any system of units can be used so long as the same system is used for all the data entered.

2.1.1.2 Defining Element Types

The element library contains more than 150 different element types. Each element type has a unique number and a prefix that identifies the element category: PLANE182, SOLID185, BEAM188, ELBOW290, and so on. The element type determines, among other things:

- The degree-of-freedom set
- Whether the element lies in 2-D or 3-D space.

BEAM188, for example, has six structural degrees of freedom (UX, UY, UZ, ROTX, ROTY, ROTZ), is a line element, and can be modeled in 3-D space. User must be in the general preprocessor to define element types. While defining the actual elements, point to the appropriate type reference number using the **Main Menu>Preprocessor>Modeling> Create> Elements> Elem Attributes**.

2.1.1.3 Defining Element Real Constants

Element real constants are properties that depend on the element type, such as the cross-sectional properties of a beam element. Not all element types require real constants, and different elements of the same type may have different real constant values.

As with element types, each set of real constants has a reference number, and the table of reference number versus real constant set is called the *real constant table*. While defining the elements, point to the appropriate real constant reference number using **Main Menu>Preprocessor>Modeling> Create> Elements> Elem Attributes**.

2.1.1.4 Defining Material Properties

Most element types require material properties. Depending on the application, material properties can be linear or nonlinear.As with element types and real constants, each set of material properties has a material reference number. The table of material reference numbers versus material property sets is called the *material table*. Within one analysis, there may be multiple material property sets. The program identifies each set with a unique reference number.

Linear Material Properties

Linear material properties can be constant or temperature-dependent, and isotropic or orthotropic. UseMain **Menu> Preprocessor> Material Props> Material Models** to define constant material properties.

Also specify the appropriate property label; for example, EX, EY, EZ for Young's modulus, KXX, KYY, KZZ for thermal conductivity, and so forth. For isotropic material it is necessary to define only the X-direction property; the other directions default to the X-direction value.

Besides the defaults for Y- and Z-direction properties (which default to the X-direction properties), other material property defaults are built in to reduce the amount of input. For example, Poisson's ratio (NUXY) defaults to 0.3 and shear modulus (GXY) defaults to EX/2(1+NUXY).

The analyst can choose constant, isotropic, linear material properties from a material library available through the GUI. Young's modulus, density, coefficient of thermal expansion, Poisson's ratio, thermal conductivity and specific heat are available for 10 materials in four unit systems.

Material Model Interface

The program GUI includes an intuitive hierarchical tree structure interface for defining many material models. A logical top-down arrangement of material categories guides in defining the appropriate model for the analysis.

Accessing the Material Model Interface

Access the material model interface via **Main Menu> Preprocessor> Material Props> Material Models**. The **Define Material Model Behavior** dialog box appears, which originally displays the top level of the tree structure, as shown in Fig. 2.2.

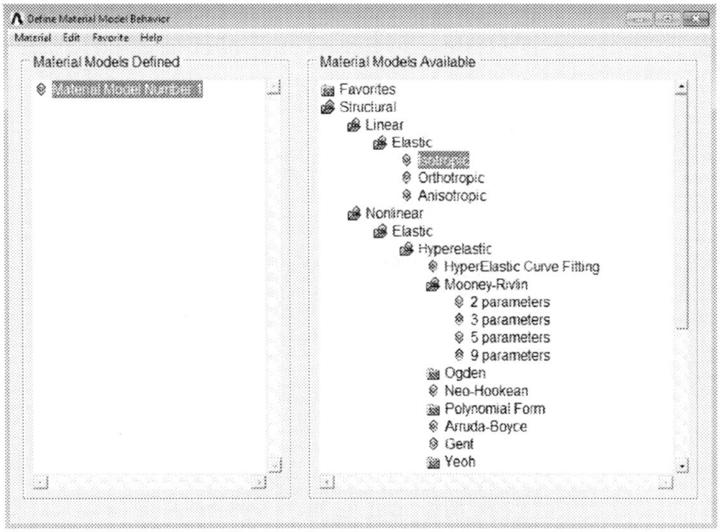

Fig. 2.2 Material model interface initial screen.

Choosing Material Behavior

The **Material Models Available** window on the right displays a list of material categories. If a category is preceded by a folder icon, there are subcategories available under the main category. On double-clicking on the category, the subcategories appear indented, and below the category as shown in Fig. 2.3.

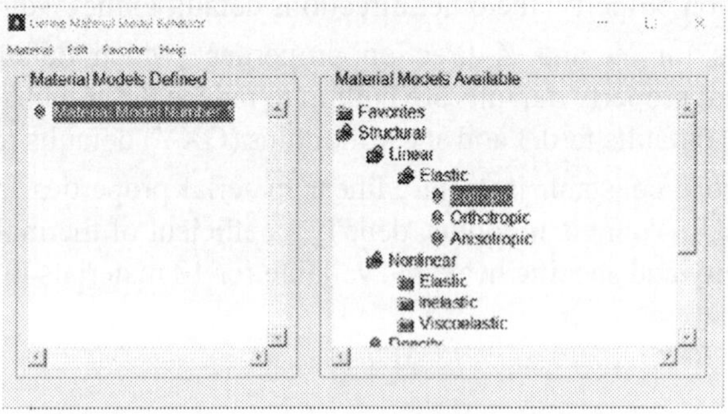

Fig. 2.3 Material model interface tree structure.

For example, under Structural are categories Linear, Nonlinear, and others. The models are further categorized so that the analyst will eventually see a vertical list of material property sets or material models that are included under that category. Once it is decided which material property set or model will be used, then choose it by double-clicking on the item. A dialog box appears that prompts for the required input data for that particular model or property set.

Entering Material Data

Included in a data input dialog box is a table whose rows and columns can be altered depending on the requirements of the specific material property or model chosen. A typical data input dialog box is shown in Fig. 2.4.

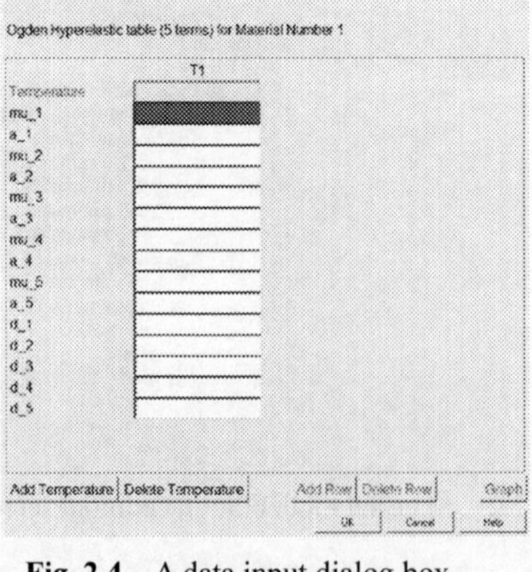

Fig. 2.4 A data input dialog box.

There are two interaction areas within a material data input dialog box: the data input table, and a series of action buttons that appear at the bottom. Depending on the material item defined, the labels in the table vary, as do the number of rows and columns that appear initially. The material item also dictates the number of rows and columns that are allowed to be added or deleted. In most cases, the columns represent temperatures, and the rows represent data values.

Temperature Dependent Data

Initially, the table is set up for temperature independent data so the temperature field is grayed out. At this point, should the analyst decide to enter data for various temperatures, s/he can quickly add columns of text fields for the data representing each temperature. The temperature dependent data can be added or deleted at any time. There is no necessity to predetermine if the data should be temperature dependent.

Adding and Deleting Columns

In order to add a column, position the text cursor in any field in the existing column, then click on the **Add Temperature** button. A new column appears to the right of the existing column, and both temperature fields become active, as shown in Fig. 2.5.

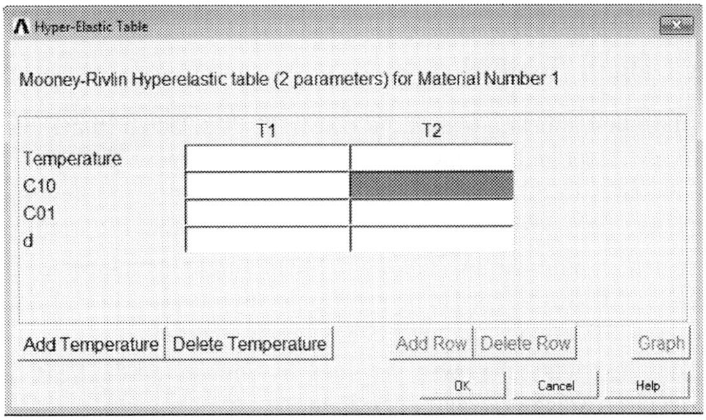

Fig. 2.5 Data input dialog box – Added column.

Then enter the two temperatures and the associated data in the rows. More temperature columns can be added, as needed, by following the same procedure. Insert columns between existing columns by clicking the text cursor in a field within a column that is to the left of where it is nesessary to insert the new column, then clicking on the **Add Temperature** button. A scroll bar appears across the bottom of the table when the number of columns exceeds the width of the dialog box.

Adding and Deleting Rows

It may be necessary to add another row of constants or other data for a specific temperature. Then add or delete rows in a similar way as is described above for adding or deleting columns. In order to add a row, click the text cursor in any field in an existing row, then click on the **Add Row** (or **Add Point**) button. A new row appears beneath the existing row, as shown in Fig. 2.6.

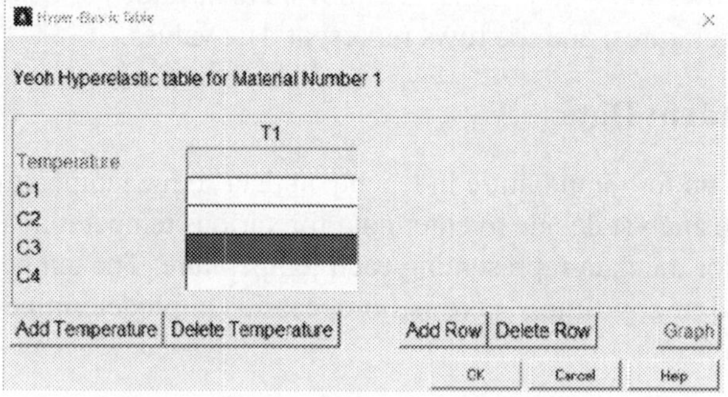

Fig. 2.6 Data input dialog box – Added row.

More rows can be added, as needed, by following the same procedure. Insert rows between the existing rows by positioning the text cursor in a field in the top row, then clicking on the **Add Row** (or **Add Point**) button. A vertical scroll bar appears in the table when the number of rows exceeds the height of the dialog box.

Entering/Editing Data in Text Fields

When a data dialog box first appears, one of the text fields is selected (black highlight), meaning that the field is ready to accept and display data as the analyst types. The arrow keys can be used to move the selection status to other text fields. Also, pressing the Tab key allows to move the selection status to the text field positioned to the right of the field that is currently selected.

When the analyst starts typing within a text field, the highlight is replaced by the characters that are typed. The left and right arrow keys can be used to position the text cursor anywhere within the field should it be necessary to replace or delete characters in that field.

In order to edit data, first select the text field either by clicking on the field, or using the arrow keys to move the selection status to the particular field.

Logging/Editing Material Data

The **Material Models Defined** window displays a log of each material model specified. After OK is chosen in the data input dialog box, this window displays a folder icon, and **Material Model Number #**, followed by the properties defined for this model. Additional models can be defined with unique numbers by choosing **Material> New Model**, then typing a new number in the **Define Material ID** dialog box. On double-clicking on any material model or property, the associated data input dialog box appears where the data can be edited, if choosen.

Using Material Library Files

Although material properties can be defined separately for each finite element analysis, the analyst can also store a material property set in an archival material library file, then retrieve the set and reuse it in multiple analyses. Each material property set has its own library file. The material library files also enable several users to share commonly used material property data.

2.1.1.5 Creating the Model Geometry

Once material properties are defined, the next step in an analysis is generating a finite element model – nodes and elements – that adequately describes the model geometry. Figure 2.7 shows some sample finite element models.

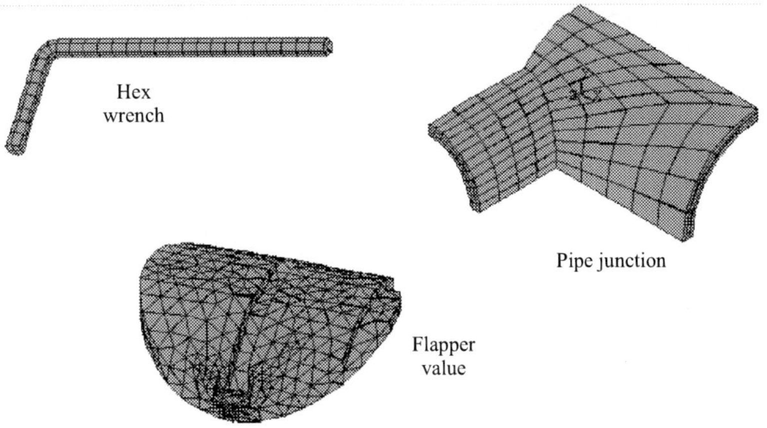

Fig. 2.7 Sample finite element models.

There are two methods to create the finite element model: solid modeling and direct generation. In solid modeling, the analyst describes the geometric shape of model, then instructs the program to automatically mesh the geometry with nodes and elements. The size and shape in the elements that the program creates can be controlled. In direct generation, s/he 'manually' defines the location of each node and the connectivity of each element. Several convenience operations, such as copying patterns of the existing nodes and elements, symmetry reflection, etc., are available.

2.1.2 Applying Loads and Obtaining the Solution

In this step, the SOLUTION processor defines the analysis type and analysis options, apply loads, specify load step options, and initiate the finite element solution. Loads can also be applied via the PREP7 preprocessor.

2.1.2.1 Specifying the Analysis Type and Analysis Options

Specify the analysis type based on the loading conditions and the response to be calculated. For example, if natural frequencies and mode shapes are to be calculated, choose a modal analysis. The following analysis types can be performed in the program: static (or steady-state), transient, harmonic, modal, spectrum, buckling, and substructuring.

Analysis options allows the analyst to customize the analysis type. Typical analysis options are the method of solution, stress stiffening on or off, and Newton-Raphson options.

Use **Main Menu> Preprocessor> Loads> Analysis Type> New Analysis** or **Main Menu> Preprocessor> Loads> Analysis Type> Restart** to define the analysis type and analysis options.

The analyst can specify either a new analysis or a restart, but a new analysis is the norm in most cases. A multiframe restart that allows to restart an analysis at any point is available for static and transient analyses. The analysis type and analysis options cannot be changed after the first solution.

After the analysis type and analysis options are defined, the next step is to apply loads. Some structural analysis types require other items to be defined first, such as master degrees of freedom and gap conditions.

2.1.2.2 Applying Loads

The word loads as used in ANSYS includes boundary conditions (constraints, supports, or boundary field specifications) as well as other externally (Fig. 2.8) and internally applied loads. Loads are divided into the following categories:

- DOF Constraints
- Surface Loads
- Inertia Loads
- Forces
- Body Loads
- Coupled -field Loads.

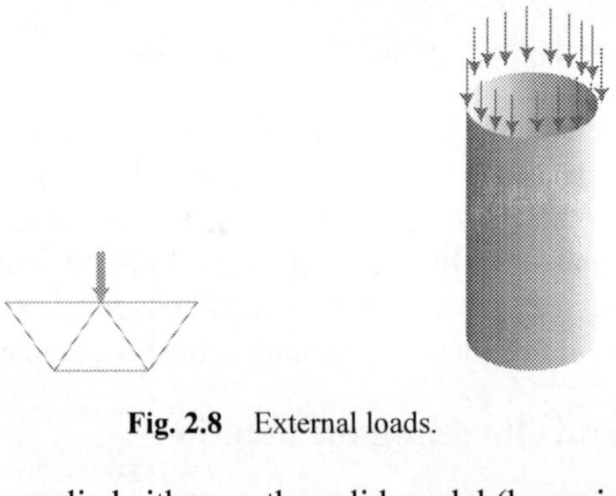

Fig. 2.8 External loads.

Most of these loads can be applied either on the solid model (keypoints, lines, and areas) or the finite element model (nodes and elements). Two important load-related terms need to be known are *load step* and *substep*. A load step is simply a configuration of loads for which a solution is to be obtained. In a structural analysis, for example, the analyst may apply wind loads in one load step and gravity in a second load step. Load steps are also useful in dividing a transient load history curve into several segments.

Substeps are incremental steps taken within a load step. They are mainly used for accuracy and convergence purposes in transient and nonlinear analyses. Substeps are also known as *time steps* – steps taken over a period of time.

The program uses the concept of time in transient analyses as well as static (or steady-state) analyses. In a transient analysis, time represents actual time, in seconds, minutes, or hours. In a static or steady-state analysis, time simply acts as a counter to identify load steps and substeps.

Loads and boundary conditions can be applied in both the Preprocessor (**Main Menu> Preprocessor> Loads > Define Loads > Apply**), and the Solution processor (**Main Menu > Solution > Define Loads > Apply**).

1. Select the kind of constraint to be applied.
2. Select the geometric entity on which it is to be applied.
3. Enter the value and direction for it.

There is no modify command for loads and B.C.'s. If a mistake is made simply apply it again with a new value (the old one will be replaced if it's on the same entity), or delete it and reapply it.

Although loads and boundary conditions can be applied to nodes or elements, it's generally better to apply all B.C.'s to the geometry. When the solve command is issued, they will be automatically transferred to the underlying nodes and elements. If B.C.'s are put on the geometry, that geometry can be remeshed without having to reapply them.

2.1.2.3 Specifying Load Step Options

Load step options are options that the analyst can change from load step to load step, such as number of substeps, time at the end of a load step, and output controls. Depending on the type of analysis s/he is doing, load step options may or may not be required. The analysis procedures in the analysis guide manuals describe the appropriate load step options as necessary.

2.1.2.4 Initiating the Solution

Use **Main Menu> Solution> Solve> Current LS** to initiate solution calculations. When this command is issued, the program takes model and loading information from the database and calculates the results. Results are written to the results file (`Jobname.RST`, `Jobname.RTH`, or `Jobname.RMG`) and also to the database. The only difference is that only one set of results can reside in the database at one time, while ANSYS can write all sets of results (for all substeps) to the results file.

Multiple load steps can be solved in a convenient manner using: **Main Menu> Solution> Solve> From LS Files**.

2.1.3 Reviewing the Results

After the solution has been calculated, use the postprocessors to review the results. Two postprocessors are available: POST1 and POST26.

Use POST1, the general postprocessor, to review results at one substep (time step) over the entire model or selected portion of the model. The GUI path for entering POST1 is **Main Menu> General Postproc**, valid only at the Begin level. The analyst can obtain contour displays, deformed shapes, and tabular listings to review and interpret the results of the analysis. POST1 offers many other capabilities, including error estimation, load case combinations, calculations among results data, and path operations.

Use POST26, the time-history postprocessor, to review results at specific points in the model over all time steps. The GUI path for entering POST26 is **Main Menu> TimeHist Postpro**, valid only at the Begin level. Graph plots of results data versus time (or frequency) and tabular listings can be obtained.

2.2 LOADING

The primary objective of a finite element analysis is to examine how a structure or component responds to certain loading conditions. Specifying the proper loading conditions is, therefore, a key step in the analysis. Loads can be applied on the model in a variety of ways. With the help of load step options, the analyst can control how the loads are actually used during solution.

2.2.1 Understanding Loads

The term *loads* includes boundary conditions and externally or internally applied forcing functions, as illustrated in Fig. 2.9. Examples of loads in different disciplines are:

Structural: displacements, velocities, accelerations, forces, pressures, temperatures (for thermal strain), gravity

Thermal: temperatures, heat flow rates, convections, internal heat generation, infinite surface

Boundary conditions, as well as other types of loading, are also shown.

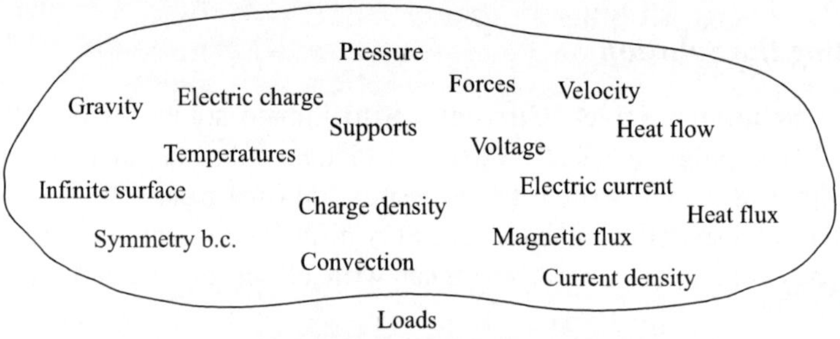

Fig. 2.9 Loads.

Loads are divided into six categories: DOF constraints, forces (concentrated loads), surface loads, body loads, inertia loads, and coupled-field loads.

- A *DOF constraint* fixes a degree of freedom (DOF) to a known value. Examples of constraints are specified displacements and symmetry boundary conditions in a structural analysis, prescribed temperatures in a thermal analysis, and flux-parallel boundary conditions.
- A *force* is a concentrated load applied at a node in the model. Examples are forces and moments in a structural analysis, heat flow rates in a thermal analysis, and current segments in a magnetic field analysis.
- A *surface load* is a distributed load applied over a surface. Examples are pressures in a structural analysis and convections and heat fluxes in a thermal analysis.
- A *body load* is a volumetric or field load. Examples are temperatures and fluences in a structural analysis, heat generation rates in a thermal analysis, and current densities in a magnetic field analysis.

- *Inertia loads* are those loads that are attributable to the inertia (mass matrix) of a body, such as gravitational acceleration, angular velocity, and angular acceleration. They are mainly used in a structural analysis.
- *Coupled-field loads* are simply a special case of one of the above loads, where results from one analysis are used as loads in another analysis. For example, magnetic forces calculated in a magnetic field analysis can be applied as force loads in a structural analysis.

A *load step* is simply a configuration of loads for which a solution is obtained. In a linear static or steady-state analysis, the analyst can use different load steps to apply different sets of loads – wind load in the first load step, gravity load in the second load step, both loads and a different support condition in the third load step, and so on. In a transient analysis, multiple load steps apply different segments of the load history curve.

The program uses the set of elements which the analyst selects for the first load step. It doesn't matter which element sets are specified for the later steps for all subsequent load steps.Use **Utility Menu> Select> Entities** to select an element set.

Figure 2.10 shows a load history curve that requires three load steps: the first load step for the ramped load, the second load step for the constant portion of the load, and the third load step for load removal.

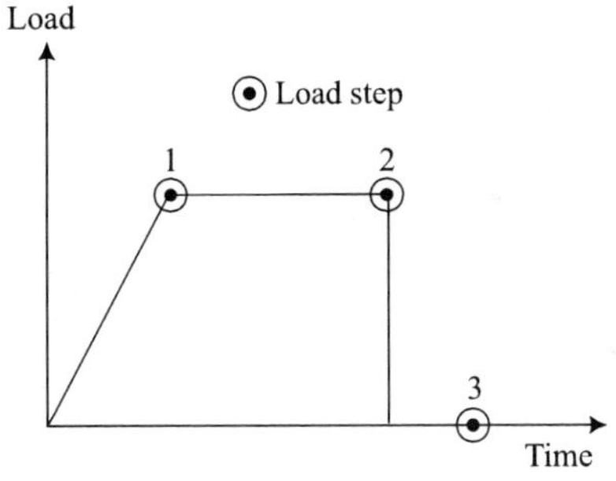

Fig. 2.10 Transient load history curve.

2.2.2 The Role of Time in Tracking

The program uses time as a tracking parameter in all static and transient analyses, whether they are or are not truly time-dependent. The advantage of this is that the analyst can use one consistent 'counter' or 'tracker' in all cases, eliminating the need for analysis-dependent terminology. Moreover, time always increases monotonically, and most things in nature happen over a period of time, however brief the period may be.

Obviously, in a transient analysis or in a rate-dependent static analysis (creep or viscoplasticity), time represents actual, chronological time in seconds, minutes, or hours. Assign the time at the end of each load step while specifying the load history curve. Use one of the following to assign time:

- **Main Menu> Preprocessor> Loads> Load Step Opts> Time/Frequenc> Time and Substps**

- **Main Menu> Preprocessor> Loads> Load Step Opts> Time/Frequenc> Time - Time Step**
- **Main Menu> Solution> Analysis Type> Sol'n Control**
- **Main Menu> Solution> Load Step Opts> Time/Frequenc> Time and Substps**
- **Main Menu> Solution> Load Step Opts> Time/Frequenc> Time - Time Step**
- **Main Menu> Solution> Load Step Opts> Time/Frequenc> Time and Substps**
- **Main Menu> Solution> Load Step Opts> Time /Frequenc> Time - Time Step**

In a rate-independent analysis, however, time simply becomes a counter that identifies load steps and substeps. By default, the program automatically assigns time = 1.0 at the end of load step 1, time = 2.0 at the end of load step 2, and so on. Any substeps within a load step are assigned the appropriate, linearly interpolated time value.

2.2.3 Stepped versus Ramped Loads

When more than one substep is specified in a load step, the question of whether the loads should be *stepped* or *ramped* arises (Fig. 2.11).

- If a load is *stepped*, then its full value is applied at the first substep and stays constant for the rest of the load step.
- If a load is *ramped*, then its value increases gradually at each substep, with the full value occurring at the end of the load step.

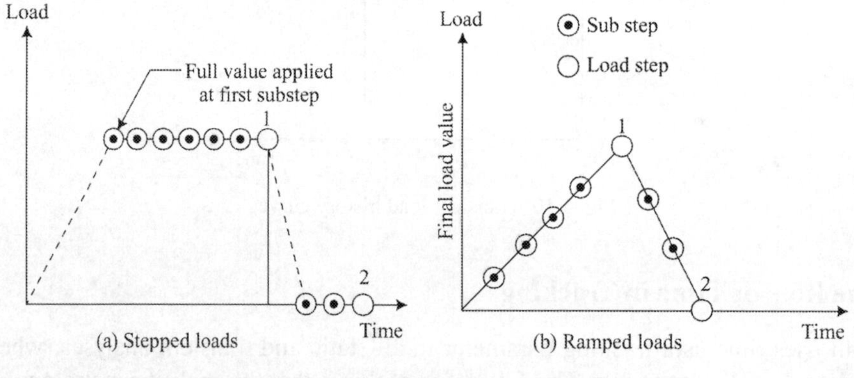

Fig. 2.11 Stepped versus ramped loads.

Use either of the following to indicate whether loads are ramped or stepped:

- **Main Menu> Solution> Load Step Opts> Time/Frequenc> Freq & Substeps: Transient Tab**
- **Main Menu> Solution> Load Step Opts> Time/Frequenc> Time and Substps**
- **Main Menu> Solution> Load Step Opts> Time/Frequenc> Time & Time Step**
- **Main Menu> Solution> Load Step Opts> Time/Frequenc> Freq & Substeps**
- **Main Menu> Solution> Load Step Opts> Time/Frequenc> Time and Substps/Main**
- **Menu> Solution> Load Step Opts> Time/Frequenc> Time & Time Step**

2.2.4 Applying Loads

Most loads can be applied either on the solid model or on the finite element model. For example, the analyst can specify forces at a keypoint or a node. Similarly, s/he can specify convections on lines and areas or on nodes and element faces. No matter how the loads are specified, the solver expects all loads to be in terms of the finite element model. Therefore, if loads are specified on the solid model, the program automatically transfers them to the nodes and elements at the beginning of the solution.

2.2.4.1 Solid-Model Loads: Advantages and Disadvantages

Advantages

Solid-model loads are independent of the finite element mesh. That is, the element mesh can be changed without affecting the applied loads. This allows to make mesh modifications and conduct mesh sensitivity studies without having to reapply loads each time.The solid model usually involves fewer entities than the finite element model. Therefore, selecting solid model entities and applying loads on them is much easier, especially with graphical picking.

Disadvantages

Elements generated by meshing commands are in the currently active element coordinate system. Nodes generated by meshing commands use the global Cartesian coordinate system. Therefore, the solid model and the finite element model may have different coordinate systems and loading directions.Applying keypoint constraints can be tricky, especially when the constraint expansion option is used. All solid-model loadscan not be displayed.

2.2.4.2 Finite-Element Loads: Advantages and Disadvantages

Advantages

There is no need to worry about constraint expansion. All desired nodes can be selected and the appropriate constraints can be specified.

Disadvantages

Any modification of the finite element mesh invalidates the loads, requiring the analyst to delete the previous loads and re-apply them on the new mesh. Applying loads by graphical picking is inconvenient, unless only a few nodes or elements are involved.

2.2.4.3 DOF Constraints

Following are some of the GUI paths that can be used to apply DOF constraints:

- **Main Menu> Preprocessor> Loads> Define Loads> Apply>** `load type` **> On Nodes**
- **Utility Menu> List> Loads> DOF Constraints> On All Keypoints (or On Picked KPs)**

- **Main Menu > Solution > Define Loads > Apply >** `load type` **> On Lines**

2.2.4.4 Transferring Constraints

Use **Main Menu> Preprocessor> Loads> Define Loads> Operate> Transfer to FE> Constraints** or **Main Menu> Solution> Define Loads> Operate> Transfer to FE> Constraints** to transfer constraints that have been applied to the solid model to the corresponding finite element model.

Use **Main Menu> Preprocessor> Loads> Define Loads> Operate> Transfer to FE> All Solid Lds** or **Main Menu> Solution> Define Loads> Operate> Transfer to FE> All Solid Lds** to transfer all solid model boundary conditions.

2.2.4.5 Forces (Concentrated Loads)

Following are examples of some of the GUI paths used for applying force loads:

- **Main Menu > Preprocessor > Loads > Define Loads > Apply >** `load type` **> On Nodes**
- **Utility Menu > List > Loads > Forces > On All Keypoints (or On Picked KPs)**
- **Main Menu > Solution > Define Loads > Apply >** `load type` **> On Line**

2.2.4.6 Surface Loads

Following are examples of some of the GUI paths used for applying surface loads:

- **Main Menu > Preprocessor > Loads > Define Loads > Apply >** `load type` **> On Nodes**
- **Utility Menu > List > Loads > Surface > On All Elements (or On Picked Elements)**
- **Main Menu > Solution > Define Loads > Apply >** `load type` **> On Lines**

The program stores surface loads specified on nodes internally in terms of elements and element faces. Therefore, if both nodal and element surface load commands are used for the same surface, only the last specification is used.

The program applies pressures on axisymmetric shell elements or beam elements on their inner or outer surfaces, as appropriate. In-plane pressure load vectors for layered shells (such asSHELL281) are applied on the nodal plane. KEYOPT(11) determines the location of the nodal plane within the shell. When using flat elements to represent doubly curved surfaces, values which should be a function of the active radius of the meridian will be inaccurate.

Applying Pressure Loads on Beams

Use **Main Menu> Preprocessor> Loads> Define Loads> Apply> Structural> Pressure> On Beams** or **Main Menu> Solution> Define Loads> Apply> Structural> Pressure> On Beams** to apply pressure loads on the lateral faces and the two ends of beam elements.

Lateral pressures, which have units of force per unit length, can be applied both in the normal and tangential directions. The pressures may vary linearly along the element length, and can be specified on a portion of the element, as shown in Fig. 2.12. The analyst can also reduce the pressure down to a force at any location on a beam element by setting the `JOFFST` field to -1. End pressures have units of force.

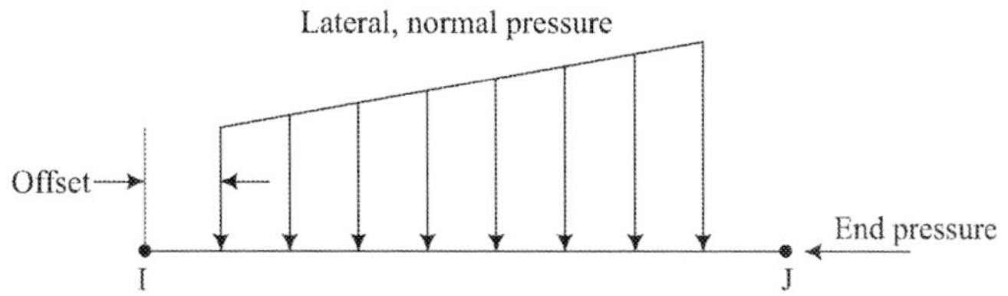

Fig. 2.12 Example of beam surface loads.

2.2.4.7 Applying Body Loads

Following are examples of some of the GUI paths used for applying body loads:

- **Main Menu> Preprocessor> Loads> Define Loads> Apply>** `load type`**> On Nodes**
- **Utility Menu> List> Loads> Body> On Picked Elems**
- **Main Menu> Solution> Define Loads> Apply>** `load type`**> On Keypoints**
- **Utility Menu> List> Loads> Body> On Picked Lines**
- **Main Menu> Solution> Define Loads> Apply>** `load type`**> On Volumes.**

2.2.4.8 Applying Coupled-Field Loads

A coupled-field analysis usually involves applying results data from one analysis as loads in a second analysis. For example, the nodal temperatures calculated in a thermal analysis can be applied as body loads in a structural analysis. Similarly, magnetic forces calculated in a magnetic field analysis can be applied as nodal forces in a structural analysis. Use **Main Menu> Preprocessor> Loads> Define Loads> Apply>** `load type`**> From source** or **Main Menu> Solution> Define Loads> Apply>** `load type`**> From source** to apply such coupled-field loads.

2.2.4.9 Axisymmetric Loads and Reactions

For constraints, surface loads, body loads, and Y-direction accelerations, loads are defined exactly as they would be for any nonaxisymmetric model. However, for concentrated forces the procedure is a little different. For these quantities, input load values of force, moment, etc., are on a '360° basis'. That is, the load value is entered in terms of *total load around the circumference.* For example, if an axisymmetric axial load of 1500 pounds per inch of circumference were applied to a 10′ diameter pipe (Fig. 2.13), the total load of 47,124 lb. ($1500*2\pi*5 = 47,124$) would be applied to node *N.*

Axisymmetric results are interpreted in the same fashion as their corresponding input loads. That is, reaction forces, moments, etc., are reported on a total load (360°) basis.

Axisymmetric harmonic elements require that their loads be supplied in a form that the program can interpret as a Fourier series. The GUI path **Main Menu> Preprocessor> Loads> Load Step Opts> Other> For Harmonic Ele** or **Main Menu> Solution> Load Step Opts> Other> For Harmonic Ele**, together with other load commands (D, F, SF, etc.) is required for these elements.

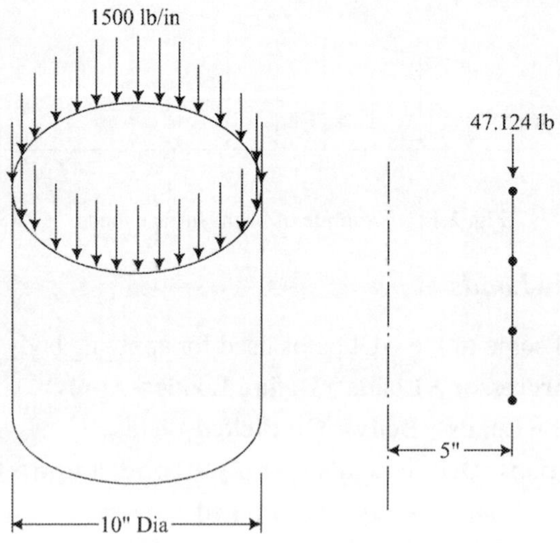

(a) 3-D Structure (b) 2-D model

Fig. 2.13 Concentrated axisymmetric loads.

2.2.5 Specifying Load Step Options

As mentioned earlier, *load step options* is a collective name for options that control how loads are used during solution and other options such as output controls, damping specifications, and response spectrum data. Load step options can vary from load step to load step. There are six categories of load step options:

- General Options
- Nonlinear Options
- Biot-Savart Options
- Dynamics Options
- Output Controls
- Spectrum Options.

These include such options as time at the end of a load step in transient and static analyses, number of substeps or the time step size, stepping or ramping of loads, and reference temperature for thermal strain calculations.

2.2.6 Creating Multiple Load Step Files

All loads and load step options put together form a load step, for which the program can calculate the solution. If there are multiple load steps, store the data for each load step on a file, called the load step file, and read it in later for solution.Use **Main Menu> Preprocessor> Loads> Load Step Opts> Write LS File** or **Main Menu> Solution> Load Step Opts> Write LS File.**

If the Solution Controls dialog box is used to set the analysis and load step options, define each load step using the Basic tab. After all load step files are written, use one action command to read in the files sequentially and obtain the solution for each load step.

2.3 SOLUTION

In the solution phase of an analysis, the computer takes over and solves the simultaneous set of equations that the finite element method generates. The results of the solution are:

- Nodal degree of freedom values, which form the primary solution
- Derived values, which form the element solution.

The element solution is usually calculated at the elements' integration points. The program writes the results to the database as well as to the results file (`.RST`, `.RTH`, or `.RMG` files).

2.3.1 Selecting a Solver

Several methods for solving the system of simultaneous equations are available in the program: sparse direct solution, Preconditioned Conjugate Gradient (PCG) solution, Jacobi Conjugate Gradient (JCG) solution, Incomplete Cholesky Conjugate Gradient (ICCG) solution, and Quasi-Minimal Residual (QMR) solution. In addition, distributed versions of the sparse, PCG, and JCG solvers are available for use in Distributed ANSYS.

Select a solver using: **Main Menu>Preprocessor> Loads> Analysis Type> Analysis Options** or **Main Menu> Solution> Load Step Options>Sol'n Control (: Sol'n Options Tab)** or **Main Menu> Solution> Analysis Options** or **Main Menu> Solution> Unabridged Menu> Analysis Options**

2.3.2 Obtaining the Solution

Use **Main Menu> Solution> Current** LS to initiate the solution. Because the solution phase generally requires more computer resources than the other phases of an analysis. It is better suited to batch mode than interactive mode.

The solver writes output to the output file (`Jobname.OUT`) and the results file. If the solution is run interactively, the output 'file' is actually the screen. By using one of the following before issuing **SOLVE**, the output can be diverted to a file instead of the screen: **Utility Menu> File> Switch Output to> File** or **Output Window**

Data written to the output file consist of the following:

- Load summary information
- Mass and moments of inertia of the model
- Solution summary information
- A final closing banner that gives total CPU time and elapsed time
- Data requested by the **OUTPR** output control command or its GUI counterpart

In interactive mode, much of the output is suppressed. The results file (`.RST, .RTH, or .RMG`) contains all results data in binary form, which can be reviewed in the postprocessors.

Another useful file produced during solution is `Jobname.STAT`, which gives the status of the solution. This file can be used to monitor an analysis while it is running. It is particularly useful in iterative analyses such as nonlinear and transient analyses.The **SOLVE** command calculates the solution for the load step data currently in the database.

2.3.3 Solving Multiple Load Steps

There are three ways to define and solve multiple load steps:

- Multiple **SOLVE** method
- Load step file method
- Array parameter method.

2.3.3.1 Using the Multiple SOLVE Method

This method is the most straightforward. It involves issuing the **SOLVE** command after each load step is defined. The main disadvantage, for interactive use, is that the analyst has to wait for the solution to be completed before defining the next load step.

2.3.3.2 Using the Load Step File Method

The load step file is a convenient method to solve problems while the analyst is away from the terminal. It involves writing each load step to a load step file (via the **LSWRITE** command or its GUI equivalent) and, with one command, reading in each file and obtaining the solution.

Issue **Main Menu> Solution> From LS Files** to solve multiple load steps. This reads in the load step files sequentially and initiates the solution for each load step.

2.3.3.3 Using the Array Parameter Method

This method, mainly intended for transient or nonlinear static analyses, requires knowledge of array parameters and do-loops, which are part of ANSYS Parametric Design Language (APDL). The array parameter method involves building tables of load versus time using array parameters.

2.4 OVERVIEW OF POSTPROCESSING

After building the model and obtaining the solution, the analyst wants answers to some critical questions: Will the design really work when put to use? How high are the stresses in this region? How does the temperature of this part vary with time? What is the heat loss across this face of my model? How does the magnetic flux flow through this device? How does the placement of this object affect fluid flow? The postprocessors in the ANSYS program can help answer these questions and others.

Postprocessing means reviewing the results of an analysis. It is probably the most important step in the analysis, because the analyst is trying to understand how the applied loads affect the design, how good the finite element mesh is, and so on.

2.4.1 Postprocessors Available

Two postprocessors are available for reviewing the results: POST1, the general postprocessor, and POST26, the time-history postprocessor. POST1 allows the analyst to review the results over the entire model at specific load steps and substeps. In a static structural analysis, for example, s/he can display the stress distribution for load step 3. Or, in a transient thermal analysis, s/he can display the temperature distribution at time = 100 seconds. Figure 2.14 is a typical example of a POST1 plot.

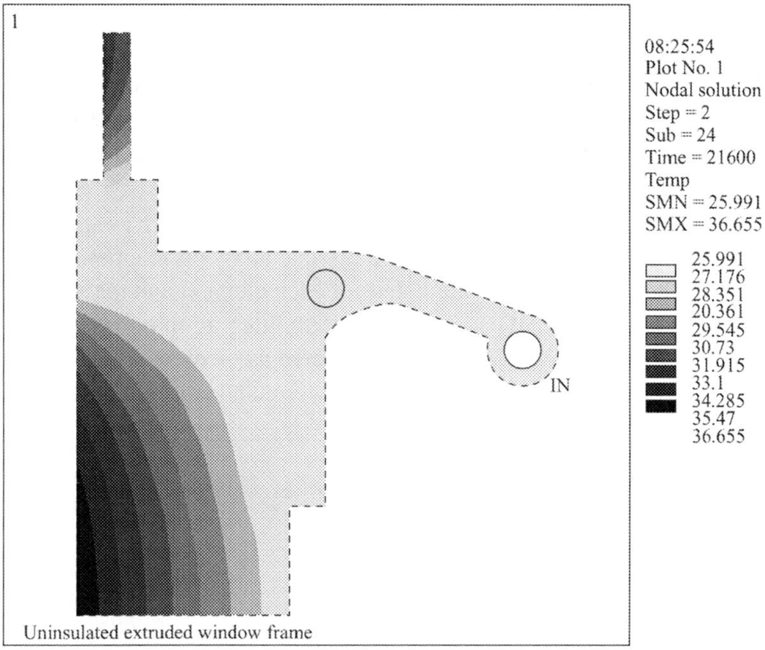

Fig. 2.14 A typical POST1 contour display.

POST26 allows the analyst to review the variation of a particular result item at specific points in the model with respect to time, frequency, or some other result item. In a transient magnetic analysis, for instance, the analyst can graph the eddy current in a particular element versus time. Or, in a nonlinear structural analysis, s/he can graph the force at a particular node versus its deflection. A typical POST26 graph is shown in Fig. 2.15.

It is important to remember that the postprocessors in ANSYS are just tools for reviewing analysis results. The analyst still needs to use engineering judgment to interpret the results. For example, a contour display may show that the highest stress in the model is 37,800 psi. It is now up to her/him to determine whether this level of stress is acceptable for the design.

2.4.2 Results Files

The analyst can use **OUTRES** to direct the ANSYS solver to append selected results of an analysis to the results file at specified intervals during solution. The name of the results file depends on the analysis discipline:

- `Jobname.RST` for a structural analysis and coupled-field analysis
- `Jobname.RTH` for a thermal and diffusion analyses
- `Jobname.RMG` for a magnetic field analysis.

For fluid analyses, the file extension is `.RST` or `.RTH`, depending on whether structural degrees of freedom are present.

2.4.3 Types of Data Available for Postprocessing

The solution phase calculates two types of results data – *primary data* and *derived data. Primary data* consist of the degree-of-freedom solution calculated at each node: displacements in a structural analysis, temperatures in a thermal analysis, magnetic potentials in a magnetic analysis, and so on. These are also known as nodal solution data.

Derived data are those results calculated from the primary data, such as stresses and strains in a structural analysis, thermal gradients and fluxes in a thermal analysis, magnetic fluxes in a magnetic analysis, and the like. They are typically calculated for each element and may be reported at any of the following locations: at all nodes of each element, at all integration points of each element, or at the centroid of each element. Derived data are also known as element solution data, except when they are averaged at the nodes. In such cases, they become nodal solution data.

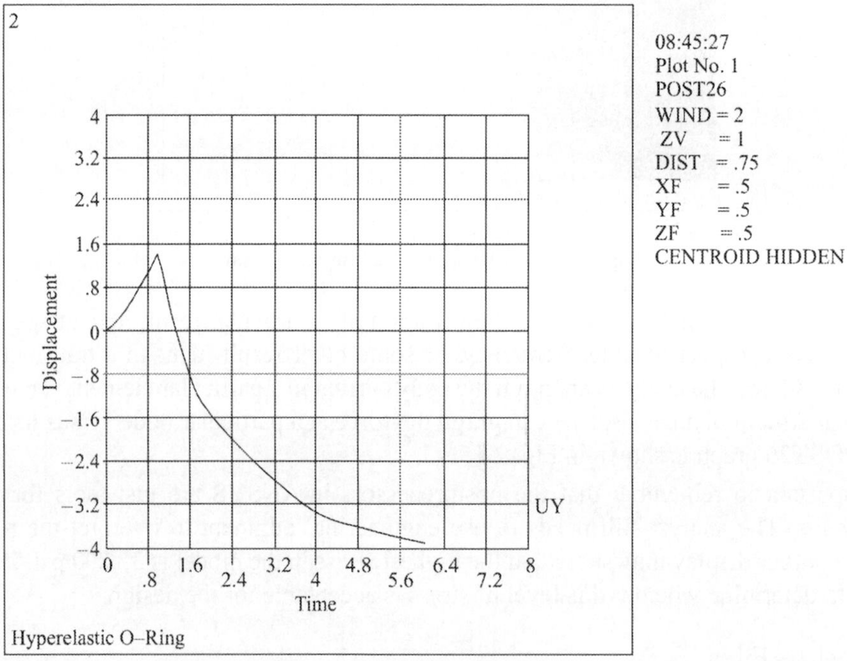

Fig. 2.15 A typical POST26 graph.

2.5 THE GENERAL POSTPROCESSOR (POST1)

Use POST1 is the general postprocessor, to review analysis results over the entire model, or selected portions of the model, for a specifically defined combination of loads at a single time. POST1 has many capabilities, ranging from simple graphics displays and tabular listings to more complex data manipulations such as load case combinations. Issue **Main Menu> General Postproc** to enter the general postprocessor.

2.5.1 Reading Results Data into the Database

The first step in POST1 is to read data from the results file into the database. Model data must exist in the database to do so. If the database does not already contain model data, issue **Utility Menu> File> Resume Jobname.db** to read the database file, `Jobname.DB`. The database should contain the same model for which the solution was calculated, including the element types, nodes, elements, element real constants, material properties, and nodal coordinate systems.

2.5.2 Reviewing Results in POST1

Once the desired results data are stored in the database, the analyst can review them through graphics displays and tabular listings. In addition, s/he can map the results data onto a path.

2.5.2.1 Displaying Results Graphically

Graphics displays are perhaps the most effective way to review results. The following types of graphics can be displayed in POST1:

- Contour displays
- Vector displays
- Reaction force displays
- Deformed shape displays
- Path plots
- Particle flow traces.

2.5.2.2 Listing Results in Tabular Form

An effective way of documenting analysis results is to produce tabular listings in POST1. Listing options are available for nodal and element solution data, reaction data, element table data, and more.

Listing Nodal and Element Solution Data

Use **Main Menu> General Postproc> List Results> Nodal Solution** to list specified nodal solution data. Use **Main Menu> General Postproc> List Results> Element Solution** to list specified results for selected elements. Specify the ELEM option with **PRESOL** to obtain line element solution printout. The program will list all applicable element results for the selected elements.

Listing Reaction Loads and Applied Loads

There are several options in POST1 for listing reaction loads and applied loads. The GUI path **Main Menu> General Postproc> List Results> Reaction Solu** lists reactions at constrained nodes in the selected set. The GUI path **Main Menu> General Postproc> List Results> Nodal Loads** lists the summed element nodal loads for the selected nodes, except for any zero values.

Another useful command is **Main Menu> General Postproc> Nodal Calcs> Total Force Sum**. It calculates and lists the force and moment summation for the selected set of nodes.

Listing Element Table Data

Use **Main Menu> General Postproc> Element Table> List Elem Table** or **Main Menu> General Postproc> List Results> Elem Table Data** to list specified data stored in the element table. Use **Main Menu> General Postproc> Element Table> Sum of Each Item** to list the sum of each column in the element table.

2.5.2.3 Using the Results Viewer to Access Results File Data

The Results Viewer (Fig. 2.16) is a compact toolbar for viewing the analysis results. Selecting the Results Viewer disables much of the standard GUI functionality. Many of these operations are not available because of PowerGraphics limitations. However, a good deal of the POST1 functionality is contained in the Result Viewer menu structure, and in the right and middle mouse button context sensitive menus that are accessible in the Results Viewer.

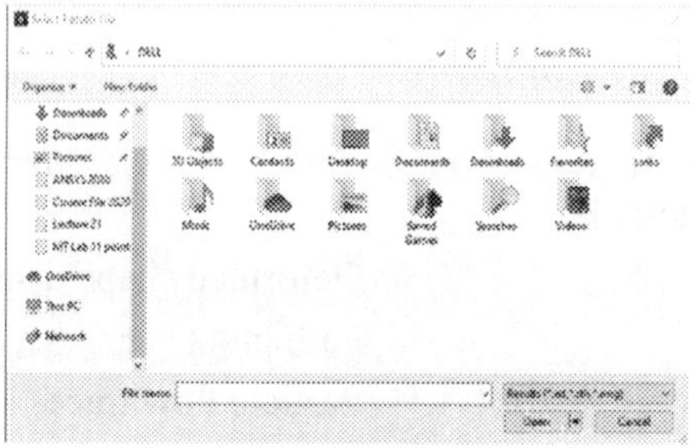

Fig. 2.16 Results Viewer.

The Results Viewer can be used to access any data stored in a valid results file (such as `*.RST`, `*.RTH`, and `*.RMG`). Because the viewer can access results data without loading the entire database file, it is an ideal location from which to compare data from many different analyses.

2.6 TIME-HISTORY POSTPROCESSOR (POST26)

Use the time-history postprocessor to review analysis results at specific locations in the model as a function of time, frequency, or some other change in the analysis parameters that can be related to time. In this mode, the analyst can process results data in many ways. S/he can construct graphics displays, chart representations or tabular listings, or perform math operations on the data sets. A typical time-history task would be to graph result items versus time in a transient analysis, or to graph force versus deflection in a nonlinear structural analysis.

Following is the general process for using the time-history postprocessor:

1. Start the time-history processor, either interactively or via the command line.
2. Define time-history variables. This involves not only identifying the variables, but also storing the variables.
3. Process the variables to develop calculated data or to extract or generate related variable sets.
4. Prepare output. This can be done via graph plots, tabular listings or file output.

2.6.1 Entering the Time-History Postprocessor

The analyst enters the time history processor to process time or frequency related results data. Once an analysis is solved, ANSYS uses the results data to create a 'Results File.' The active results file (`*.RST`, `*.RTH`, `*.RMG`, etc.) is automatically loaded when the analyst enters postprocessing. If the current analysis contains no results file, the program queries for one. The file option can also be used to load any other results file for processing.

2.6.2 Importing Data

This feature allows the analyst to read in set(s) of data from a file into time history variable(s). This enables the user, for instance, to display and compare test results data against the corresponding ANSYS results data.

2.6.3 Exporting Data

This feature allows the analyst to write out selected time history variable(s) to an ASCII file or to APDL array/table parameter. This enables to perform several functions such as pass data on to another program for further processing or to archive data in an easily retrievable format.

2.6.4 Reviewing the Variables

Once the variables are defined, they can be reviewed via graph plots or tabular listings.

2.6.4.1 Plotting Result Graphs

The description for graph plotting, both with the variable viewer and from the command line follows:

Interactive

The 'Graph Data' button in the variable viewer allows to plot all the selected variables. A maximum of 10 variables can be plotted on a single graph. By default, the variable used for the X-axis of the graphs is TIME for static and transient analyses or FREQUENCY for harmonic analysis. The analyst can select a different variable for the X-axis of the graph using the radio button under the column X-AXIS in the list of variables.

When plotting complex data such as from a harmonic analysis, use the 'results to view' drop-down list on the right top corner of the variable viewer to indicate whether to plot Amplitude (default), Phase angle, Real part or Imaginary part.

The variable viewer stores all the time points available on the results file. A portion of this data can be displayed by selecting a range for the X-axis value. This is useful when it is necessary to focus on the response around a certain time point, e.g., around the moment of impact in a drop test analysis. This is available in the 'Data Properties' dialog under the X-AXIS tab. Note that this is a global setting, i.e., this setting is used for all subsequent graph plots.

2.6.4.2 Listing Results in Tabular Form

Use the following procedures to create tabular data lists, interactively:

Interactive

The 'List Data' button of the variable viewer can be used to list up to six variables in the variable viewer.

When listing complex data such as from a harmonic analysis, use the 'results to view' drop-down list on the right top corner of the variable viewer to indicate whether to printout 'amplitude and phase angle' or 'real and imaginary parts' in the listing. Select amplitude or phase to list 'Amplitude and Phase Angle' results. Select real or imaginary to list 'Real and Imaginary' results.

Data being listed can be restricted to a range of time or frequency. This and other listing controls are available through the 'Lists' tab under Data Properties dialog. In addition to setting the range of time or frequency, this dialog also allows to:

- Control the number of lines before repeating headers on the listings.
- Additionally, print the extreme values of the selected variables.
- Specify printing every 'n'th data point.

2.7 SELECTING AND COMPONENTS

2.7.1 Selecting

Selecting (Fig. 2.17) is an important and fundamental concept in ANSYS. Selected entities are the active entities. All operations (including Solving) are performed on the selected set. In many operations items are selected on the fly; ANSYS prompts for what volumes to mesh, for example, analyst picks them with the mouse, and ANSYS does the meshing. However, there are many times when it is necessary to select things in more sophisticated ways. Also, in an ANSYS input file or batch file things cannot be selected with the mouse! Examples, where this would be useful:

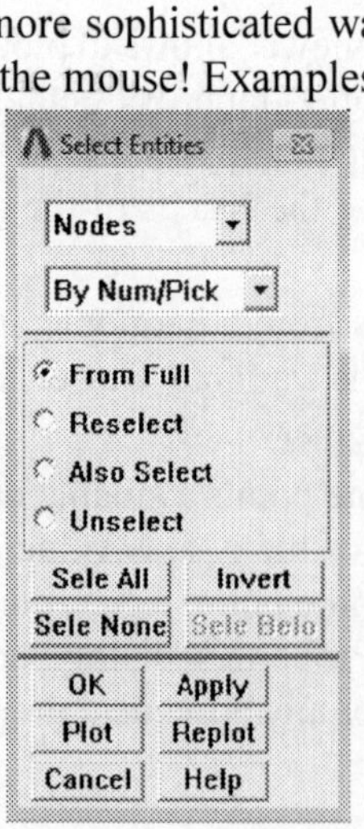

Fig. 2.17

- There are many different areas at Z = 0 an analyst would want to constrain. He selects them all one by one when applying the constraint, or selects 'By Location' beforehand, then says 'Pick All' in the picking dialog.
- There is a structure with many fastener holes that analyst wants to constrain. Again, he could select them all one by one when applying the constraint, or select lines 'By Length/ Radius', type in the radius of the holes to select all of them in one shot, then 'Pick All' in the picking dialog when applying the constraint.

After working with the selected set, use **Utility Menu> Select> Everything** to make the whole model active again. Select Entities Dialog Box Terminology (Fig. 2.18):

- *From Full*: Select from the entire set of entities in the model.
- *Reselect:* Select a subset from the currently selected entities.
- *Also Select:* Select in addition to (from the whole model) the set analyst has currently selected.
- *Unselect:* Remove items from the selection set.
- *Select All:* This is not the same as **Utility Menu> Select> Everything.** This selects all of whatever entity analyst has specified at the top of the dialog.
- *Invert:* Reverses the selected and unselected entities (just the entities specified at the top of the dialog).
- *OK:* This does the select operation (or brings up a picker dialog so that analyst can pick with the mouse) and then dismisses the dialog.
- *Apply:* This does the operation but keeps the dialog box. Typically use this so the dialog stays active.
- *Replot:* Replots whatever is active in the graphics window.
- *Plot:* Plots only the entity specified at the top of the dialog.

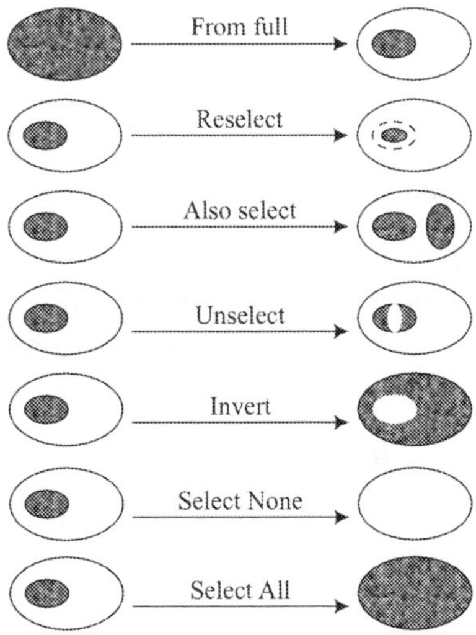

Fig. 2.18

2.7.2 Organizing the Model using Components

If a group of entities is selected and it might be necessary to use that selection set again, create a component out of it. Components are groups of entities but hold only one kind of entity at a time. Components can themselves be grouped into Assemblies, so this is the way to group different types of entities together. Use **Utility Menu > Select > Comp/Assembly > Create Component…** to create a component. The Component Manager (Fig. 2.19) makes it very easy to manage and manipulate groups and select/plot what analyst wants to see to the screen. This is found under **Utility Menu > Select > Component Manager**.

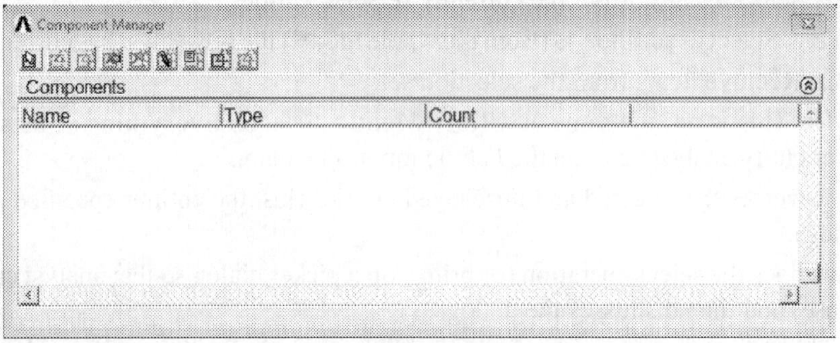

Fig. 2.19 Component Manager.

2.8 GETTING STARTED WITH GRAPHICS

The program enables the analyst to portray almost any aspect of the model in pictures or graphs that s/he can view on the terminal screen, store on a file, or plot out as hard copy. The program has numerous features to help customize or enhance the graphics displays.

2.8.1 Interactive versus External Graphics

Any discussion of graphics assumes that the analyst is running the program interactively and viewing graphics images on the terminal screen. For the most part, this section is written for such a scenario. However, the program can be run in either interactive or batch mode and graphics images stored on a file for later viewing and processing. This process is called creating *external graphics*.

2.8.2 Specifying the Graphics Display Device Type

For Windows users, the program supports these drivers and capabilities:

- A window device
- Hot keyboard/mouse
- Two- or three-button mouse
- Hidden line removal
- Light source shading

On a two-button mouse, the shift-right button functions like the middle button of a three-button mouse. While running the program on Windows platforms, there are three alternatives for specifying the graphics device type:

- Double-click on the Interactive icon in the Program Folder. Click on the down arrow next to **Graphics device name** and choose the appropriate device.
- Within the program, issue **Utility Menu> PlotCtrls> Device Options**.
- Include the device type on the program execution command line. The command option - **d** or **-D** must precede the device type, as shown below:

```
Ansys201 -d device_type
```

The device type is one of the following:

- WIN32
- WIN32c
- 3D

A colour setting higher than 256 colours is recommended.

Specifying an invalid device type causes the program to divert the graphics to a disk file and inhibits the opening of the menu system, even if the **-g** option is included on the program execution command.

2.8.3 Creating Graphics Displays

Many types of graphics displays can be created: geometry displays (nodes, elements, keypoints, etc.), results displays (temperature or stress contours, etc.), and graphs (stress-strain curves, time-history displays, etc.). Creating any display is a two-step process:

1. Use graphics *specification* functions to establish specifications (such as the viewing direction, number and color controls, etc.) for the display.
2. Use graphics *action* functions to actually produce the display.

The analyst can perform both types of graphics functions either by using menu functions in the GUI or by typing in commands directly.

2.8.3.1 GUI-Driven Graphics Functions

When running the program interactively, most users will prefer to use the GUI. As they use the GUI functions, they execute commands without actually seeing or editing them. (The program will record all underlying executed commands in your `Jobname.LOG` file.) Graphics specification functions can be accessed via **Utility Menu> PlotCtrls**. Graphics action functions reside under **Utility Menu> Plot**.

2.8.3.2 Immediate Mode Graphics

By default in the GUI, the model will immediately be displayed in the Graphics Window as new entities (such as areas, keypoints, nodes, elements, local coordinate systems, boundary conditions, etc.) are created. This is called *immediate mode* graphics. Anything drawn immediately in this way, however, will be destroyed if the analyst brings up a menu or dialog box on top of it. Or, if the GUI is iconified, the immediate mode graphics image will not be shown when the analyst restores the GUI icon.

An immediate image will also be automatically scaled to fit nicely within the Graphics Window, a feature called *automatic scaling*. Periodically, though, the analyst may need to issue an explicit plot function because new entities which lie 'outside' the boundaries of the scaled image are created already in the Graphics Window and are thus not captured with immediate mode graphics. The plot function will rescale and redraw the image.

To obtain a more 'permanent' image, execute one of the plot functions (such as **Utility Menu> Plot> Volumes**) or a graphics action command (such as VPLOT). An image generated in this way will not be destroyed by menu pop-ups or by iconifying the GUI. Also note that symbols (such as keypoint or node numbers, local coordinate systems, boundary conditions, etc.) are also shown immediately but will not be present on a 'permanent' display unless the analyst first 'turns on' the appropriate symbol using the functions under **Utility Menu> PlotCtrls** or the appropriate graphics specification command.

If the analyst prefers not to see things immediately as s/he defines them, s/he can use **Utility Menu>PlotCtrls> Erase Options> Immediate Display** to turn off immediate mode. When the program is run interactively without using the GUI, immediate mode is off by default.

2.8.3.3 *Replotting the Current Display*

The GUI path **Utility Menu> Plot>Replot** re-executes the last display action command that was executed. However, the program can execute that command only if it is valid in the current routine.

2.8.3.4 *Erasing the Current Display*

The current graphics display can be cleared by issuing the **Utility Menu> PlotCtrls> Erase Options> Erase Screen.** GUI menus will not be erased, however.

2.8.3.5 *Aborting a Display in Progress*

Invoke the system 'break' to terminate a display before it is completed. Typically, this means moving the mouse pointer to the Output Window and typing Ctrl+C. However, the specific procedure varies from system to system. This break **must be executed** while the display is visibly in progress, or else the entire session will terminate.

2.8.4 Multi-Plotting Techniques

The multi-plotting capabilities within the program enable to display both multiple entities within a window and multiple windows with varying entity types. Defining each window's composition is a four-step process:

1. Define the window layout.
2. Choose the entities to be displayed in each window.
3. Choose the type of element or graph display used for plots.
4. Display the entities selected.

2.8.4.1 *Defining the Window Layout*

First define how many windows are to be used for plotting and how those windows appear on the screen. The following layout options are available:

- One window
- Two windows (left and right of the screen, or top and bottom)
- Three windows (two at the top of the screen and one at the bottom, or one window at the top and two windows at the bottom)
- Four windows (two at the top of the screen and two at the bottom)

Issue the **Utility Menu> PlotCtrls> MultiWindow Layout** to define the window layout. The program displays a dialog box, in which the analyst can click on the layout s/he prefers. That dialog box also contains a **Display upon OK/Apply** field, where the analyst also can specify what the program displays next. Choices for this field are **Multi-Plots, Replot**, and **No redisplay**. When specifying the layout design is finished, click on **Apply** or **OK**.

2.8.4.2 *Choosing What Entities Each Window Displays*

Once the window layout is designed, choose what entities each window will display. To do so, use: **Utility Menu> PlotCtrls> Multi-Plot Controls**. A dialog box appears. In its **Window to edit** field, click on either **All window** or a specific window number (default is window 1). In the **Display type** field, choose either **Entity plots** or **Graph plots**. Then, click on **OK**. If **Entity plots** is choosen, another dialog box appears, listing the types of entities available for display. For Label, specify any of these entity types:

- NODE (nodes)
- ELEM (elements)
- KEYP (keypoints)
- LINE (lines)
- AREA (areas)
- VOLU (volumes)
- GRPH (graph displays)

2.8.4.3 *Choosing the Display Used for Plots*

When displaying either the ELEM or GRPH entity type, the analyst can control the type of element or graph display used for plots. Use: **Utility Menu>PlotCtrls> Multi-Plot Controls**. 'ALL'can be specified to have all windows use the selected display type, or that display type can be applied only to a specific window (default is window 1).

2.8.4.4 *Displaying Selected Entities*

Issue **Utility Menu> PlotCtrls> Multi-Plots** or **Utility Menu> Plot> Replot** to display the entities selected.

2.9 CREATING GRAPHS

A graph is used to review the material property curves, trace the time-history response of the system, or examine the relationship between any two items in the analysis. ANSYS graphs can be either 2-D (X-Y) or 3-D (X-Y-Z, where Z must always be *TIME*). Figure 2.20 shows two typical graphs.

The most convenient way to create and control graph displays is by using the GUI operations available under **Utility Menu> Plot** and **Utility Menu> PlotCtrls**. Alternatively, graphics action and control commands can be used.

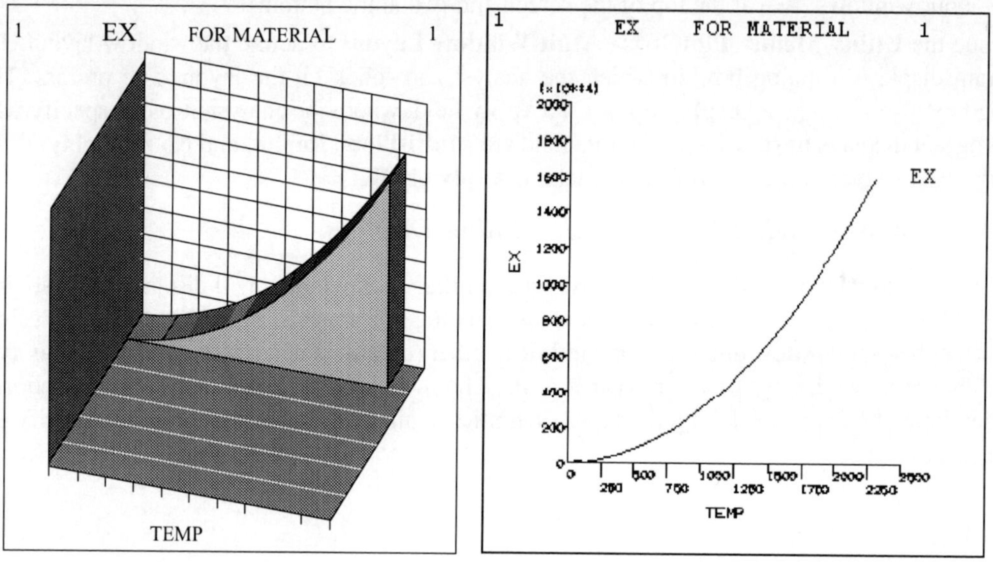

Fig. 2.20 Typical ANSYS graphs.

2.9.1 Graph Display Actions

Use **Utility Menu> Plot> Materials** to display linear material properties as a function of temperature. Use **Utility Menu> Plot> Data Tables** to display nonlinear data curves family of commands. Use **Utility Menu> Plot> Array Parameters** to display column vectors of array parameters.Use **Main Menu> General Postproc> Fatigue> Store Stresses> Plot Stresses** to display a stress item associated with a particular location and event versus loading number.

2.9.2 Changing the Specifications for Graph Displays

2.9.2.1 Changing the Type, Style, and Color of the Graph Display

The appearance of the graph display can be altered as follows:

Turning axis divisions (tick marks) on or off. This feature can be controlled using the command **Utility Menu> PlotCtrls> Style> Graphs.**

Setting line thickness for axes, grid lines or graph curve lines. Graph items can be accentuated by increasing their line thickness, using the AXIS, GRID, and CURVE labels using **Utility Menu> Plot Ctrls> Style> Graphs.**

Turning the grid on or off (in the XY plane). A grid can be added to the graph displays, using **Utility Menu> Plot Ctrls> Style> Graphs.** A grid can be either a full grid or a partial grid.

Producing a dashed tolerance curve about the displayed curve. A range of data spread, tolerance, or uncertainty can be indicated on the graph curves using **Main Menu> TimeHist Postpro> Settings> Graph**.

Changing the color of curves (and color-filled areas under curves). The CURVE label on the **Utility Menu> PlotCtrls> Style> Colors>** `color type` allows controlling to control the color of each curve in the graph.

2.9.2.2 Labeling the Graph

Labeling the axes. The X and Y axes can be labeled using **Utility Menu> PlotCtrls> Style> Graphs**.

Labeling the curves. For POST26 plotted-variable graphs, the labels applied to the curves are established choosing: **Main Menu> TimeHist Postpro> Define Variables** or **Main Menu> TimeHist Postpro> Elec&Mag> Circuit> Define Variables.**

Adding user-defined graphics and text. Extra graphics and text can be added to the displays using the annotation functions by choosing **Utility Menu> PlotCtrls> Annotation**.

2.10 ANNOTATION

A common step in the analysis process is presenting model and results data with additional notations applied, such as dimensions, comments, highlights, or other text or artwork. The standard ANSYS display can be enhanced with a variety of annotation primitives including text, dimensions, polygons, symbols, and even pie charts. The '!' and '$' characters are not available for text annotation.

2-D text and graphics annotations are formed as a 2-D overlay on the graphics screen. Because this overlay exists as an imaginary plane, when the analyst transforms the model, the constructed annotation will not move with the model. Because of this, 2-D annotation should be used primarily for finalized reports and printouts. Access 2-D annotation functions through **Utility Menu> PlotCtrls> Annotation> Create 2D Annotation**.

The following annotation primitives are available from the 2-D annotation dialog box:

- Text
- Dimensions
- Polygons
- Rectangles
- Symbols
- Arrows
- Arcs
- Lines
- Pies
- Wedges
- Circles

On choosing **Utility Menu> PlotCtrls> Annotation> Create 2D Annotation**, the text annotation dialog box appears (Fig. 2.21). Text annotation can be applied either as stroke text or as bitmap fonts. The '!' and '$' characters can't be used in ANSYS text annotation.

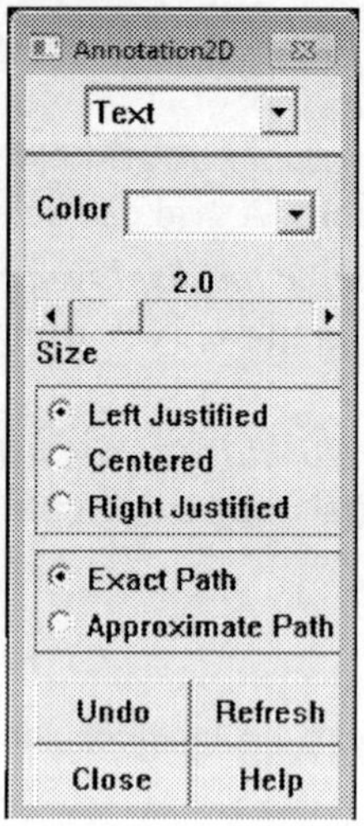

Fig. 2.21 Stroke text annotation dialog box.

The fields and buttons presented in the annotation dialog box change when the annotation entity type is reset. For example, if the annotation entity is reset to arcs, the dialog box shown, changes to display the options available for annotation arcs. Regardless of which annotation entity is chosen, the annotation dialog box always displays four action buttons:

- **Undo** – Erases the last annotation entity created.
- **Refresh** – Redisplays the annotation, which is useful after move and delete operations.
- **Close** – Closes the annotation dialog box.
- **Help** – Displays online help for the dialog of the currently selected annotation entity.

Once annotations are created, the analyst can control their display by selecting **Utility Menu> PlotCtrls> Annotation> Display Annotation.** Accessing this menu pick toggles annotation display on and off.

2.11 ANIMATION

Animation is a valuable tool for graphically interpreting many analysis results, especially nonlinear or time-dependent behavior. The ANSYS program provides tools that enable the analyst to animate any type of display.

2.11.1 Creating Animated Displays within ANSYS

The easiest way to perform animation in ANSYS is to use the functions available under **Utility Menu> PlotCtrls> Animate**. These GUI functions allow to achieve'push-button animation' effects in ANSYS. The GUI functions internally execute ANSYS animation commands, which the analyst can type in directly if prefered.

2.11.2 Basic Animation Commands

Several frames can be displayed in rapid succession to achieve an animation effect, via these commands:

- **Utility Menu> PlotCtrls> Redirect Plots> Delete Segments**
- **Utility Menu> PlotCtrls> Redirect Plots> Segment Status**
- **Utility Menu> PlotCtrls> Redirect Plots> To Animation File** (Windows)
- **Utility Menu> PlotCtrls> Animate> Replay Animation**
- **Utility Menu> PlotCtrls> Animate> Replay Animation**

2.12 DOCUMENTING THE ANALYSIS

ANSYS has a Report Generator (Fig. 2.22) available from the main toolbar, which can help to put together an HTML report by capturing images, window listings, etc.

Fig. 2.22 Report generator.

To print a hardcopy of the graphics window: **Utility Menu > Plot Ctrls> Hard Copy > To Printer**. There are several ways of capturing a graphic image for use in Microsoft Word, Powerpoint or some other software.

Exact screen shot of the graphics window: **Utility Menu> Plot Ctrls> Capture Image** will pop up another window with a screen shot of the graphics window. It can be kept available for later reference or save the image to a bitmap (`.bmp`) file. Note that although a windows bitmap file is not compressed, when it is inserted into Word it does get compressed automatically so analyst doesn't end up with a huge bloated document.

Output a vector image: **Utility Menu> Plot Ctrls> Redirect Plots> To PSCR File....** A Postscript file is a vector file, which means that it is a 2D representation of all of the entities in the graphics window in an editable format. Because it is not a bitmap, it can be scaled to any size without losing any resolution, and is always very crisp looking. It can also be imported into a technical illustration program and manipulated very easily: change the colors, add annotations, change or resize fonts, etc. All this can be done in ANSYS but it can be quicker in an illustration package. One caution about Postscript files! Since they actually write out every entity in the model, if the model is large (say a tet mesh of a CAD model) this file can be huge. It is best suited for getting very crisp images of smallish models or wireframe displays. Microsoft Word will not display the image until it is printed.

Output a bitmap image: **Utility Menu> Plot Ctrls> Redirect Plots> To xyz File...,** where *xyz* is JPEG, TIFF, PNG, etc. These file formats produce good images with reasonably small file sizes. The size of the image file for these formats is not dependent on the size of the model like Postscript.

Controlling the Way the Model Looks

All of the visual aspects of what is seen in the graphics window are controlled from the 'Plot' and 'Plot Ctrls' pull downs from the Utility menu. Use **'Utility Menu> Plot'** to plot different types of entities to the screen. Use **'Utility Menu> Plot Ctrls'** to control the characteristics of what is to be plotted.

'Utility Menu> Plot Ctrls> Numbering': Entity Numbers on and off.

'Utility Menu> Plot Ctrls> Symbols': Turn various markers and symbols on and off (Fig. 2.23).

'Utility Menu> Plot Ctrls> Style': Change hidden line, element edges, element shrink, etc.

'Utility Menu> Plot Ctrls> Device Options': Change between solid shaded and wireframe display.

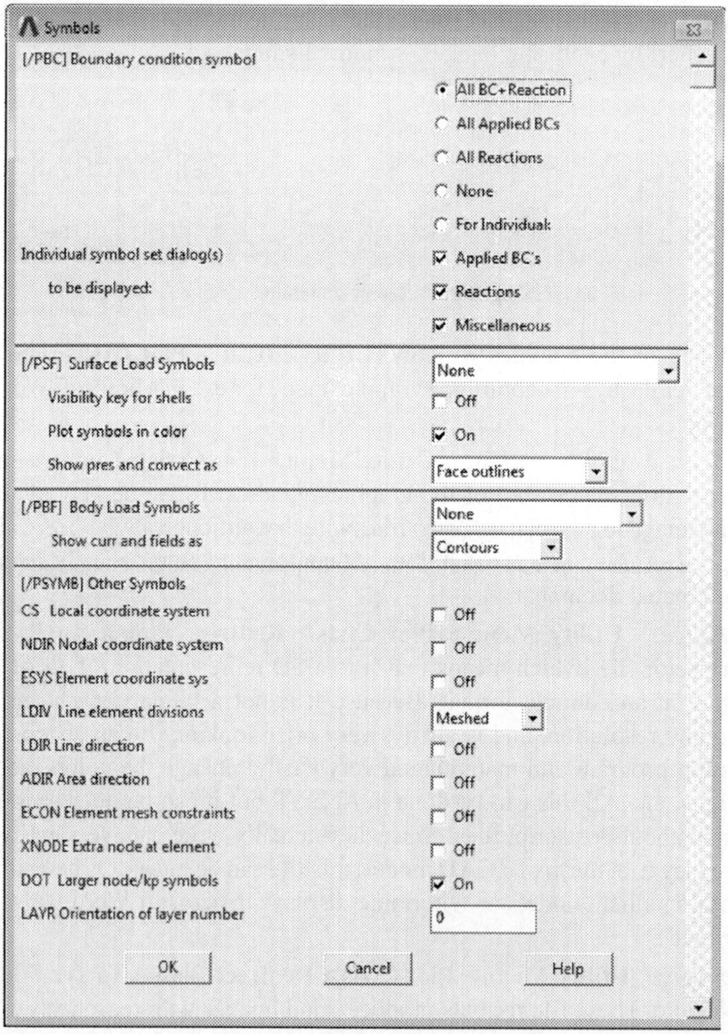

Fig. 2.23

2.13 TYPICAL MODELING DIFFICULTIES

Certain modeling problems can be considered 'typical'. A typical analyst modeling problem is the case of keypoints, lines, areas, volumes, nodes, and elements that are identical and occupy the same space. This can lead to erroneous models. Proper use of the merge command can eliminate many instances of these problems. The merge can fail if, for example, two elements share the same space, but were defined via alternative sequences of nodes (e.g., elements in the same place, one numbered by nodes selected clockwise, the other counterclockwise).

Another problem is failure of keypoints, or lines, or areas to be shared by higher geometric modeling entities. When this happens, the higher entities are not 'fused' or 'welded' together as intended. Consequently, the elements will not share nodes along what should have been the common boundary. The analyst must always use caution and double-check everything while developing a model.

A problem most analysts encounter is to make a change to a model in error, long after the database was saved. The analyst will have to learn to use a text editor on the log file, to extract that portion of the log file after the last time the database was saved, or retrieved (whichever was most recent). Remove the offending command. That portion of the log file will have to be run on the model as it was the last time it was saved or retrieved. Make sure to be in the correct part of ANSYS (usually /PREP7) when reading in the instructions with /INPUT. The same method can apply if the computer is subjected to a power failure, or if ANSYS crashes without leaving an '`ansabort.db`' file. After re-starting, take a text editor to the log file, and re-run the appropriate instructions on the model database file as it was when last saved or retrieved.

The most common of all errors in Finite Element Modeling is the incorrect application of loads and boundary conditions. This must be thought about very carefully. Most models (not all) are prevented from undergoing free body motion in 2-D or 3-D space, by eliminating at least a minimal number of degrees of freedom (2 translations plus 1 rotation in 2-D, and 3 translations plus 3 rotations in 3-D). Rotations can be prevented either by having constraints on translations at enough distinct nodes in space, or by directly constraining a rotational degree of freedom at a node. A common check on results is to see whether the sums of the reaction forces at the constrained nodes equal the sums of the applied forces and gravity loads.

3 CHAPTER MODELING AND MESHING

3.1 MODEL GENERATION

The function of finite element analysis is to rebuild mathematically the behavior of a real engineering system. The analysis must be a precise mathematical model of the system. This model comprises the nodes, elements, material properties, real constants, boundary conditions, and other features that are used to represent the physical system.

3.1.1 What is Model Generation?

Model generation usually takes on the narrower meaning of generating the nodes and elements that represent the spatial volume and connectivity of the real system. Thus, *model generation* involves the process of *defining the geometric configuration of the model's nodes and elements*. ANSYS offers the following approaches to model generation:

- Creating a solid model using ANSYS
- Using direct generation
- Importing a model created in a computer-aided design (CAD) system.

3.1.2 Steps Involved in Model Generation

A general modeling session might follow this common outline:

- Start by *planning the approach*. Find out the objectives, choose what basic form the model will take, decide appropriate element types, and consider how to set up an appropriate mesh density. Carry out general planning before initiating the session.
- Enter the preprocessor (PREP7) to begin the model-building session. Most often, the model is built using *solid modeling* procedures.
- Establish a *working plane*.
- Create basic geometric features using *geometric primitives* and *Boolean operators*.
- Activate a suitable *coordinate system*.
- Generate other solid model features from the *bottom up*. That is, create *keypoints*, and then define *lines, areas,* and *volumes* as needed.
- Employ more *Boolean operators* or *number controls* to join separate solid model regions together as appropriate.

- Create tables of element attributes.
- Set *element attribute pointers*.
- Set *meshing controls* to set up the desired mesh density if desired. This step is not always required because default element sizes exist.
- Generate nodes and elements by *meshing* the solid model.
- After nodes and elements are generated, add features such as *surface-to-surface contact elements, coupled degrees of freedom,* and *constraint equations*.
- Save the model data to `Jobname.DB`.
- Exit the preprocessor.

3.1.3 Solid Modeling vs Direct Generation

Two different methods can be used to generate a model: *solid modeling* and *direct generation*. With *solid modeling*, describe the geometric boundaries of the model, establish controls over the size and desired shape of the elements, and then instruct the ANSYS program to generate the nodes and elements automatically. In the *direct generation method*, the analyst determines the location of every node and the size, shape, and connectivity of every element prior to defining these entities.

The direct generation method is essentially a hands-on, 'manual' method that requires to keep track of all the node numbers as the finite element mesh is developed. This detailed accounting can become tiresome for large models, leading to the potential for modeling errors. Solid modeling is usually more dominant and flexible than direct generation, and is by and large the preferred choice for generating the model.

The relative advantages and disadvantages of the two approaches are listed here.

Solid Modeling

On the optimistic side, solid modeling is generally more apt for large or complex models, especially 3-D models. This method permits to work with a comparatively small number of entities and permits geometric operations that cannot be completed with nodes and elements. It supports the use of 'primitive' areas and volumes. It also facilitates Boolean operations for 'top down' construction and readily permits modifications to geometry.

However, solid modeling can necessitate large amounts of system time. It can be more taxing, requiring more data entries than direct generation and can 'fail' under certain situations.

Direct Generation

On the positive side, direct generation is handy for small or simple models. It offers complete control over the geometry and numbering of every node and every element.

However, direct generation is usually more time consuming; the volume of data one must work with can become vast. It cannot be used with adaptive meshing. Direct generation can become monotonous, requiring the analyst to pay more concentration to every detail of the mesh.

3.1.4 Importing Solid Models Created in CAD Systems

Instead of generating the solid models using ANSYS, the analyst can produce them in her/his favorite CAD system and then import them into ANSYS, by saving them in the IGES file format. ANSYS comes with IGES support by default. There are Geometry Interfaces available for Pro/E, CATIA, UG, Solidworks, Parasolid, etc. All of these geometry interfaces on the ANSYS Traditional side perform a conversion of the geometry into an ANSYS Neutral File (`.anf`) format, which it then reads in.

Import CAD geometry using '**File > Import**'. Models imported from CAD systems may require general repair if they are not of appropriate quality. Generating a model using a CAD package has the following advantages:

- Repetition of effort can be avoided by using existing CAD models to produce solid models for analysis.
- More familiar tools can be used to create models.

3.2 PLANNING THE APPROACH

As the analyst begins to create a model, s/he will make a number of decisions that determine how the real system will be simulated mathematically. For example: What are the objectives of the analysis? Will s/he model all, or just a portion, of the real system? How much detail will be included in the model? What kind of elements will be used? How dense should the finite element mesh be?

An attempt will be made to balance computational expense against precision of results. The decisions made in the planning stage of the analysis will mainly decide the success or failure of the analysis.

This first step of the analysis depends not on the capabilities of the ANSYS program, but on knowledge, experience, and professional expertize of the analyst. Only s/he can determine what the objectives must be. The objectives established at the start will control the rest of the choices as the model is created.

3.2.1 Choosing a Model Type

The ANSYS model may be categorized as 2-D or 3-D, and as composed of point elements, line elements, area elements, or solid elements. Of course, the analyst can mix different kinds of elements (Fig. 3.1) as required. For example, the analyst might model a stiffened shell structure using 3-D shell elements to represent the skin and 3-D beam elements to represent the ribs.

LINE models can represent 2-D or 3-D beam or pipe structures, as well as 2-D models of 3-D axisymmetric shell structures. Solid modeling usually does not offer much benefit for generating line models; they are more often created by direct generation methods.

2-D SOLID analysis models are used for thin planar structures (plane stress), 'infinitely long' structures having a constant cross section (plane strain), or axisymmetric solid structures. Although many 2-D analysis models are relatively easy to create by direct generation methods, they are usually easier to create with solid modeling.

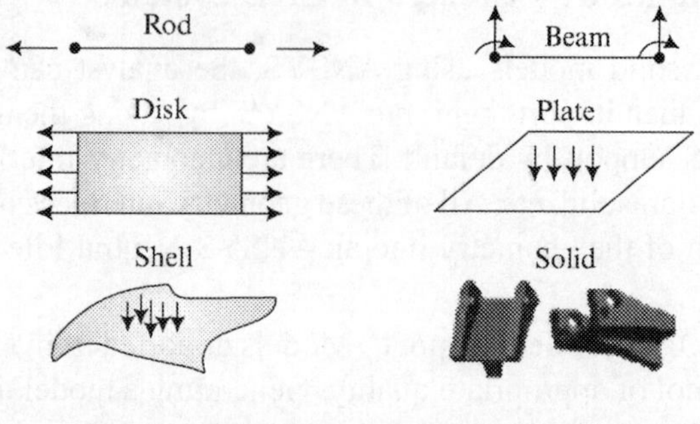

Fig. 3.1 Different types of elements.

3-D SHELL models are used for thin structures in 3-D space. Although some 3-D shell analysis models are relatively easy to create by direct generation methods, they are usually easier to create with solid modeling.

3-D SOLID analysis models are used for thick structures in 3-D space that have neither a constant cross section nor an axis of symmetry. Creating a 3-D solid analysis model by direct generation methods usually requires considerable effort. Solid modeling nearly always makes the job easier.

3.2.2 Choosing btween Linear and Higher Order Elements

The ANSYS program's element library includes two basic types of area and volume elements: *linear* and *quadratic*. These basic element types are represented schematically in Fig. 3.2. Look at some of the points involved in selecting between these two basic element types.

(a) Linear isoparametric (b) Linear isoparametric with extra shapes (c) Quadratic

Fig. 3.2 Area and volume types.

3.2.2.1 Linear Elements

For structural analyses, the corner node elements with extra shape functions will often yield an accurate solution. When using these elements, it is essential to avoid their degenerate forms. That is, avoid using the triangular form of 2-D linear elements and the wedge or tetrahedral forms of 3-D linear elements. Also take care to avoid using excessively distorted linear elements. In nonlinear structural analyses, better accuracy can be obtained at less cost by using a fine mesh rather than a comparable coarse mesh of quadratic elements. Examples of (a) linear and (b) quadratic elements are shown in Fig. 3.3.

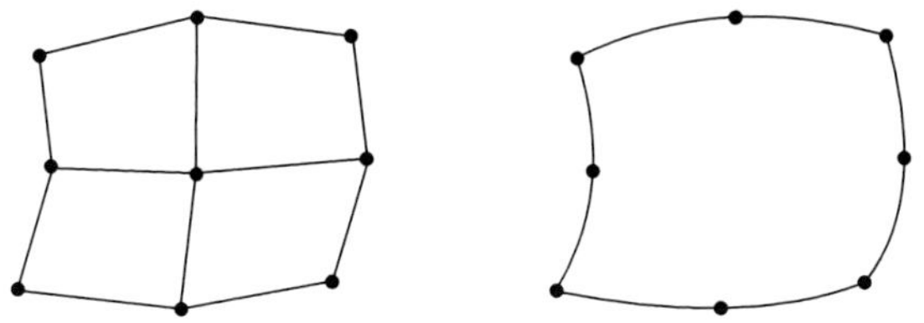

(a) (b)

Fig. 3.3 Comparable grids.

When modeling a curved shell, choose between using curved or flat shell elements. Each option has its own pros and cons.The majority of problems can be solved to a high degree of accuracy in a minimum amount of computer time with flat elements. Take care to ensure that enough flat elements are used to model the curved surface satisfactorily. Obviously, the smaller the element, the higher will be the accuracy.

For most non-structural analyses, the linear elements are nearly as good as the higher order elements, and are less expensive to use. Degenerate elements usually produce accurate results in non-structural analyses.

3.2.2.2 Quadratic Elements (Mid-side Nodes)

For linear structural analyses with degenerate element shapes, the quadratic elements usually give better results at less expense than the linear elements.

3.3 COORDINATE SYSTEMS

The ANSYS program uses several types of coordinate systems, each used for a different purpose:

- Global and local coordinate systems are used to locate geometry items in space.
- The display coordinate system determines the system in which geometry items are listed or displayed.
- The nodal coordinate system defines the degree of freedom directions at each node and the orientation of nodal results data.
- The element coordinate system determines the orientation of material properties and element results data.
- The results coordinate system is used to convert nodal or element results data to a particular coordinate system for listings, displays, or general postprocessing operations (POST1).

The working plane, which is separate from the coordinate systems, is for locating geometric primitives during the modeling process.

3.3.1 Global and Local Coordinate Systems

Global and local coordinate systems are used to locate geometry items. When a node or a keypoint is defined, its coordinates are interpreted in the global Cartesian system by default. For some models, however, it may be more convenient to define the coordinates in a system other than global Cartesian. Input the geometry in any of three predefined (global) coordinate systems, or in any number of user defined (local) coordinate systems.

3.3.1.1 Global Coordinate Systems

A global coordinate system (Fig. 3.4) can be thought of as an absolute reference frame. Three predefined global systems are available: *Cartesian*, *cylindrical*, and *spherical*. All these systems are right-handed and, by definition, share the same origin. They are identified by their coordinate system (CS) numbers: 0 for Cartesian, 1 and 5 for cylindrical, and 2 for spherical.

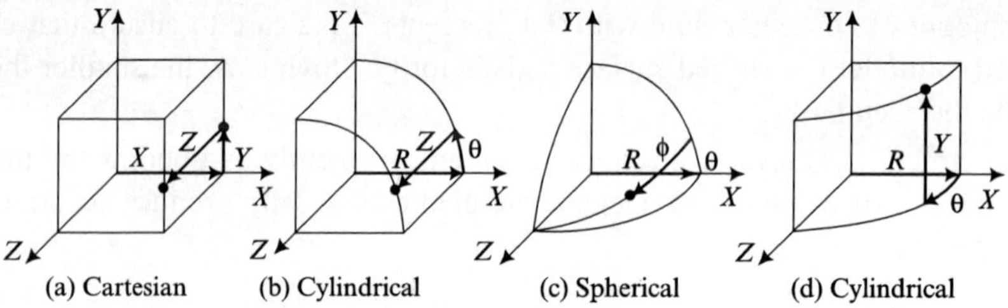

(a) Cartesian (X, Y, Z components) coordinate system 0 (C.S.0)

(b) Cylindrical (R, θ, Z components) coordinate system 1 (C.S.1)

(c) Spherical (R, θ, φ components) coordinate system 2 (C.S.2)

(d) Cylindrical (R, θ, Y components) coordinate system 5 (C.S.5)

Fig. 3.4 Global coordinate systems.

3.3.1.2 Local Coordinate Systems

In many cases, it may be necessary to establish own coordinate system, whose origin is offset from the global origin, or whose orientation differs from that of the predefined global systems. Such user defined coordinate systems, known as *local* coordinate systems, can be created in the following ways:

- Use **Utility Menu> WorkPlane> Local Coordinate Systems> Create Local CS> At Specified Loc** to define the local system in terms of global Cartesian coordinates.
- Use **Utility Menu> WorkPlane> Local Coordinate Systems> Create Local CS> By 3 Nodes** to define the local system in terms of existing nodes.
- Use **Utility Menu> WorkPlane> Local Coordinate Systems> Create Local CS> By 3 Keypoints** to define the local system in terms of existing keypoints.
- Use **Utility Menu> WorkPlane> Local Coordinate Systems> Create Local CS> At WP Origin** to define the local system to be centered at the origin of the presently defined working plane.

When a local coordinate system is defined, it becomes an active coordinate system. As a local system is created, assign it a CS identification number (which must be 11 or greater). Local coordinate systems can be created or deleted in any phase of the ANSYS session. Use **Utility Menu> WorkPlane> Local Coordinate Systems> Delete Local CS** to delete a local system. Use **Utility Menu> List> Other> Local Coord Sys** to view the status of all global and local coordinate systems.

The local coordinate systems can be Cartesian, cylindrical, or spherical, similar in form to the three predefined global systems. The analyst may define local cylindrical and spherical coordinate systems in either circular or elliptical configuration. Additionally, a toroidal local coordinate system can be defined, as illustrated in Fig. 3.5.

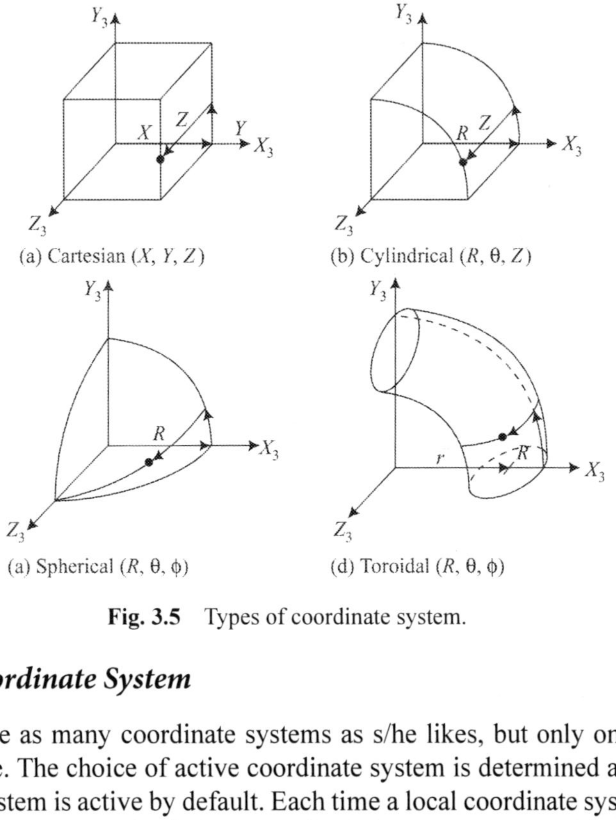

Fig. 3.5 Types of coordinate system.

3.3.1.3 Active Coordinate System

The analyst may define as many coordinate systems as s/he likes, but only one of these systems may be active at a time. The choice of active coordinate system is determined as follows: Initially, the global Cartesian system is active by default. Each time a local coordinate system is defined, that newly-defined system then automatically becomes the active one. Use one of the following methods to activate one of the global coordinate systems or some other previously defined coordinate system:

- **Utility Menu> WorkPlane> Change Active CS to> Global Cartesian**
- **Utility Menu> WorkPlane> Change Active CS to> Global Cylindrical**
- **Utility Menu> WorkPlane> Change Active CS to> Global Spherical**
- **Utility Menu> WorkPlane> Change Active CS to> Specified Coord Sys**
- **Utility Menu> WorkPlane> Change Active CS to> Working Plane**

A coordinate system can be activated in any stage of the ANSYS session. That same coordinate system will remain active in all later stages until it is changed intentionally.

3.3.2 Display Coordinate System

By default, a listing of nodes or keypoints always shows their global Cartesian coordinates, even if they were defined in a different coordinate system. The display coordinate system used in such listings can be changed by one of the following methods:

- **Utility Menu> WorkPlane> Change Display CS to> Global Cartesian**
- **Utility Menu> WorkPlane> Change Display CS to> Global Cylindrical**
- **Utility Menu> WorkPlane> Change Display CS to> Global Spherical**
- **Utility Menu> WorkPlane> Change Display CS to> Specified Coord Sys**

Changing the display coordinate system will also affect the graphical displays. Unless a specific effect is advantageous in the displays, reset the display coordinate system to the global Cartesian system before issuing any graphics display action commands.

3.3.3 Element Coordinate Systems

Every element has its own coordinate system, the *element coordinate system*, that determines the direction of orthotropic material properties, applied pressures, and results for that element. All element coordinate systems are right-handed orthogonal systems.

The default orientations for most elements' coordinate systems fit the following patterns:

- *Line elements* usually have the element X-axis directed from their node I toward their node J.
- *Shell elements* usually have the element X-axis similarly directed (from I toward J), the Z-axis normal to the shell surface, and the Y-axis perpendicular to the X and Z axes.
- For *2-D and 3-D solid elements*, the element coordinate system is usually parallel to the global Cartesian system.

Many element types have key options that permit to change the default element coordinate system orientation. For area and volume elements, the analyst can also change the orientation to align the element coordinate system with a previously defined local system by using: **Main Menu> Preprocessor> Meshing> Mesh Attributes> Default Attribs** or **Main Menu> Preprocessor> Modeling> Create> Elements> Elem Attributes**.

3.3.4 Results Coordinate System

Results data are calculated during solution and consist of displacements (UX, UY, ROTX, etc.), gradients (TGX, TGY, etc.), stresses (SX, SY, SZ, etc.), strains (EPPLX, EPPLXY, etc.), etc. These data are stored in the database and on the results file in either the nodal coordinate system or the element coordinate system. However, results data are generally rotated into the active results coordinate system for displays, listings, and element table data storage.

The active results coordinate system can be changed to another system, or to the coordinate systems used during solution. On listing, displaying, or operating on the results data, they are rotated to this results coordinate system first. Use **Main Menu> General Postproc> Options for Output** or **Utility Menu> List> Results> Options** to change the results coordinate system.

3.4 WORKING PLANES

Although the cursor appears as a point on the screen, it actually represents a line through space, normal to the screen. In order to be able to pick a point with the cursor, it is necessary to define an imaginary plane that, when intersected by the normal line of the cursor, will yield a unique point in space. This imaginary plane is called a *working plane* (Fig. 3.6). Another way to consider the interaction between the cursor and the working plane is to picture the cursor as a point that moves around on theworking plane. The working plane, then, acts as a 'tablet' on which the analyst writes with the cursor.

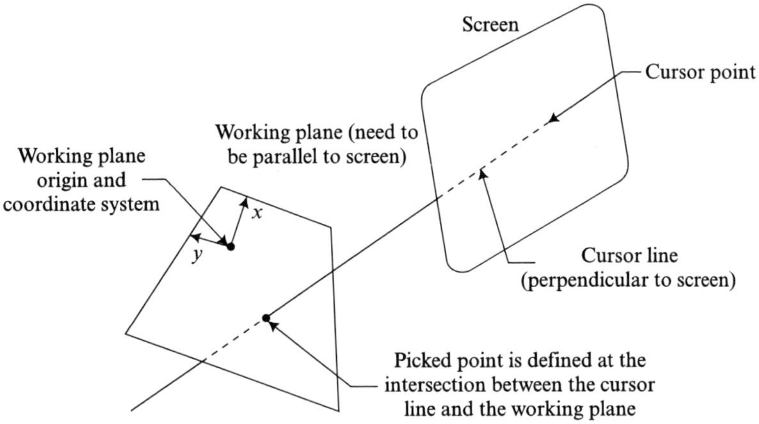

Fig. 3.6 Relationships among display screen, cursor, working plane, and picked point.

A working plane is an infinite plane with an origin, a 2-D coordinate system, a snap increment, and a display grid. Only one working plane can be defined at a time. Creating a new working plane eliminates the existing working plane. The working plane is separate from the coordinate systems; for example, the working plane can have a different point of origin and rotation from the active coordinate system.

3.4.1 Creating a Working Plane

By default, when the analyst initiates the ANSYS session, there is a working plane located on the global Cartesian X-Y plane, with its x and y axes are colinear with the global Cartesian X and Y axes.

3.4.1.1 Defining a New Working Plane

A new working plane (Fig. 3.7) can be defined by using any of the following methods:

- Use **Utility Menu> WorkPlane> Align WP with> XYZ Locations** to define a working plane by three points, or by locating it on the plane normal to a viewing vector at a specified point.

- Use **Utility Menu> WorkPlane> Align WP with> Nodes** to define a working plane by three nodes, or by locating it on the plane normal to a viewing vector at a specified node.
- Use **Utility Menu> WorkPlane> Align WP with> Keypoints** to define a working plane by three keypoints, or by locating it on the plane normal to a viewing vector at a specified keypoint.

3.4.1.2 *Controlling the Display and Style of the Working Plane*

Use **Utility Menu> List> Status> Working Plane** to obtain the current status of the working plane.

3.4.1.3 *Moving the Working Plane*

A working plane can be moved to a new location using any of the following methods:

- Use **Utility Menu> WorkPlane> Offset WP to> Keypoints** to move the working plane origin to the average location of keypoints.
- Use **Utility Menu> WorkPlane> Offset WP to> Nodes** to move the working plane origin to the average location of nodes.
- Use **Utility Menu> WorkPlane> Offset WP by Increments** to offset the working plane.

3.4.1.4 *Rotating the Working Plane*

The working plane can be rotated to a new orientation in two ways: by rotating the working plane's x-y coordinate system within the plane, or by rotating the entire plane to a new position. Use **Utility Menu> WorkPlane> Offset WP by increments** to rotate the working plane.

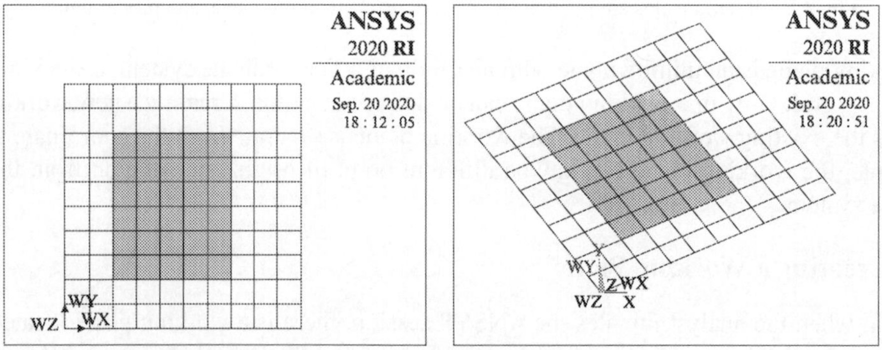

Fig. 3.7 Working plane.

3.4.2 Working Plane Enhancements

The working plane can be enhanced with a *snap increment*, a *display grid*, *retrieval tolerance*, and *coordinate type* using GUI path described earlier. Then, the coordinate system can be forced to follow the working plane as the working plane is moved using:

- **Utility Menu> WorkPlane> Change Active CS to> Global Cartesian**
- **Utility Menu> WorkPlane> Change Active CS to> Global Cylindrical**
- **Utility Menu> WorkPlane> Change Active CS to> Global Spherical**
- **Utility Menu> WorkPlane> Change Active CS to> Specified Coordinate Sys**
- **Utility Menu> WorkPlane> Change Active CS to> Working Plane**
- **Utility Menu> WorkPlane> Offset WP to> Global Origin**.

All geometry is created with respect to the working plane, which by default is aligned with the global Cartesian coordinate system. The 'Working Plane' is actually the XY plane of the working coordinate system. The working coordinate system ID is coordinate system 4 in ANSYS. Global Cartesian is ID 0, Global Cylindrical is ID 1, and Global Spherical is ID 2.

Turn on the working plane using **Utility Menu> Work Plane> Display Working Plane**. Change the way the working plane looks or adjust the snap settings under **Utility Menu> Work Plane> WP Settings...**. Move the working plane around using **Utility Menu> Work Plane> Offset WP to...**. Align the working plane with various parts of the model using **Utility Menu> Work Plane> Align WP with...**. If more than one node or keypoint is selected to offset the working plane to, it will go to the average location of the selected entities. The working plane can be used to slice and dice the model. For example, to cut an area in pieces use **Main Menu> Modeling> Operate> Booleans> Divide> Area by WrkPlane**.

3.5 SOLID MODELING

A solid model relieves the analyst of the time-consuming task of building a complicated finite element model by direct generation. Some solid modeling and meshing operations can help to speed up the creation of the final analysis model.

3.5.1 Solid Modeling Operations

The points that define the vertices of the model are called keypoints and are the 'lowest-order' solid model entities. If, in building the solid model, the analyst first creates the keypoints, and then uses those keypoints to define the 'higher-order' solid model entities, the model is said to be built 'from the bottom up' (Fig. 3.8). Models built from the bottom up are defined within the currently active coordinate system.

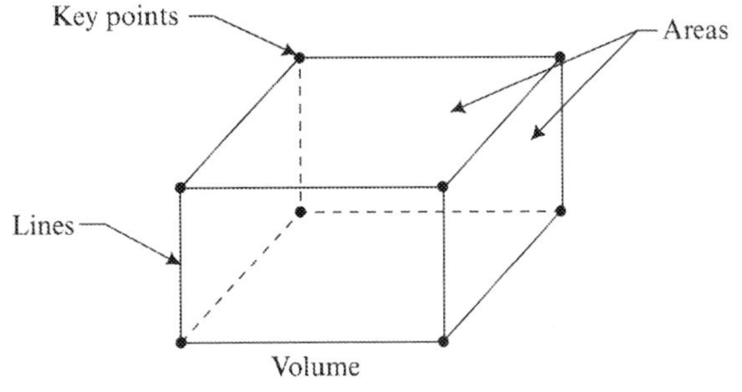

Fig. 3.8 Bottom up construction.

The ANSYS program also gives the analyst the ability to assemble the model using *geometric primitives*, which are fully defined lines, areas, and volumes. As a primitive is created, the program automatically creates all the 'lower' entities associated with it. If the modeling effort begins with the 'higher' primitive entities, the model is said to be built 'from the top down.' Bottom up and top down modeling techniques can be freely combined, as appropriate, in any model. Remember that geometric primitives (Fig. 3.9) are built within the working plane while bottom up techniques are defined against the active coordinate system.

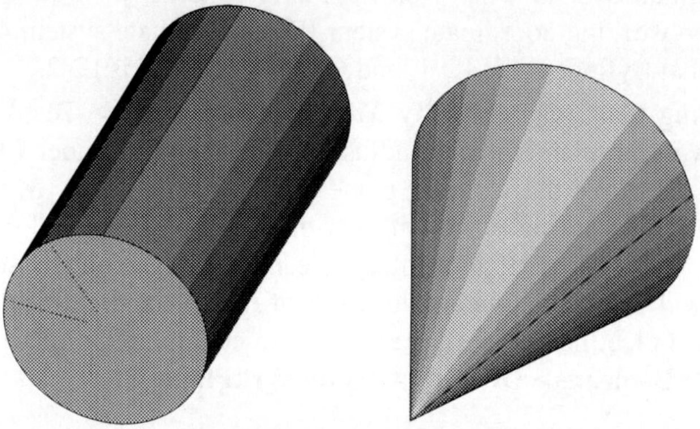

Fig. 3.9 Top down constructions (primitives).

Using Boolean operators: The analyst can 'sculpt' the solid model using intersections, subtractions, and other Boolean operations (Fig. 3.10). Booleans allow to work directly with higher solid model entities to create complex shapes. Both bottom up and top down creations can be used in Boolean operations.

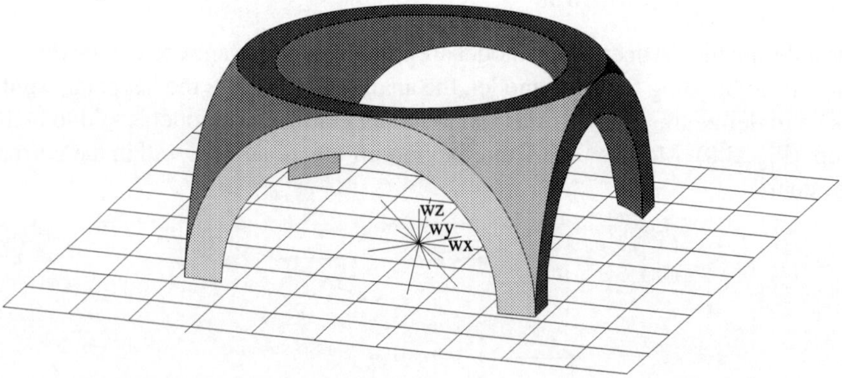

Fig. 3.10 Complex shape created with Boolean operations.

Dragging and rotating: Boolean operators, although convenient, can be computationally costly. Sometimes a model can be constructed more efficiently by dragging (Fig. 3.11) or rotating.

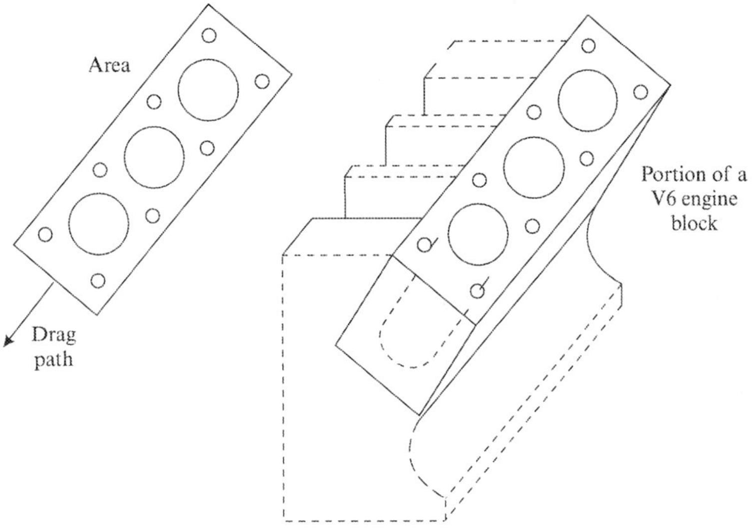

Fig. 3.11 Dragging an area to create a volume.

Moving and copying solid model entities: A complicated area or volume that appears repetitively in the model needs only be constructed once; it can then be moved, rotated, and copied (Fig. 3.12) to a new location on the model. It might be more convenient to place geometric primitives in their proper locations by moving them, rather than by changing the working plane.

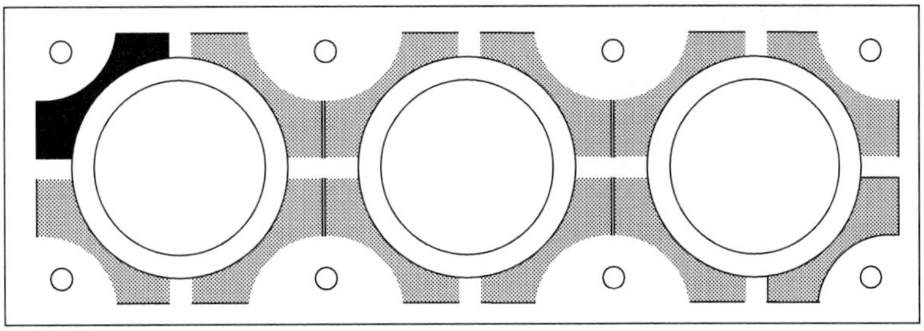

A complicated area (black) can be copied as many times as needed (gray)

Fig. 3.12 Copying an area.

Meshing: The ultimate objective of building a solid model is to mesh (Fig. 3.13) that model with nodes and elements. First complete the solid model, set element attributes, and establish meshing controls.Then turn the ANSYS program free to generate the finite element mesh. A'mapped' mesh containing all quadrilateral, all triangular, or all brick elements can be requested by taking care to meet certain requirements.

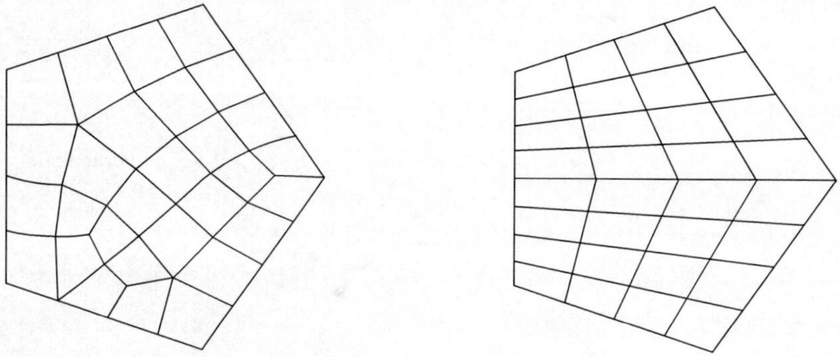

Fig. 3.13 Free and mapped meshes.

Moving and copying nodes and elements: Automatic meshing is a vast improvement over direct generation of nodes and elements, but it can sometimes be computationally time consuming. If the model contains repetitive features, the most efficient approach to model generation would be to model and mesh a pattern region of the model and then generate copies (Fig. 3.14) of that meshed region.

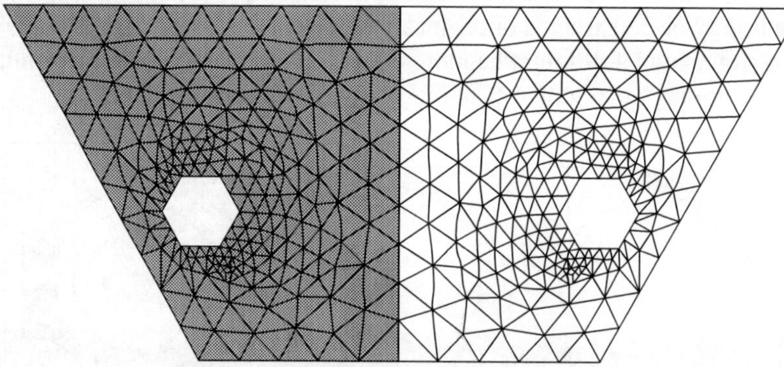

Fig. 3.14 Copying a meshed area.

Solid model loads: In the ANSYS program, loads are normally associated with nodes and elements. However, using solid modeling, it is inconvenient to define loads at nodes and elements. Fortunately, loads may be assigned directly to the solid model.When the solution calculations are initiated, the program will automatically transfer these solid model loads to the finite element model.

Revising the model (clearing and deleting): Before revising the model, it is necessary to be aware of the hierarchy of solid model and finite element model entities. A lower order entity cannot be deleted if it is attached to a higher-order entity. Thus, a volume cannot be deleted if it has been meshed, a line cannot be deleted if it is attached to an area, and so forth. If an entity is attached to any loads, deleting or redefining that entity will delete the attached loads from the database. The hierarchy of modeling entities is as listed below:

Highest Elements (and Element Loads)
Nodes (and Nodal Loads)
Volumes (and Solid-Model Body Loads)
Areas (and Solid-Model Surface Loads)
Lines (and Solid-Model Line Loads)
Lowest Keypoints (and Solid-Model Point Loads)

If it is necessary to revise (Fig. 3.15) a solid model after it has been meshed, first delete all the nodes and elements in the portion of the model to be revised, using **Main Menu> Preprocessor> Meshing> Clear**. Solid model entities can be deleted and redefined after clearing the solid model. As an alternative to clearing, deleting, and redefining, consider modifying the keypoints directly, using: **Main Menu> Preprocessor> Modeling> Move/Modify>Keypoints> Set of KPs** or **Main Menu> Preprocessor> Modeling> Move/Modify>Keypoints> Single KP**.

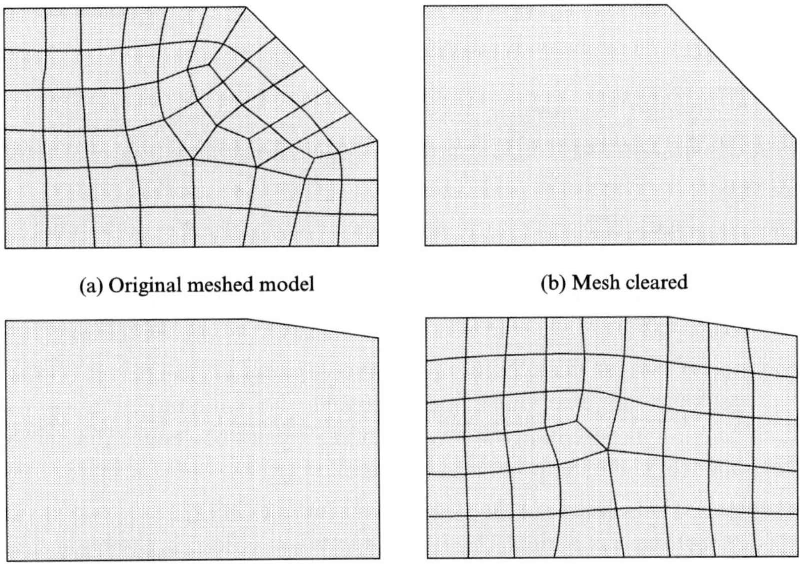

(a) Original meshed model (b) Mesh cleared

(c) Solid model modified (d) Revised remeshed model

Fig. 3.15 Revising a meshed solid model.

3.5.2 Creating the Solid Model from the Bottom Up

Any solid model, whether assembled from the bottom up or from the top down, is defined in terms of keypoints, lines, areas, and volumes. Figure 3.16 illustrates these entities.

Geometry in ANSYS is created from **Main Menu> Preprocessor> Modeling> Create** and has the following terminology (Fig. 3.16).

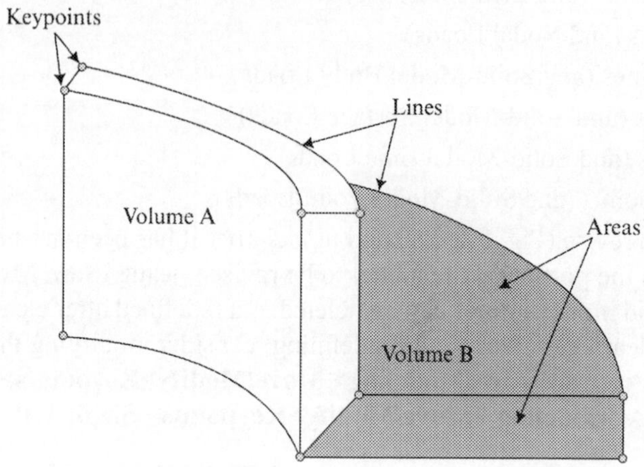

Fig. 3.16 Basic solid model entities.

Keypoints: These are points, locations in 3D space.

Lines: This includes straight lines, curves, circles, spline curves, etc. Lines are typically defined using existing keypoints.

Areas: This is a surface. When an area is created, its associated lines and keypoints are automatically created to border it.

Volumes: This is a solid. When a volume is created, its associated areas, lines and keypoints are automatically created.

Solid model: In most packages this would refer to the volumes only, but in ANSYS this refers to the geometry. Any geometry, for example, a line is considered a 'solid model'. A child entity cannot be deleted without deleting its parent, in other words a line cannot be deleted if it's a part of an area, can't delete a keypoint if it's the end point of a line, etc.

Keypoints are the vertices, lines are the edges, areas are the faces, and volumes are the interior of the object. Notice that there is a hierarchy in these entities: volumes, the highest-order entities, are bounded by areas, which are bounded by lines, which, in turn, are bounded by keypoints.

In bottom up construction, first create keypoints and use those keypoints to define higher-order solid model entities (Fig. 3.17).

Fig. 3.17 Bottom up construction.

3.5.2.1 *Keypoints*

When building the model from the bottom up, begin by defining the lowest-order solid model entities, *keypoints*. Keypoints are defined within the currently active coordinate system. Then define lines, areas, and volumes connecting these keypoints. It is not always necessary to explicitly define all entities in ascending order to create higher-order entities: define areas and volumes directly in terms of the keypoints at their vertices. The intermediate entities will then be generated automatically as needed. For example, a brick-like volume is defined in terms of the eight keypoints at its corners, the program will automatically generate the bounding areas and lines.

Use **Main Menu> Preprocessor> Modeling> Create> Keypoints> In Active CS or Main Menu> Preprocessor> Modeling> Create> Keypoints> On Working Plane** to define individual keypoints in the active coordinate system. Use **Main Menu> Preprocessor> Modeling> Create> Keypoints> On Line or Main Menu> Preprocessor> Modeling> Create> Keypoints> On Line w/Ratio** to define individual keypoints at a given location on an existing line.

Once an initial pattern of keypoints is created, generate additional keypoints and work with existing keypoints:

- Use **Main Menu> Preprocessor> Modeling > Create> Keypoints> KP between KPs** to create a keypoint between two existing keypoints
- Use **Main Menu> Preprocessor> Modeling > Create> Keypoints> Fill between KPs** to generate keypoints between two keypoints.
- Use **Main Menu> Preprocessor> Modeling > Create> Keypoints> KP at Center** to create a keypoint at the center of a circular arc defined by three locations.
- Use **Main Menu> Preprocessor> Modeling> Copy> Keypoints** to generate additional keypoints from a pattern of keypoints.

The keypoints can be maintained using the methods listed below:

- Use **Utility Menu> List> Keypoint> Coordinates +Attributes** or **Utility Menu> List> Keypoint> Coordinates +Attributes** or **Utility Menu> List> Keypoint> Coordinates only** or **Utility Menu> List> Keypoint> Hard Points** or **Utility Menu> List> Keypoint> Coordinates +Attributes** or **Utility Menu> List> Keypoint> Coordinates only** or **Utility Menu> List> Keypoint> Hard Points** or **Utility Menu> List> Keypoint> Coordinates only** or **Utility Menu> List> Keypoint> Hard Points** to list defined keypoints.
- Use **Utility Menu> Select> Entities** to select keypoints.

3.5.2.2 *Lines*

Lines are mainly used to represent the edges of an object. As with keypoints, lines are defined within the currently active coordinate system. It is not always necessary to define all lines explicitly, because the program will generate the necessary lines in many instances when an area or volume is defined. Lines are required if it is necessary to generate line elements or to create areas from lines.

Copy a pattern of lines to generate additional lines using any of the methods described below:

Use **Main Menu > Preprocessor > Modeling > Copy > Lines or Main Menu > Preprocessor > Modeling > Move/Modify > Lines** to generate additional lines from a pattern of lines. Use **Main Menu > Preprocessor > Modeling > Reflect > Lines** to generate lines from a line pattern by symmetry reflection.

An existing line can be modified by redefining it by using one of the methods described below:

- Use **Main Menu > Preprocessor > Modeling > Operate > Booleans > Divide > Line into 2 Ln's** or **Main Menu > Preprocessor > Modeling > Operate > Booleans > Divide > Line into N Ln's** or **Main Menu > Preprocessor > Modeling > Operate > Booleans > Divide > Line into 2 Ln's** or **Main Menu > Preprocessor > Modeling > Operate > Booleans > Divide > Line into N Ln's** or **Main Menu > Preprocessor > Modeling > Operate > Booleans > Divide > Lines w/Options** or **Main Menu > Preprocess** or**> Modeling > Operate > Booleans > Divide > Lines w/ Options** to divide a line into smaller lines.

Lines can be maintained using the methods listed below:

- Use **Utility Menu > Plot > Lines or Utility Menu > Plot > Specified Entities > Lines** to display lines.
- Use **Main Menu > Preprocessor > Modeling > Delete > Line and Below or Main Menu > Preprocessor > Modeling > Delete > Lines Only to** delete lines.

3.5.2.3 Areas

Flat areas are used to represent 2-D solid objects. Curved as well as flat areas are used to represent 3-D surfaces, such as shells, and the faces of 3-D solid objects. Areas are required to use area elements or to create volumes from areas. Most commands that create areas will also automatically generate the necessary lines and keypoints; similarly, many areas can be conveniently generated by defining volumes.

Any of the methods described below can be used to explicitly define areas. Several geometric primitives and Boolean commands can also be used to generate or modify areas.

- Use **Main Menu> Preprocessor> Modeling> Create> Areas> Arbitrary> Through KPs** to define an area in terms of its vertices.
- Use **Main Menu> Preprocessor> Modeling> Create> Areas> Arbitrary> By Lines** to define an area in terms of its boundaries.
- Use **Main Menu> Preprocessor> Modeling> Operate> Extrude> Along Lines** to define an area by 'sweeping' a line along a characteristic path.
- Use **Main Menu> Preprocessor> Modeling> Operate>Extrude> About Axis** to define an areaby rotating a line pattern about an axis.

Existing areas can be copied to generate additional areas using the methods described below:

- Use **Main Menu> Preprocessor> Modeling> Copy> Areas** or **Main Menu> Preprocessor> Modeling> Move/Modify> Areas> Areas** to generate additional areas from a pattern of areas.
- Use **Main Menu> Preprocessor> Modeling> Reflect> Areas to** generate areas from an area pattern by symmetry reflection.

Use the methods listed below to maintain areas. Only unmeshed areas that are not attached to a volume can be redefined or deleted.

- Use **Utility Menu> List> Areas or Utility Menu> List> Picked Entities> Areas** to list defined areas.
- Use **Utility Menu> Plot> Areas or v Utility Menu> Plot> Areas or Utility Menu> Plot> Specified Entities> Areas or Utility Menu> Plot> Specified Entities> Areas to** display areas.
- Use **Main Menu> Preprocessor> Modeling> Delete> Area and Below** or **Main Menu> Preprocessor>Modeling> Delete> Areas Only** to delete areas.

3.5.2.4 Volumes

Volumes are used to represent 3-D objects, and are required only to use volume elements. Most commands that create volumes will also automatically generate the necessary lower-order entities.

Adopt any of the methods described below to define volumes. Several geometric primitives and Boolean commands can also be used to generate or modify areas.

- Use **Main Menu> Preprocessor> Modeling> Create> Volumes> Arbitrary> Through KPs** to define a volume in terms of its vertices.
- Use **Main Menu> Preprocessor> Modeling> Create> Volumes> Arbitrary> By Areas** to define a volume in terms of its boundaries.
- Use **Main Menu> Preprocessor> Modeling> Operate> Extrude> About Axis** to define a volume by rotating an area pattern about an axis.

Volumes can be maintained using the methods described below. Note that only unmeshed volumes can be redefined or deleted.

- Use: **Utility Menu> List> Picked Entities> Volumes or Utility Menu> List> Volumes** to list volumes.
- Use **Utility Menu> Plot> Specified Entities> Volumes or Utility Menu> Plot> Volumes** to display volumes.
- Use **Utility Menu> Select> Entities** to select volumes.

3.5.3 Creating the Solid Model from the Top Down

In top down construction, geometric primitives are used to assemble the model. As a primitive is created, the program automatically creates all the 'lower' entities associated with it. A *geometric primitive* is a commonly used solid modeling shape that can be created with a single ANSYS command.

Because primitives are higher-order entities that can be constructed without first defining any keypoints, model generation that uses primitives is sometimes referred to as 'top down' modeling. Geometric primitives are created within the working plane.

Bottom up and top down modeling techniques can be freely combined, as appropriate, in any model. Remember that geometric primitives are built within the working plane while bottom up techniques are defined against the active coordinate system.

3.5.3.1 *Creating Area Primitives*

Any area primitives created will lie flat on the working plane and will be oriented according to the working plane coordinate system. Area primitives must have surface areas greater than zero.

3.5.3.2 *Creating Volume Primitives*

Volume primitives are positioned relative to the working plane.

3.5.4 Sculpting the Model with Boolean Operations

Boolean operations can be applied to almost any solid model construction, whether it was created from the top down or from the bottom up. The only exceptions are that Boolean operations are not valid for entities created by concatenation and that some Boolean operations cannot always be performed on entities that contain degeneracies.

Also, all solid-model loads and element attributes should be defined after the Boolean operations are completed. Take care to redefine the element attributes and solid-model loads when using Booleans to modify an existing model.

3.5.4.1 *Intersect*

An *intersection* defines a new set of entities which is common to every original entity included in the operation (Fig. 3.18). In other words, an intersection represents the region of overlap of two or more entities. The new set can be of the same or lower dimension as the original entities. For instance, the intersection of two lines can be a keypoint, or it can be a line. The Boolean intersect commands are as follows:

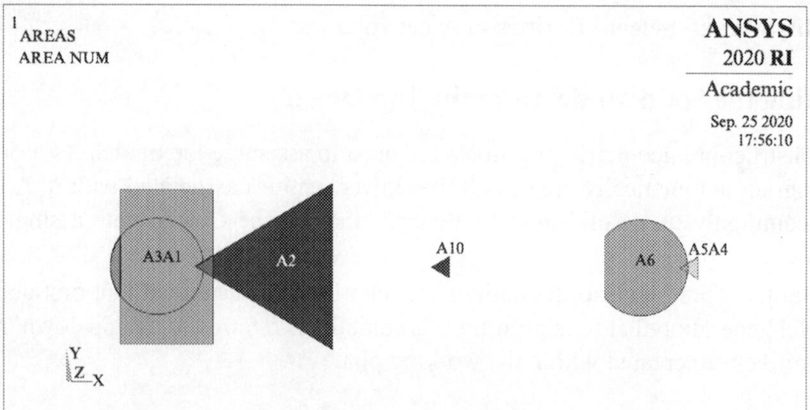

Fig. 3.18 Intersect operation.

- Use **Main Menu> Preprocessor> Modeling> Operate> Booleans> Intersect> Common> Lines** to create the intersection of lines.
- Use **Main Menu> Preprocessor> Modeling> Operate> Booleans> Intersect> Common> Areas** to create the intersection of areas.
- Use **Main Menu> Preprocessor> Modeling> Operate> Booleans> Intersect> Common> Volumes** to create the intersection of volumes.

3.5.4.2 Add

An *addition* of entities defines a new entity that includes all parts of the originals (Fig. 3.19). The resulting entity is a single seamless whole, containing no internal divisions. Only volumes or coplanar 2-D areas can be added in the ANSYS program. Areas added may contain holes within the area; that is, internal loops. The Boolean add commands are as follows:

- Use **Main Menu> Preprocessor> Modeling> Operate> Booleans> Add> Areasto add separate areas** to create a single area.
- Use **Main Menu> Preprocessor> Modeling> Operate> Booleans> Add> Volumes** to add separate volumes to create a single volume.

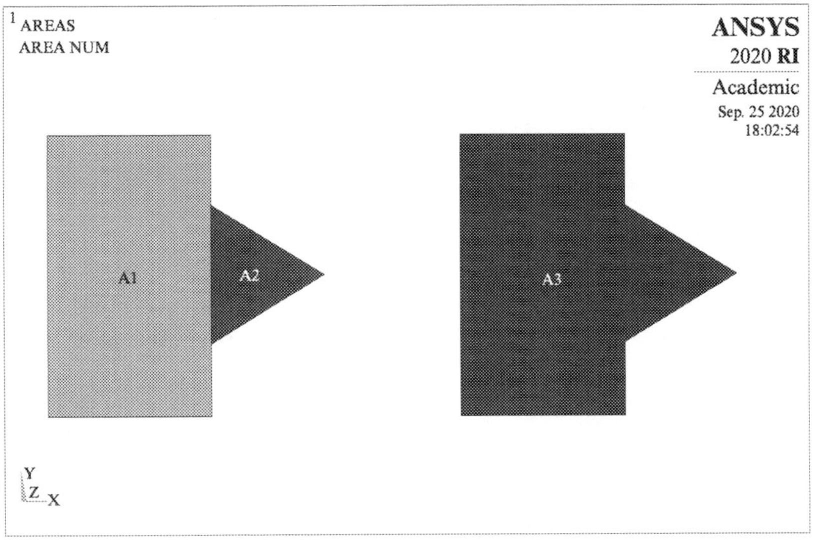

Fig. 3.19 Add operation.

3.5.4.3 Subtract

Subtracting (Fig. 3.20) one entity (E2) from another (E1), gives one of two results: Either the analyst creates a new entity or entities (E1 – E2 ≥ E3), that is, of the same dimensionality as E1 and that contains no overlap with E2, or, if the overlap is of a lower dimensionality, simply E1 will be divided into two or more new entities (E1 – E2 ≥ E3 and E4).The Boolean subtract GUI paths are as follows:

- Use **Main Menu> Preprocessor> Modeling> Operate> Booleans> Subtract> Lines or Main Menu> Preprocessor> Modeling> Operate> Booleans> Subtract> With Options> Lines** to subtract lines from lines.
- Use **Main Menu> Preprocessor> Modeling> Operate> Booleans> Subtract> Areas or Main Menu> Preprocessor> Modeling> Operate> Booleans> Subtract> With Options> Areas** to subtract areas from areas.

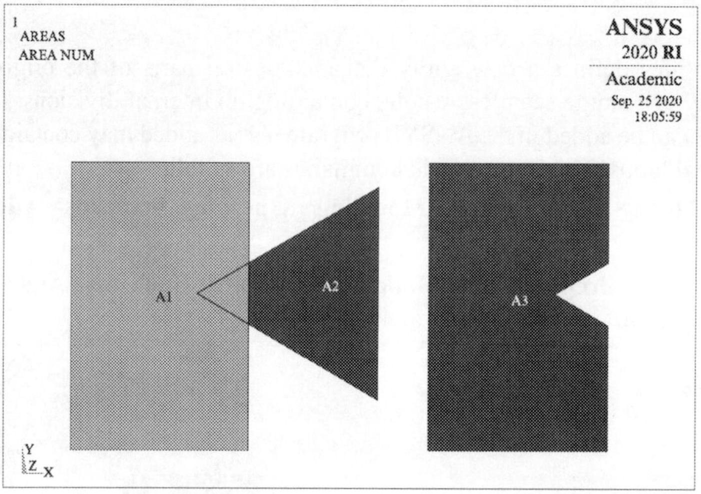

Fig. 3.20 Subtract operation.

3.5.4.4 *Overlap*

The *overlap* commands will join two or more entities to create three or more new entities that encompass all parts of the originals (Fig. 3.21). The end result is similar to an 'add' operation,

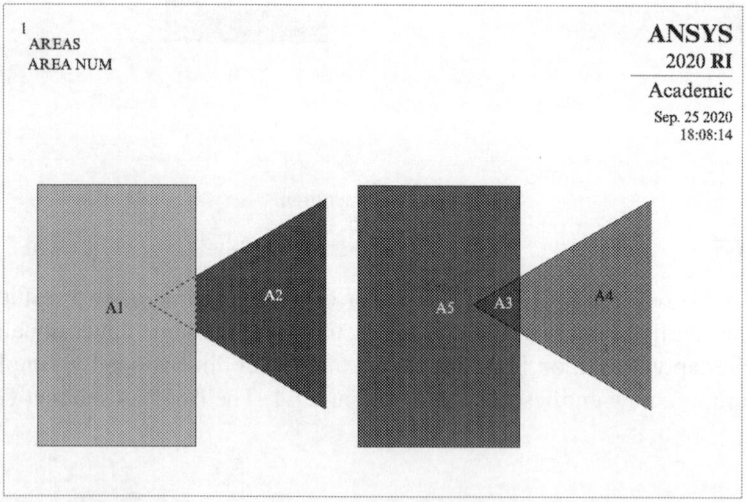

Fig. 3.21 Overlap operation.

except that boundaries will be created around the overlap zone. Thus, the overlap operation produces a number of relatively uncomplicated regions, as compared to the single relatively complicated region created by the add operation. For this reason, overlapped entities will often mesh better than added entities.

Overlapping is valid only if the overlap region has the same dimensionality as the original entities. The Boolean overlap GUI paths are as follows:

- Use **Main Menu> Preprocessor> Modeling> Operate> Booleans> Overlap> Lines** to overlap lines
- Use **Main Menu> Preprocessor> Modeling> Operate> Booleans> Overlap> Areas** to overlap areas.

3.5.4.5 Partition

The *partition* commands will join two or more entities to create three or more new entities that encompass all parts of the originals (Fig. 3.22). The end result is similar to an 'overlap' operation if the overlap is of the same dimensionality as the original entities. However, unlike the *overlap* operations, non-overlapping input entities will not be deleted. The Boolean partition GUI paths are as follows:

- Use **Main Menu> Preprocessor> Modeling> Operate> Booleans> Partition> Lines** to partition lines.
- Use **Main Menu> Preprocessor> Modeling> Operate> Booleans> Partition> Areas** to partition areas.

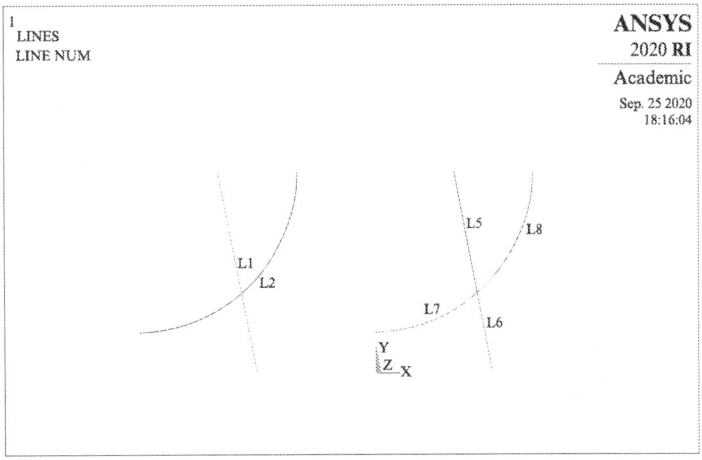

Fig. 3.22 Partition operation.

3.5.4.6 Glue (or Merge)

Glue is similar to overlap, except that it applies only to cases in which the intersection between entities occurs at a boundary, and is one dimension lower than the original entities (Fig. 3.23). The entities maintain their individuality, but they become connected at their intersection, as shown in the illustrations below. The Boolean glue GUI paths are as follows:

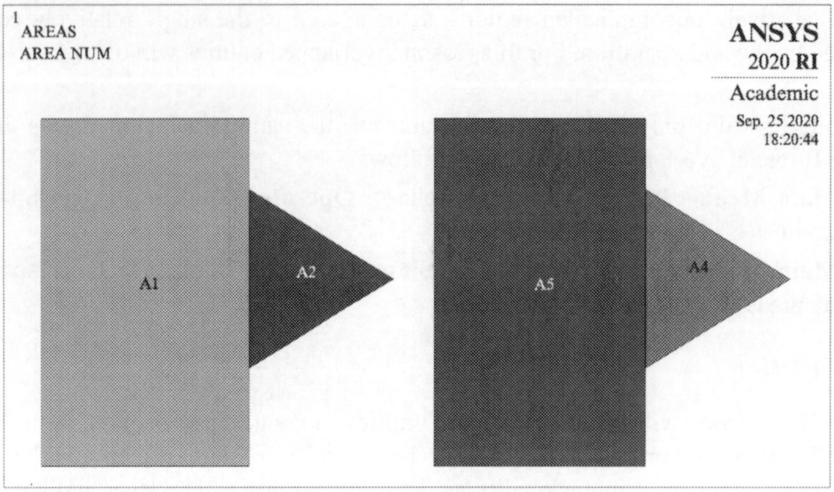

Fig. 3.23 Glue operation.

- Use **Main Menu> Preprocessor> Modeling> Operate> Booleans> Glue> Lines** to glue lines by 'gluing' lines.
- Use **Main Menu> Preprocessor> Modeling> Operate> Booleans> Glue> Areas** to glue areas by 'gluing' areas.

3.5.4.7 Divide

Divides an entity into two or more entities (Fig. 3.24).

Fig. 3.24 Divide operation.

3.5.5 Updating After Boolean Operations

Some Boolean commands will automatically update entities after the Boolean operation is performed on attached lower-order entities. For instance, if the GUI path **Main Menu> Preprocessor> Modeling> Operate> Booleans> Add> Areas** is used to add several areas together that are attached to a single volume, that volume will be updated by replacing the original areas with the newly produced area. This releases the analyst from the work of deleting the higher-order entity and rebuilding it with bottom up techniques.

3.5.6 Moving and Copying Solid Model Entities

If the model repetitively uses a relatively complicated area or volume, it is necessary to construct that part only once. Then copies of that part can be generated in new locations and new orientations as needed.

Move translates or rotates an entity by specifying DX, DY, DZ offsets (Fig. 3.25). DX, DY, DZ are interpreted in the active CS. To translate an entity, make the active CS Cartesian. To rotate an entity, make the active CS cylindrical or spherical. Another option is to transfer coordinates to a different system.

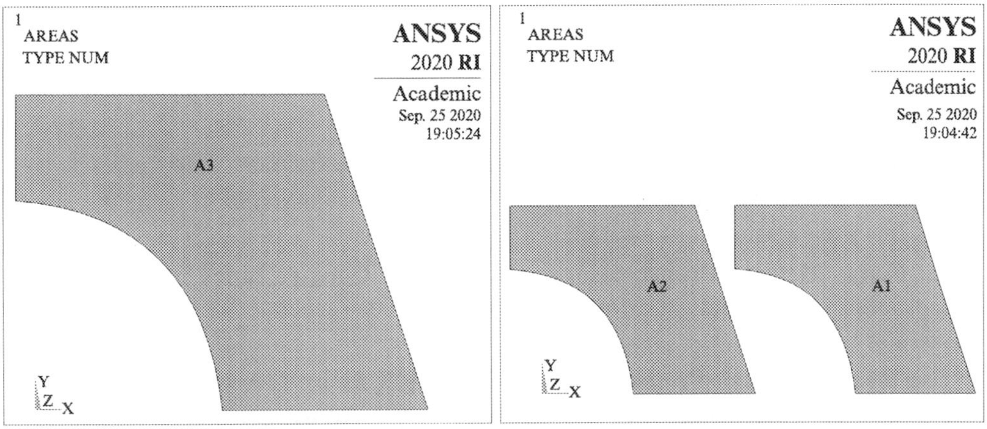

Fig. 3.25 Moving an area.

Geometric primitives can also be considered to be 'parts'. As geometric primitives are created, their location and orientation will be determined by the current working plane. Because it is not always particularly convenient to redefine the working plane for each new primitive that is created, it is more practical to allow a primitive to be created at the 'wrong' location, and then *move* that primitive to its correct position. Of course, this operation is not limited to geometric primitives: any solid model entity can be copied or moved.

Copy generates multiple copies of an entity. Specify the number of copies (2 or more) and the DX, DY, DZ offset for each copy. DX, DY, DZ are interpreted in the active CS. It is useful to create multiple holes, ribs, protrusions, etc. (Fig. 3.26).

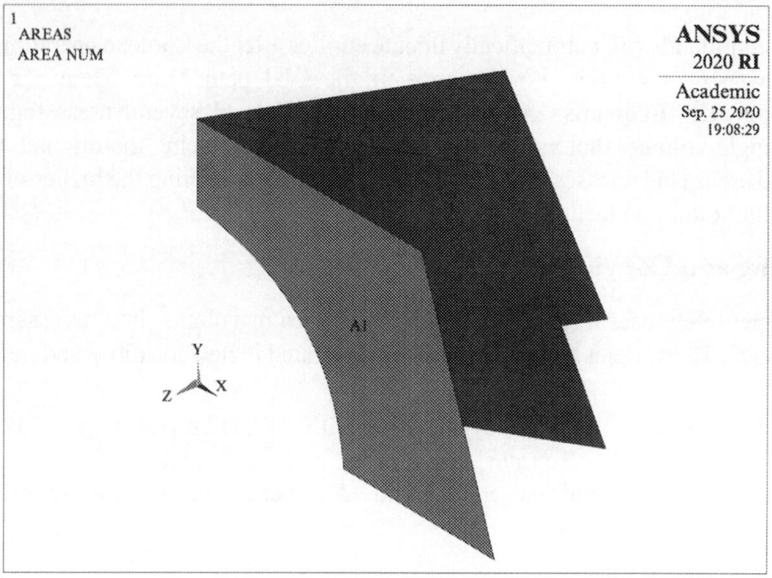

Fig. 3.26 Copying an area.

3.5.6.1 Generating Entities from a Pattern

ANSYS provides the following GUI paths for generating entities from a pattern:

- Use **Main Menu> Preprocessor> Modeling> Copy> Keypoints** to generate additional keypoints from a pattern of keypoints.
- Use **Main Menu> Preprocessor> Modeling> Copy> Lines** or **Main Menu> Preprocessor> Modeling> Move/Modify> Lines** to generate additional lines from a pattern of lines.
- Use **Main Menu> Preprocessor> Modeling> Copy> Areas** or **Main Menu> Preprocessor> Modeling> Move/Modify> Areas> Areas** to generate additional areas from a pattern of areas.
- Use **Main Menu> Preprocessor> Modeling> Copy> Volumes** or **Main Menu> Preprocessor> Modeling> Move/Modify> Volumes** to generate additional volumes from a pattern of volumes.

3.5.6.2 Generating Entities by Symmetry Reflection

ANSYS provides the following GUI paths for generating entities by symmetry reflection:

- Use **Main Menu> Preprocessor> Modeling> Reflect> Keypoints** to generate a reflected set of keypoints.
- Use **Main Menu> Preprocessor> Modeling> Reflect> Lines** to generate lines from a line pattern by symmetry reflection.

- Use **Main Menu> Preprocessor> Modeling> Reflect> Areas** to generate areas from an area pattern by symmetry reflection.
- Use **Main Menu> Preprocessor> Modeling> Reflect> Volumes** to generate volumes from a volume pattern by symmetry reflection.

3.5.6.3 Transferring a Pattern of Entities to a Coordinate System

ANSYS provides the following GUI paths:

- Use **Main Menu> Preprocessor> Modeling> Move/Modify> Transfer Coord> Keypoints** to transfer a pattern of keypoints to another coordinate system.
- Use **Main Menu> Preprocessor> Modeling> Move/Modify> Transfer Coord> Lines** to transfer a pattern of lines to another coordinate system.
- Use **Main Menu> Preprocessor> Modeling> Move/Modify> Transfer Coord> Areas** to transfer a pattern of areas to another coordinate system.
- Use **Main Menu> Preprocessor> Modeling> Move/Modify> Transfer Coord> Volumes** to transfer a pattern of volumes to another coordinate system.

3.5.7 Other Operations

Besides Booleans, many other operations are available like Extrude, Scale, Reflect, Merge, Fillet.

3.5.7.1 Extrude

Extrude (Fig. 3.27) quickly creates volumes from the existing areas (or areas from lines, and lines from keypoints). If the area is meshed, the elements can be extruded along with the areas. There are four ways to extrude areas:

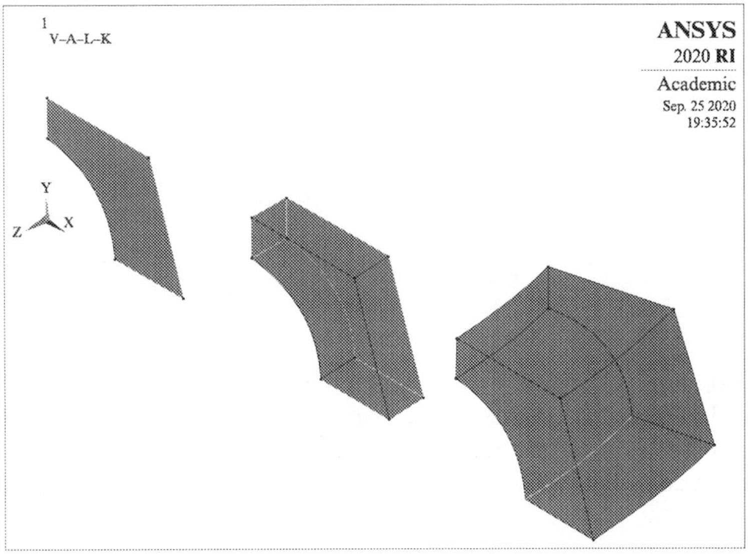

Fig. 3.27 Extrude operation.

Along normal — creates volume by normal offset of areas (Fig. 3.27b)

By XYZ offset — creates volume by a general x-y-z offset. Allows tapered extrusion.

About axis — creates volume by revolving areas about an axis (specified by two keypoints).

Along lines — creates volume by 'dragging' areas along a line or a set of contiguous lines.

3.5.7.2 *Reflect*

Reflect facilitates reflecting entities about a plane (Fig. 3.28). Specify the direction of reflection:

- X for reflection about the YZ plane
- Y for XZ plane
- Z for XY plane

All directions are interpreted in the active CS, which must be a Cartesian system.

Fig. 3.28 Reflect operation.

3.5.7.3 *Merge*

Merge attaches two entities together by removing coincident keypoints (Fig. 3.29). Observe the removal of coincident keypoints on the centre line. Merging keypoints will automatically merge coincident higher-order entities, if any. This is usually required after a reflect, copy, or other operation that causes coincident entities.

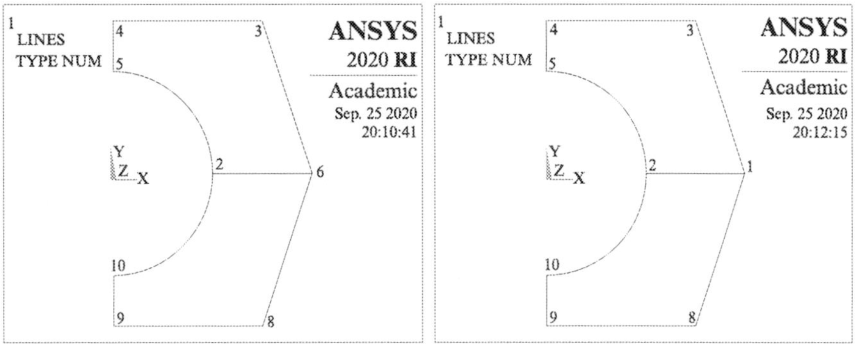

Fig. 3.29 Merge operation.

3.5.7.4 *Fillet*

Line fillet requires two intersecting lines with a common keypoint at the intersection (Fig. 3.30). If the common keypoint does not exist, do a partition operation first. ANSYS does not update the underlying area (if any), so either add or subtract the fillet region. Area filleting is similar.

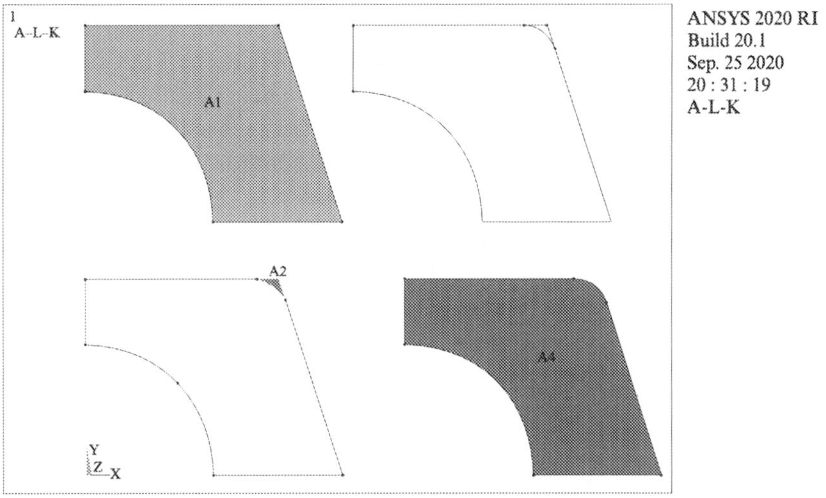

Fig. 3.30 Fillet operation.

3.6 IMPORTING SOLID MODELS

A solid model can be created in a favorite CAD system, saved as an IGES file, and then imported to ANSYS. Once successfully imported, the model can be meshed similar to any model created within the program.

The Initial Graphics Exchange Specification (IGES) is a vendor-neutral standard format used to exchange geometric models between various CAD and CAE systems. The filter permits import of partial files. Multiple files can also be imported into the same model.

ANSYS provides the SMOOTH method for importing IGES files. This method uses the standard ANSYS geometry database and has no capabilities for automatically creating volumes.

3.7 GENERATING THE MESH

The process for generating a mesh of nodes and elements consists of three general steps:

- Setting the element attributes
- Setting mesh controls
- Meshing the model.

It is not always necessary to set mesh controls because the default mesh controls are appropriate for many models. If no controls are specified, the program will use the default settings to produce a free mesh. Alternatively, the SmartSize feature can be used to produce a better quality free mesh.

3.7.1 Free or Mapped Mesh

Before meshing the model, and even before building the model, it is important to think about whether a free mesh or a mapped mesh (Fig. 3.31) is appropriate for the analysis. A free mesh has no restrictions in terms of element shapes, and has no specified pattern applied to it. A mapped mesh is restricted in terms of the element shape it contains and the pattern of the mesh. A mapped area mesh contains either only quadrilateral or only triangular elements, while a mapped volume mesh contains only hexahedron elements. In addition, a mapped mesh typically has a regular pattern, with obvious rows of elements. To generate this type of mesh, build the geometry as a series of fairly regular volumes and/or areas that can accept a mapped mesh.

A mapped mesh generates a very regular grid of elements. This can only be used on rectangular shaped areas or volumes. A free mesh will mesh any entity, regardless of shape.

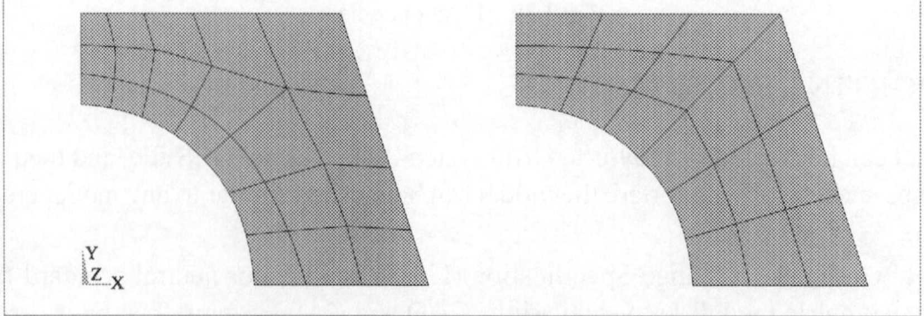

Fig. 3.31 Free and mapped meshes.

3.7.2 Setting Element Attributes

Before generating a mesh of nodes and elements, the analyst must first define the appropriate *element attributes*. That is, s/he must specify the following:

- The element type
- Real constant set
- Material properties set
- Element coordinate system
- Section ID

3.7.2.1 Creating Tables of Element Attributes

In order to assign attributes to the elements, the analyst must first build tables of element attributes. Typical models include element types (menu path **Main Menu> Preprocessor> Element Type> Add/Edit/Delete**), real constants (menu path **Main Menu> Preprocessor> Real Constants**), and material properties (menu path **Main Menu> Preprocessor> Material Props>** `material option`).

A table of coordinate systems can also be assembled using **Utility Menu> WorkPlane> Local Coordinate Systems> Create Local CS>** ***option***). This table can be used to assign element coordinate systems to elements.

3.7.2.2 Assigning Element Attributes before Meshing

Once the attribute tables are assembled, the analyst can assign element attributes to different parts of the model by 'pointing' to the appropriate entries in the tables. The pointers are simply a set of reference numbers that include a material number (MAT), a real constant set number (REAL), an element type number (TYPE), a coordinate system number (ESYS) and, for beam meshing with BEAM188, or BEAM189, a section ID number (SECNUM). The analyst can either assign the attributes directly to selected solid model entities, or define a default set of attributes that will be used for elements created in subsequent meshing operations.

3.7.3 Mesh Controls

The default mesh controls that the program uses may produce a mesh that is adequate for the model being analyzed. In this case, it is not necessary to specify any mesh controls. However, if mesh controls are used, set them before meshing the solid model.

Mesh controls allow establishing such factors as the element shape, mid-side node placement, and element size to be used in meshing the solid model. This step is one of the most important steps of the entire analysis, for the decisions made at this stage in the model development will profoundly affect the accuracy and economy of the analysis.

3.7.3.1 *MeshTool*

The MeshTool (**Main Menu> Preprocessor> Meshing> MeshTool**) provides a convenient path to many of the most common mesh controls, as well as to the most frequently performed meshing operations. The MeshTool (Fig. 3.32) is an interactive 'tool box', not only because of the numerous functions that it contains, but also because once opened, it remains open until the analyst either closes it or exits PREP7.

Although all of the functions available via the MeshTool are also available via the traditional commands and menus, using the MeshTool is a valuable shortcut.

How fine of a mesh should be? Part of this is common sense. Put a finer mesh in areas of higher stress gradient, like near a stress concentration. Do simple experiments along with hand calculations; for example, about 30 nodes are needed around a hole to get a good value of the peak stress. The stress convergence should follow a pattern as shown in Fig. 3.33. When the result is not changing much for a finer mesh, the mesh is converged.

If global attributes are set, that material, element type, real constant, beam section, etc., will be used for all elements in the model. It's better to use the drop down to assign different attributes to different geometric entities in the model. Then mesh the whole model at one time.

Smart Size is used to automatically pick the element edge length depending on the sizes of features in the geometry. It makes a finer mesh around smaller features in order to capture them adequately.

3.7.3.2 *Element Shape*

The analyst should set the allowable element shapes if s/he plans on meshing with an element type that can take on more than one shape. For instance, many area elements can both be triangular and quadrilateral shaped within the same meshed area. Volume elements can often be either hexahedral or tetrahedral shaped, but a mixture of the two shapes in the same model is not recommended.

3.7.3.3 *Choosing Free or Mapped Meshing*

In addition to specifying element shape, the analyst may also want to specify the type of meshing that should be used to mesh the model. This is done by setting the meshing key: **Main Menu> Preprocessor> Meshing> MeshTool** or **Main Menu> Preprocessor> Meshing> Mesher Opts**.

Use the MeshTool (**Main Menu> Preprocessor> Meshing> MeshTool**) to specify meshing type. The MeshTool is the recommended method. Together, the settings for element shape and meshing type affect the resulting mesh.

3.7.3.4 *Controlling Placement of Mid-side Nodes*

When meshing with quadratic elements, the placement of mid-side nodes can be controlled. The choices for mid-side node placement are:

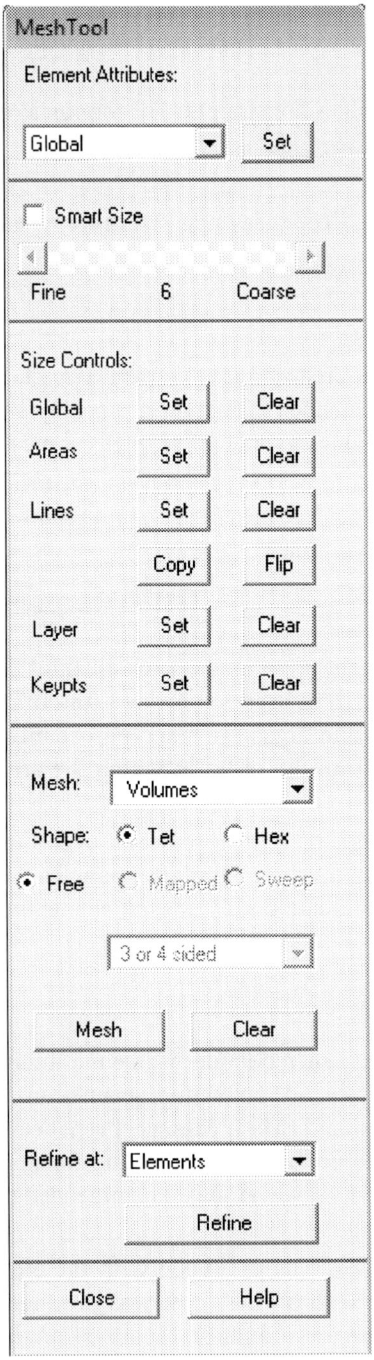

Fig. 3.32 Mesh tool.

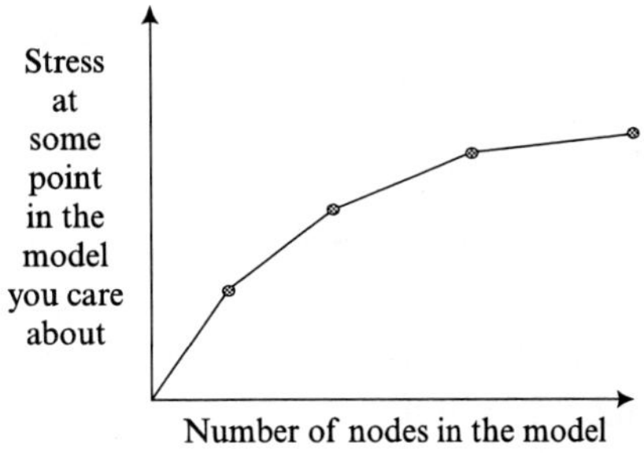

Fig. 3.33 Stress convergence with mesh fineness.

- Mid-side nodes of elements on a region boundary follow the curvature of the boundary line or area. This is a default.
- Place mid-side nodes of all elements so that element edges are straight. This option allows a coarse mesh along curves. However, the curvature of the model is not matched.
- Do not create mid-side nodes.

To control mid-side node placement, use: **Main Menu> Preprocessor> Meshing> Mesher Opts.**

3.7.3.5 Smart Element Sizing for Free Meshing

Smart element sizing (SmartSizing) is a meshing feature that creates initial element sizes for free meshing operations. SmartSizing gives the mesher a better chance of creating reasonably shaped elements during automatic mesh generation. This feature provides a range of settings for meshing.

By default, the default element size method of element sizing will be used during free meshing. However, it is recommended that SmartSizing be used instead for free meshing. To turn SmartSizing on, simply specify an element size level on the SmartSizing command.

Advantages of SmartSizing

The SmartSizing algorithm first computes estimated element edge lengths for all lines in the areas or volumes being meshed. The edge lengths on these lines are then refined for curvature and proximity of features in the geometry. Since all lines and areas are sized before meshing begins, the quality of the generated mesh is not dependent on the order in which the areas or volumes are meshed.

If quadrilateral elements are used for area meshing, SmartSizing tries to set an even number of line divisions around each area so that an all-quadrilateral mesh is possible. Triangles will be included in the mesh only if forcing all quadrilaterals would create poorly shaped elements, or if odd divisions exist on boundaries.

3.7.3.6 Default Element Sizes for Mapped Meshing

The default element size command allows the analyst to modify such defaults as: the minimum and maximum number of elements that will be attached to an unmeshed line, maximum spanned angle per element, and minimum and maximum edge length. This command (**Main Menu> Preprocessor> Meshing> Size Cntrls> Global> Other**) is always used to control element sizing for mapped meshing. The settings are also used by default for free meshing. However, it is recommended to use SmartSizing instead for free meshing operations.

As an example, the mapped mesh on the left in Fig. 3.34 was produced with the element size defaults that exist when entering the program. The mesh on the right was produced by modifying the minimum number of elements (*MINL*) and the maximum spanned angle per element (*ANGL*).

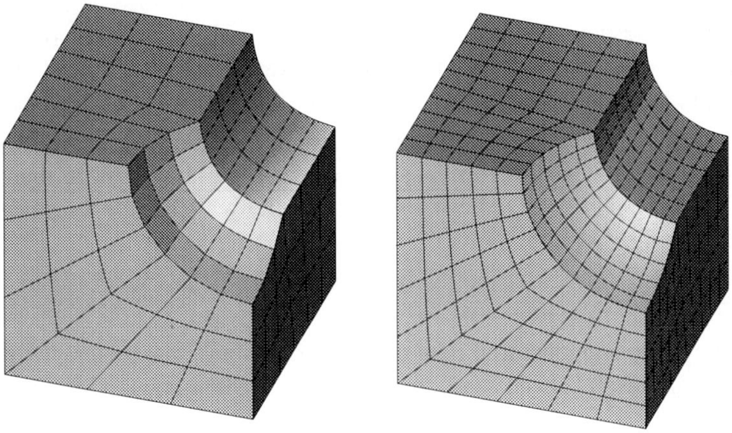

Fig. 3.34 Changing default element sizes.

3.7.3.7 *Local Mesh Controls*

In many cases, the mesh produced by default element sizes is not appropriate due to the physics of the structure. Examples include models with stress concentrations or singularities. In these cases, the analyst will have to get more involved with the meshing process. More control can be taken by using the following element size specifications:

- Use **Main Menu> Preprocessor> Meshing> Size Cntrls> Global> Size** to control the global element size in terms of the element edge length used on surface boundaries or the number of element divisions per line.
- Use **Main Menu> Preprocessor> Meshing> Size Cntrls> Keypoints> All KPs or Main Menu> Preprocessor> Meshing> Size Cntrls> Keypoints> Picked KPs** or **Main Menu> Preprocessor> Meshing> Size Cntrls> Keypoints> Clr Size** to control the element sizes near specified keypoints.
- Use **Main Menu> Preprocessor> Meshing> Size Cntrls> Lines> All Lines or Main Menu> Preprocessor> Meshing> Size Cntrls> Lines> Picked Lines or Main Menu> Preprocessor> Meshing> Size Cntrls> Lines> Clr Size** to control the number of elements on specified lines.

3.7.3.8 *Interior Mesh Controls*

The mesh can also be controlled on the interior of an area where there are no lines to guide the size of the mesh. To do so, use: **Main Menu> Preprocessor> Meshing> Size Cntrls> Global> Area Cntrls.**

3.7.4 Controls Used for Free and Mapped Meshing

3.7.4.1 *Free Meshing*

In *free meshing* operations, no special requirements restrict the solid model. Any model geometry, even if it is irregular, can be meshed. The element shapes used will depend on whether areas or volumes are being meshed. For area meshing, a free mesh can consist of only quadrilateral elements, only triangular elements, or a mixture of the two. For volume meshing, a free mesh is usually restricted to tetrahedral elements.

If the chosen element type is strictly triangular or tetrahedral, the program will use only that shape during meshing. However, if the chosen element type allows more than one shape (for example, PLANE183 or SOLID186), the analyst can specify which shape to use by: **Main Menu> Preprocessor> Meshing> Mesher Opts.**

Also specify that free meshing should be used to mesh the model, using: **Main Menu> Preprocessor> Meshing> Mesher Opts.**

For area elements that support more than one shape, a mixed shape mesh will be produced by default. An all triangle mesh can be requested, but is not recommended if lower-order elements are being used.

In order to achieve a free volume mesh, choose an element type that allows only a tetrahedral shape, or use an element that supports multiple shapes and set the shape option to tetrahedral only.

3.7.4.2 Mapped Meshing

The analyst can specify that the program uses all quadrilateral area elements, all triangle area elements, or all hexahedral volume elements to generate a *mapped mesh*. Mapped meshing requires that an area or volume be 'regular', that is, it must meet certain criteria.

3.7.5 Meshing the Solid Model

Once solid model is built, element attributes are established, and meshing controls are set, ANSYS is ready to generate the finite element mesh. First, however, it is usually good practice to save the model before initiating mesh generation: **Utility Menu> File> Save as Jobname.db.**

It may also be necessary to turn on the 'mesh accept/reject' prompt by picking **Main Menu> Preprocessor> Meshing> Mesher Opts**. This feature allows to easily discard an undesirable mesh.

If multiple volumes or areas are being meshed at one time, consider using the meshing option **By Size** so the mesh is created in the smallest volume or area first. This helps ensure that the mesh is adequately dense in smaller volumes or areas and that the mesh is of a higher quality.

3.7.6 Changing the Mesh

If it is decided that the generated mesh is not appropriate, the mesh can easily be changed by one of the following methods:

- Remesh with new element size specifications
- Use the accept/reject prompt to discard the mesh, then remesh
- Clear the mesh, redefine mesh controls, and remesh
- Refine the mesh locally
- Improve the mesh.

3.8 REVISING THE MODEL

ANSYS offers various methods for revising and refining the model.

3.8.1 Refining a Mesh Locally

There are generally two situations in which it may be necessary to refine a mesh in a local region:

- a model has been meshed and the analyst would like a finer mesh in specific regions of the model, or
- the analyst has completed the analysis and, based on the results, would like a more detailed solution in a region of interest.

For all area meshes and for volume meshes composed of tetrahedra, the ANSYS program allows to refine the mesh locally around specified nodes, elements, keypoints, lines, or areas. Meshes composed of volume elements other than tetrahedra cannot be locally refined.

3.8.1.1 Refining a Mesh

Follow these two steps to refine a mesh:

1. Select the entity around which refinement will be done.
2. Specify the level of refinement to be done. The refined elements will always be smaller than the original elements; the local mesh refinement process does not provide mesh coarsening (*LEVEL*).

3.8.1.2 Refinement Commands and Menu Paths

Use the following menu paths to select entities for refinement and to set refinement controls.

- Use **Main Menu> Preprocessor> Meshing> Modify Mesh> Refine At> Nodes** to refine around a selected set of nodes.
- Use **Main Menu> Preprocessor> Meshing> Modify Mesh> Refine At>Elements** or **Main Menu > Preprocessor> Meshing> Modify Mesh> Refine At> All** to refine around a selected set of elements.
- Use **Main Menu> Preprocessor> Meshing> Modify Mesh> Refine At> Keypoints** to refine around a selected set of keypoints.
- Use **Main Menu> Preprocessor> Meshing> Modify Mesh> Refine At> Lines** to refine around a selected set of lines.
- Use **Main Menu> Preprocessor> Meshing> Modify Mesh> Refine At> Areas** to refine around a selected set of areas.

Figure 3.35 shows examples of mesh refinement around a node, element, keypoint, and line. Figure 3.36 illustrates the use of the command to perform tetrahedral mesh refinement around an area.

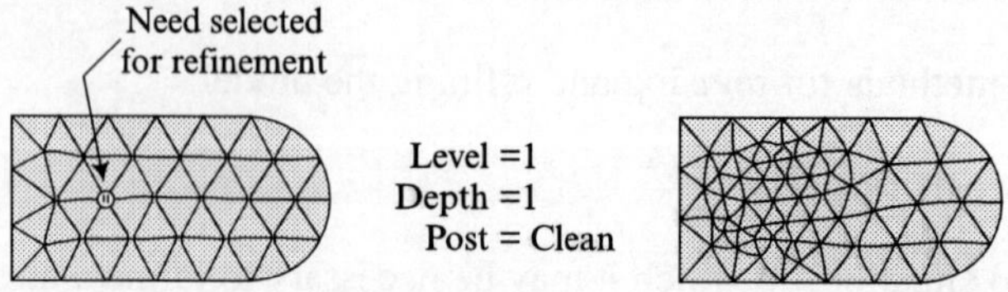

Fig. 3.35 Mesh refinement around a node.

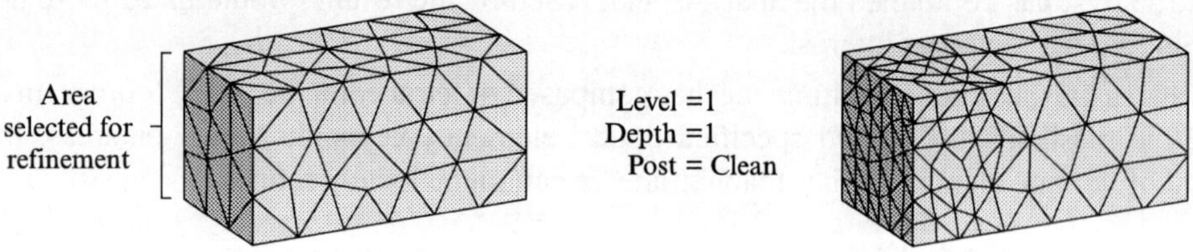

Fig. 3.36 Tetrahedral mesh refinement around an area.

3.8.1.3 *Transfer of Attributes and Loads*

Element attributes associated with the 'parent' element are automatically transferred to all of the 'child' elements. These attributes include element type, material properties, real constants, and element coordinates.

Loads and boundary conditions applied to the solid model are transferred to nodes and elements when the solution is initiated. Therefore, solid model loads will be correctly applied to the new nodes and elements created during refinement. However, loads and boundary conditions applied at the node and element level cannot be transferred to new nodes and elements created during refinement. If there are such loads in a region selected for refinement, the program will not allow refinement to take place unless the loads are first deleted. Therefore, it is recommended that loads should be applied only to the solid model rather than directly to nodes and elements if using mesh refinement is anticipated.

3.8.2 Keeping Track of Element Faces and Orientations

If the model contains shell elements, and if surface loads are applied, the analyst needs to keep track of the element faces in order to be able to define the proper direction for the loads. In general, shell surface loads will be applied to element face one, and will be positive in accordance with the right-hand rule (following the I, J, K, L nodal sequence, as illustrated in Fig. 3.37). If shell elements are created by meshing a solid model area, the normal direction of the elements will be consistent with the normal direction of the area. The area's normal direction can be determined by executing menu path **Utility Menu > List > Areas**; the direction of the sequence of lines defining that area will define the normal direction by the right-hand rule.

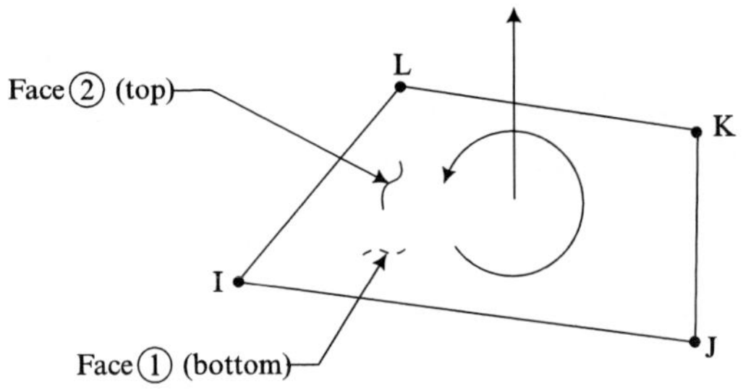

Fig. 3.37 Positive normal direction as defined by the right-hand rule.

3.8.3 Revising a Meshed Model: Clearing and Deleting

Because of the solid modeling cross-reference checking that the ANSYS program performs, meshed solid model entities cannot be deleted. In order to revise the model, the analyst needs to clear solid model entities of their meshes by using the mesh clearing commands. These clearing commands can be thought of as the inverse of the meshing commands. After clearing the model, the analyst can proceed to modify the solid model as desired.

3.8.3.1 Clearing a Mesh

The mesh clearing commands delete the nodes and elements associated with the corresponding solid model entity. When a higher level entity is cleared, all lower level entities will be automatically cleared, unless those lower entities are themselves meshed with elements. Nodes on the boundary of an entity shared by an adjoining meshed entity are not deleted as a result of clearing.

- Use **Main Menu> Preprocessor> Meshing> Clear> Keypoints** to delete nodes and point elements associated with selected keypoints.
- Use **Main Menu> Preprocessor> Meshing> Clear> Lines** to delete nodes and line elements associated with selected lines.
- Use **Main Menu> Preprocessor> Meshing> Clear> Areas** to delete nodes and area elements associated with selected areas.
- Use **Main Menu> Preprocessor> Meshing> Clear> Volumes** to delete nodes and volume elements associated with selected volumes.

The program will report how many of each kind of entity have been cleared after a mesh clearing operation. An entity is considered to have been 'cleared' if either its elements or its nodes have been cleared. If the elements/nodes being cleared are at the end of the element/node lists, then the next available element/node ID is reset accordingly.

3.8.3.2 Deleting Solid Model Entities

Solid model entities can be deleted with the entity deletion commands described below. Lower level entities cannot be independently deleted if they are attached to a higher level entity. Thus, if a block is created using a geometric primitive command, the analyst cannot selectively delete a keypoint that is associated with that block, unless he/she first deletes, in descending hierarchical order, all the higher level entities that are attached to that keypoint.

- Use **Main Menu> Preprocessor> Modeling> Delete> Area and Below** or **Main Menu> Preprocessor> Modeling> Delete> Areas Only** to delete unmeshed areas.
- Use **Main Menu> Preprocessor> Modeling> Delete> Keypoints** to delete unmeshed keypoints.
- Use **Main Menu> Preprocessor> Modeling> Delete> Line and Below** or **Main Menu> Preprocessor> Modeling> Delete> Lines Only** to delete unmeshed lines.
- Use **Main Menu> Preprocessor> Modeling> Delete> Volume and Below** or **Main Menu> Preprocessor> Modeling> Delete> Volumes Only** to delete unmeshed volumes.

3.8.3.3 Modifying Solid Model Entities

The geometry of a solid model can be modified by changing the position of its keypoints using: **Main Menu> Preprocessor> Modeling> Move/Modify> Keypoints> Set of KPs** or **Main Menu> Preprocessor> Modeling> Move/Modify> Keypoints> Single KP**

Any meshed regions attached to modified keypoints will be automatically cleared of nodes and elements. All lines, areas, and volumes attached to the modified keypoint will then be automatically redefined using the active coordinate system.

Unmeshed lines can be modified using the operations described below. These operations will also update attached unmeshed areas, even if these areas are attached to volumes.

- Use **Main Menu> Preprocessor> Modeling> Operate> Booleans> Divide> Line into 2 Lns** or **Main Menu> Preprocessor> Modeling> Operate> Booleans> Divide> Line into N Lns** or **Main Menu> Preprocessor> Modeling> Operate> Booleans> Divide> Lines w/ Options** to divide a single line into two or more lines.
- Use **Main Menu> Preprocessor> Modeling> Operate> Booleans> Add> Lines** to combine adjacent lines into one line.
- Use **Main Menu> Preprocessor> Modeling> Create> Lines> Lines> Line Fillet** to generate a fillet line between two intersecting lines.

3.8.4 Understanding Solid Model Cross-Reference Checking

Several conditions restrict the analyst as s/he modifies the meshed solid model. These restrictions arise due to the cross-reference checking that has been incorporated into the ANSYS program to prevent contamination of the solid model and finite element model data. They may be summarized as follows:

- Meshed keypoints, lines, areas, or volumes may not be deleted or moved.
- Nodes or elements associated with keypoints, lines, areas, or volumes may not be moved. They can be deleted only with the mesh clearing commands.
- Areas contained in volumes may not be deleted or changed.
- Lines contained in areas may not be deleted or changed.
- Keypoints contained in lines may not be deleted.

The basic reasoning behind these rules might be visualized from Fig. 3.38. Schematically, the completed model can be thought of as a stack of blocks, in which the bottommost block represents the keypoints, the next block represents lines, and so forth. If a lower entity was to be disturbed, the analyst would disturb the entities that are stacked on top of it. This illustration somewhat oversimplifies the dependence of higher level entities on lower entities.

Fig. 3.38 Solid model cross-reference checking.

3.9 DIRECT GENERATION

Direct generation is the approach in which the nodes and elements of a model are defined directly. Many convenience commands allow the analyst to copy, reflect, scale, etc., a given pattern of nodes or elements. However, direct generation requires about ten times as many data entries to define a model as compared to solid modeling. Planning, coordinate systems, and working planes apply to direct generation as well as to solid modeling.

A model assembled by direct generation is defined in terms of nodes and elements. Even though node and element generation operations can be interspersed, no one element can be defined until after all of its nodes have been created.

3.9.1 Nodes

This section describes various GUI paths related to the direct generation of nodes. Use **Main Menu> Preprocessor> Modeling> Create> Nodes> In Active CS** or **> On Working Plane** to define individual nodes in the active coordinate system. Use **Main Menu> Preprocessor> Modeling> Create> Nodes> On Keypoint** to define a node at an existing keypoint.

3.9.2 Elements

3.9.2.1 Prerequisites for Defining Element Attributes

Define the minimum number of nodes required for that element and specify the proper element attributes before directly generating elements. Assemble tables of element attributes.

Once the element attribute tables are in place, the analyst can 'point' to various entries in the element tables. The values of these pointers, that are in effect at the time of creating the elements, are used by the program to assign attributes from the tables to the elements.

3.9.2.2 Defining Elements

After defining the necessary nodes and setting the element attributes, the analyst can proceed to define the elements. An element can be defined by identifying its nodes using **Main Menu>**

Preprocessor> Modeling> Create> Elements> Auto Numbered> Thru Nodes or Main Menu> Preprocessor> Modeling> Create> Elements> User Numbered> Thru Nodes. The number of nodes required for each element and the order in which they must be input are determined by the element type. The order in which nodes are defined determines the element normal direction.

3.9.2.3 Reading and Writing Text Files that Contain Element Data

ANSYS program can read or write a text file that contains element data. These capabilities can be useful for transferring data to and from another program. Normally, it will not be necessary to use these capabilities in a standard ANSYS model generating session.

3.9.2.4 Overlapping Elements

If overlapping elements are created, various ANSYS features such as graphics, surface loads, selecting logic, etc., might not function as expected. It is best to avoid the use of overlapping elements altogether; if this is not possible, use extreme caution whenever overlapping elements are employed.

3.10 NUMBER CONTROL AND ELEMENT REORDERING

The numbering of keypoints, lines, areas, volumes, elements, nodes, element types, real constant sets, materials, coupled DOF sets, constraint equations, and coordinate systems can be controlled as per requirement. Control of the numbering is also useful when combining separately modeled portions of a model into one.

3.10.1 Number Control

The GUI paths described here give the analyst control of the numbering. Control of the numbering is useful when combining separately modeled portions of a model into one.

3.10.1.1 Merging Coincident Items

If two separate entities have the same location, these entities can be merged together into a single entity using **Main Menu> Preprocessor> Numbering Ctrls> Merge Items.**

For example, suppose that there are two separate but coincident nodes. If the nodes are merged, the higher numbered node will be deleted and will be replaced with the lower numbered coincident node. Two coincident nodes will thus be replaced by a single node.

3.10.1.2 Compressing Item Numbers

As the model is built, the analyst creates unused slots in the numbering sequence for various items by deleting, clearing, merging, or performing other operations. These slots will remain empty for some items but will be filled in for other items as new items are created. These gaps can be eliminated by 'compressing' the numbering to save data storage space or to preserve desired sequencing. Use **Main Menu> Preprocessor> Numbering Ctrls> Compress Numbers.** The compression operation may be selectively re-executed for chosen item groups or may be simultaneously applied to all valid items.

3.10.1.3 *Setting Starting Numbers*

When creating new, automatically-numbered items, it is suggested to set the starting number of the new series of items higher than the greatest number used by the existing items. By doing so, it is ensured that the newly created entities will be consecutively numbered, and will thus be prevented from occupying gaps in the existing numbering sequence. Also, while creating portions of the model independently of one another, a set of starting numbers is specified to avoid numbering conflicts when they are combined together into one model. Such user-defined starting numbers can be specified using: **Main Menu> Preprocessor> Numbering Ctrls> Set Start Number.**

3.10.1.4 *Adding Number Offsets*

In order to combine two independently created portions of a model and to prevent numbering conflicts, all selected items can be renumbered by *adding* an offset value to their existing numbers. Use **Main Menu> Preprocessor> Numbering Ctrls> Add Num Offset.** Re-execute this operation for each item group that is to be renumbered.

3.10.2 Element Reordering

At the start of a solution, the ANSYS program will automatically reorder the elements so that they can be accessed in close proximity to each other. This reordering scheme could improve the memory usage and, hence, the speed of the global assembly. This operation does not affect the element numbers used to identify elements in pre- and postprocessing of the database. In most cases, allowing the ANSYS program to reorder elements automatically is the preferred approach. The original element ordering can be reestablished using: **Main Menu> Preprocessor> Numbering Ctrls> Element Reorder> Reset Elem Order.**

3.11 PROBLEMS

3.11.1 1-D Problems

W1: Consider a 2 m long steel bar of 50 mm^2 cross-sectional area as shown in figure below. Use two-element mesh to model this problem.

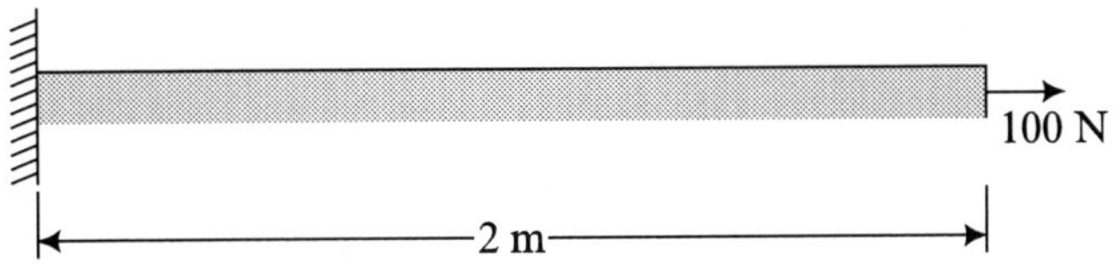

Solution:

Preprocessing

1. Change jobname

   File> Change Jobname

   Enter 'Bar with tip load', and click on 'OK'.

2. Define element types

   Preprocessor> Element Type> Add/Edit/Delete

   Click on 'Add..', highlight 'Link', then '3D finit stn 180', click on 'OK', then 'Close'.

3. Define the real constants for the Link180, which are cross-sectional area and initial strain:

   Preprocessor> Real Constants> Add

   Click 'OK' for 'Type 1 LINK180'

   In this problem, there is no initial strain (leave blank), and the area is 50. After filling in the area value, click on 'OK', then 'Close'.

4. Create nodes in the bar

   Preprocessor> Modeling> Create> Nodes> In Active CS

   Enter 1 for node number (ANSYS would automatically number nodes if this column is left blank).

   Enter the location as (X, Y, Z) = (0, 0, 0). Click on 'Apply'. Continue defining nodes 2&3 using the locations based on the sketch of the truss, but after entering the node 3 location, click on 'OK' instead of 'Apply'. So, node 2 is at (X,Y) = (500,100) and node 3 is at (X,Y) = (1000, 0).

   As a check to ensure all nodes were entered correctly, list the nodes.

   Utility Menu> List> Nodes

   Turn on node numbering.

   Utility Menu> PlotCtrls> Numbering

5. Create link elements between nodes

   Preprocessor> Create> Elements> Auto Numbered> Thru Nodes

A picking menu appears. Pick node 1, then node 2, and click on 'Apply' in the Picking Menu. Pick node 2, then node 3, and click on 'OK' in the Picking Menu.

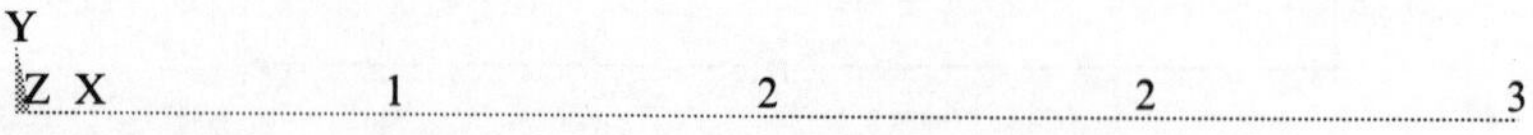

3.11.2 2-D Problems

W2: Mesh the three-sided area shown below with quad elements.

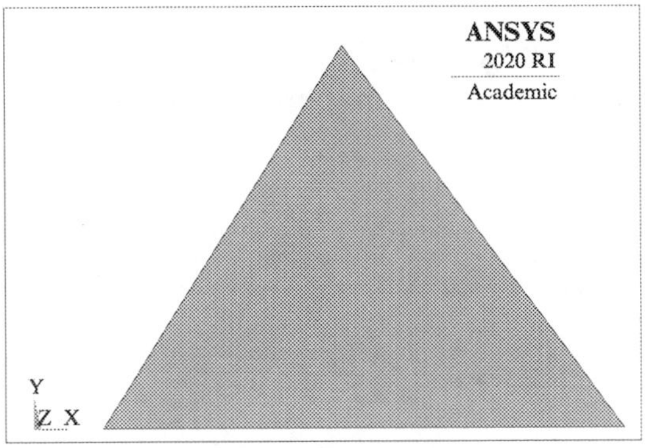

Solution:

1. Open preprocessor menu
   Main Menu> Preprocessor
2. Define keypoints
   Preprocessor> Modeling> Create> Keypoints> On Working Plane..
   Pick 3 keypoints as shown in the above figure.
3. Create areas
   Preprocessor> Modeling> Create> Areas> Arbitrary> Through KPs
   Define area through keypoints 1, 2, 3
   Select keypoints 1, 2 and 3 and then select 'OK'.
4. Define the Type of Element
   Element Type> Add/Edit/Delete> Add> Solid> Quad 4 node 182>
   Select 'OK', select 'Close'
5. Mesh the patterns
   Meshing> Mesh> Areas> Mapped> 3 or 4 sided or
   Meshing> MeshTool> Shape (Quad)> Mapped> Mesh
   Pick the area shown

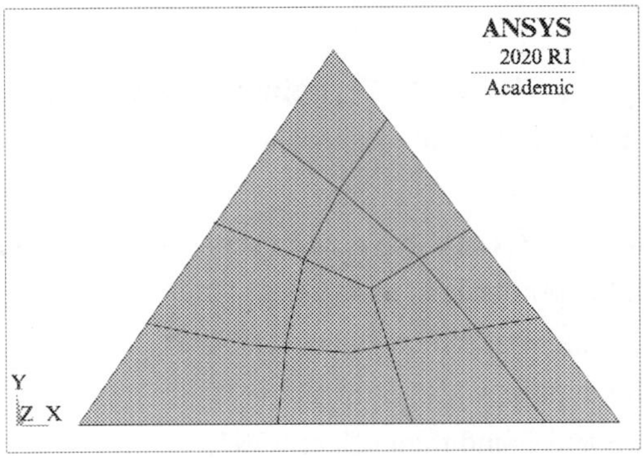

Triangular elements can also be used to create the mesh.
Meshing> MeshTool> Shape (Tri)> Mapped> Mesh

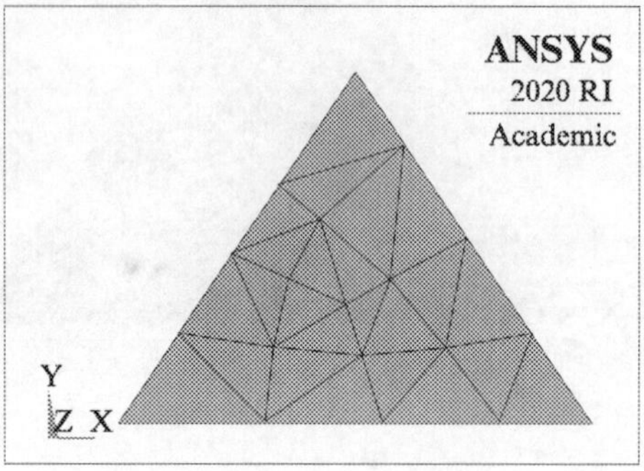

W3: Mesh the region shown below with two adjacent, but distinct areas with quad elements.

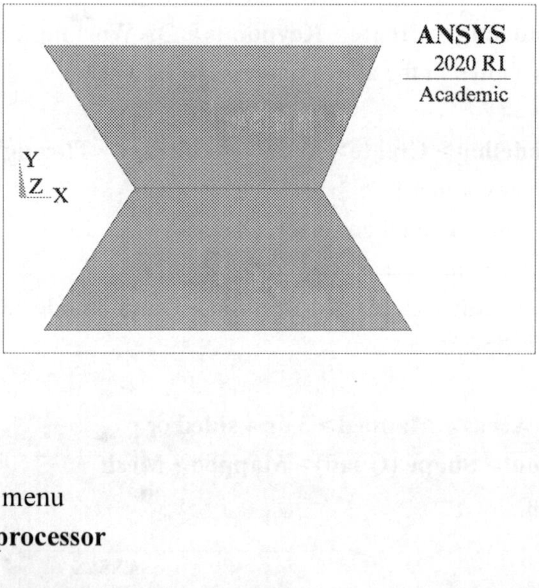

Solution:

1. Open preprocessor menu
   Main Menu> Preprocessor
2. Define Keypoints
   Preprocessor> Modeling> Create> Keypoints> On Working Plane..
   Pick 6 keypoints as shown in the above figure.
3. Create Areas
   Preprocessor> Modeling> Create> Areas> Arbitrary> Through KPs
   Define area-1 through keypoints 1, 2, 3, 4.
   Select keypoints 1, 2, 3 and 4 and then select 'OK'.
   Define area-2 through keypoints 1, 2, 5, 6.
   Select keypoints 1, 2, 5 and 6 and then select 'OK'.

4. Define the Type of Element

   Element Type> Add/Edit/Delete> Add> Solid> Quad 4 node 182

5. Mesh the areas

   Meshing> Mesh> Areas> Mapped> 3 or 4 sided

   Pick the two areas shown.

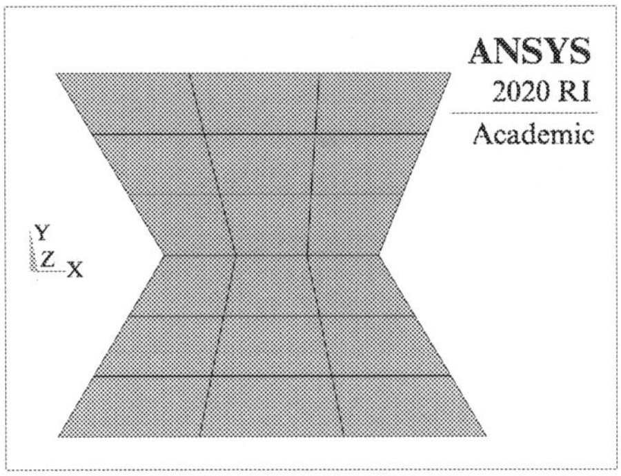

W4: Draw a rectangle of 50 mm × 10 mm. Mesh (1×5) the area. Choose the element type as Solid Quad 4 node 182.

Solution:

1. Create the area

   Preprocessor> Create> Rectangle> By Dimensions

   Enter (0, 50) for (X1, X2) and (0, 10) for (Y1, Y2).

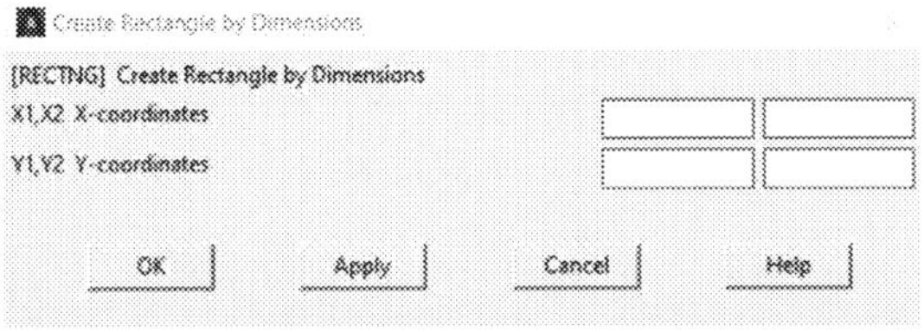
Create Rectangle by Dimensions
[RECTNG] Create Rectangle by Dimensions
X1,X2 X-coordinates
Y1,Y2 Y-coordinates
OK Apply Cancel Help

2. Mesh the area

   Preprocessor> Mesh> Areas

   Pick the area just created and click 'OK'

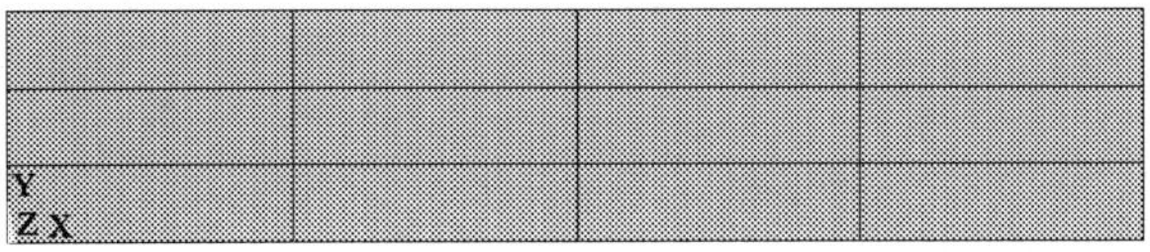

Notice that since element size and shape are not specified, ANSYS used its default values to generate the mesh. Since it is desired to use five elements, first clear the mesh and regenerate it after specifying '**Element shape and size**'.

3. Clear the area

   Preprocessor>Clear> Areas

   Pick the meshed area. Click 'OK'

4. Remesh the area

   Set element shape

   Preprocessor> Element Type> Add/Edit/Delete> Add

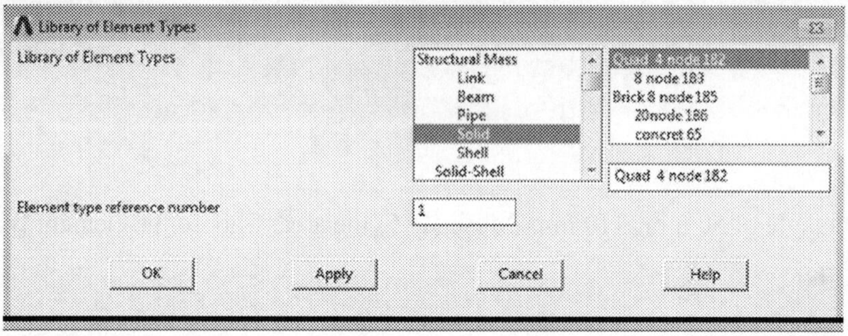

Set element size

Preprocessor> Shape & Size> For Picked Lines

Not For All Lines because different number of divisions are necessary for different lines (i.e., 5 for horizontal lines and 1 for vertical lines). Pick two horizontal lines (lines 1 and 3). Enter 5 for number of element divisions. Click 'Apply'.

Pick two vertical lines (lines 2 and 4). Enter 1 for No. of element divisions.

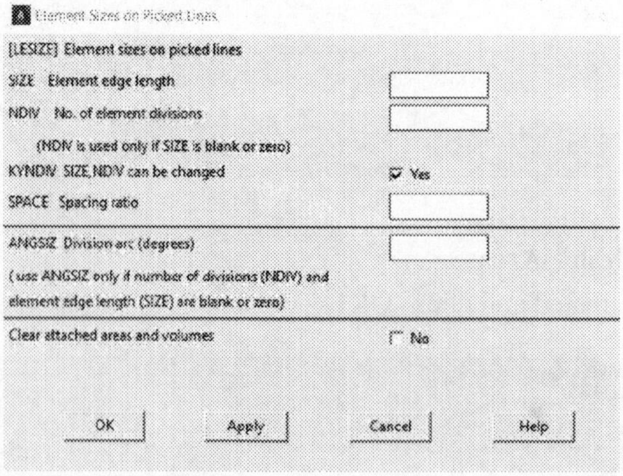

The screen should look something like this:

Plot Elements

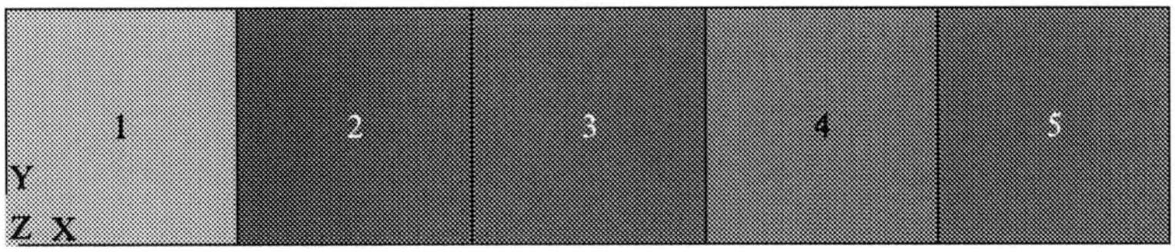

W5: Create the insignia shown below.

Solution:

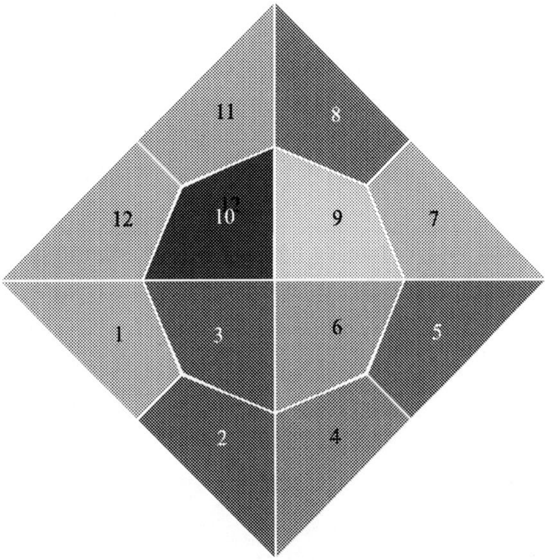

I. Cutlines Method

1. Give example a Title
   Utility Menu> File> Change Title ...
   'Meshing a plate using cutlines'

2. Open prepreocessor menu
   Main Menu > Preprocessor
3. Define Keypoints
   Preprocessor> Modeling> Create> Keypoints> In Active CS...
   Define 4 keypoints as given in the following table:

Keypoints	Coordinates (x, y)
1	(0, 50)
2	(50, 0)
3	(100, 50)
4	(50, 100)

4. Create Area
   Preprocessor> Modeling> Create> Areas> Arbitrary> Through KPs
   Define an area through keypoints 1, 2, 3, 4. Select keypoints 1, 2, 3 and 4 and then select 'OK'.
5. Divide the area into 4 parts using 2 diagonal lines
   (a) Create a line
   Preprocessor> Modeling> Create> Lines> Lines> Straight Line
   (b) Select the middle left keypoint and draw a line up to the middle right keypoint by clicking on that keypoint
6. Select the bottom middle keypoint and draw a line up to the top middle keypoint by clicking on that keypoint
   (a) Now divide the area into 4 areas using the lines by selecting
   Preprocessor> Modeling> Operate> Booleans> Divide> Area by Line
   (b) Select the area and click 'OK' in the 'Divide Area by Line' window
   (c) Now select the lines and click 'OK' in the 'Divide Area by Line' window
   (d) The area is now divided into 4 areas as shown in the figure below.

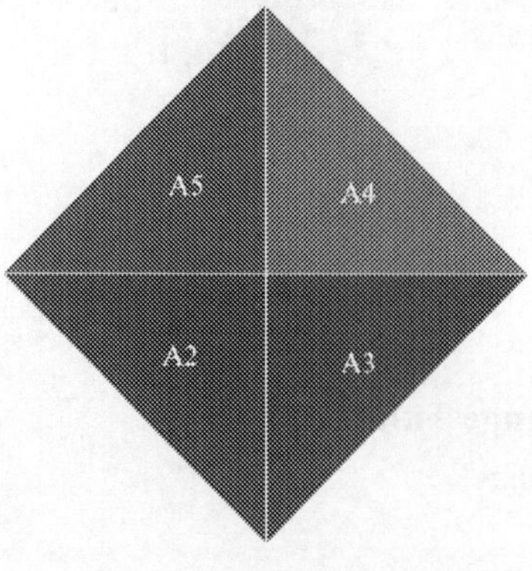

Any time the analyst deletes/modifies items like elements, nodes, areas and tries to create some more, s/he will see that ANSYS doesn't number the entities starting from one. The concept of 'Compressing the numbers' will be used to tackle this problem. Then ANSYS is forced to start numbering from the next available number.

Example: Suppose 1 area is created and for some reason the first one is split into 4. If the analyst tries to create more areas, most probably next one will be numbered as 2 NOT 1. This problem will be avoided if the following concept is used.

7. Compress the numbers

Preprocessor> Numbering Controls> Compress Numbers ...> ALL

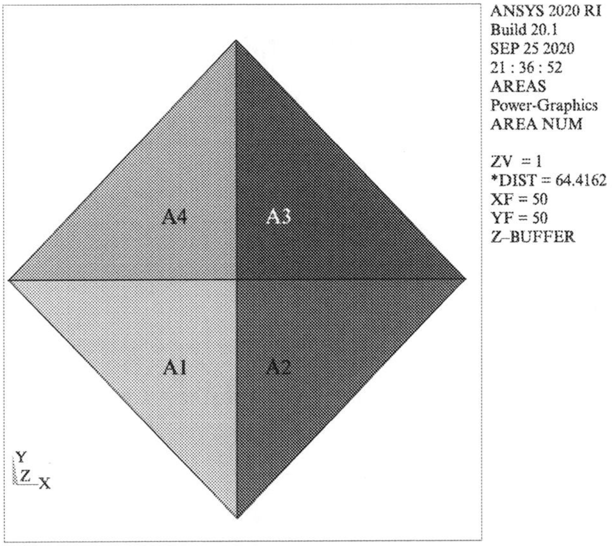

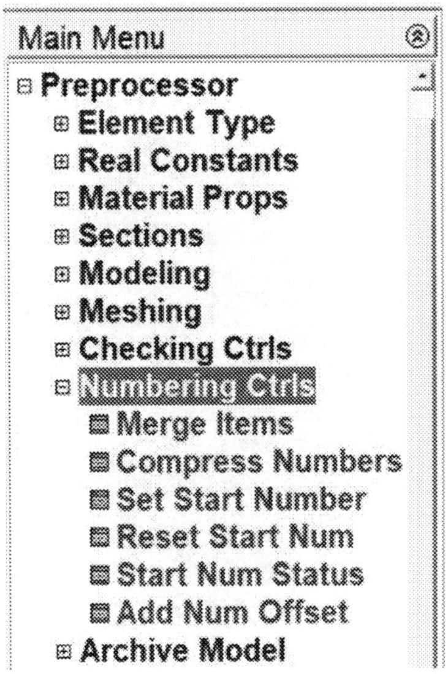

8. Define the Type of Element

   Preprocessor> Element Type> Add/Edit/Delete...> Add...> Solid> Quad 4 node 182

   For this problem use the PLANE182 (plane stress, plane strain or generalized plane strain) element. It is defined by four nodes having two degrees of freedom at each node: translations in the nodal x and y directions.

9. Select Plane Stress with Thickness

   In the Element Types window, select 'Options'... and in Element behavior select 'Plane strs w/thk'

10. Define Real Constants

    Preprocessor> Real Constants> Add/Edit/Delete> Add...> OK

    In the 'Real Constants for PLANE182' window, enter the thickness: 10

11. Define Element Material Properties

    Preprocessor> Material Props> Material Models> Structural> Linear> Elastic> Isotropic

    In the window that appears, enter the following geometric properties for steel:

    (a) Young's modulus EX: 2e5

    (b) Poisson's ratio PRXY: 0.3

12. Define Mesh Size

    Preprocessor> Meshing> Size Cntrls> ManualSize> Lines> All Lines...

    Set No. of element divisions = 2

13. Mesh the frame

    Preprocessor > Meshing > Mesh > Areas > Pick All

    The mesh then appears as shown below.

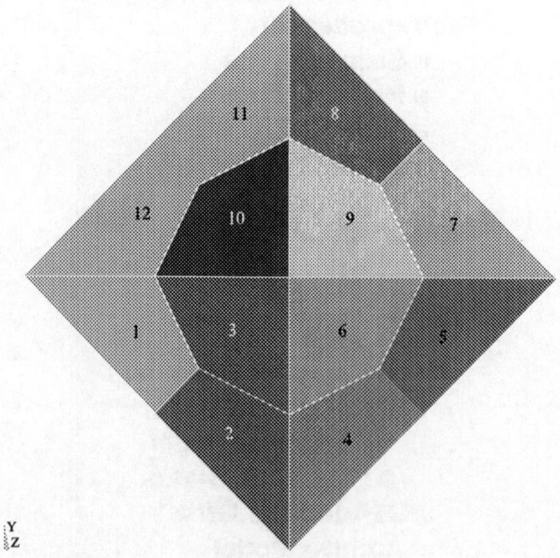

II. Merging Objects Method

1. Clear the memory and start a new model
   Utility Menu> File> Clear & Start New ...
2. Give example a Title
   Utility Menu> File> Change Title ...
   'Meshing a plate by copying elements'
3. Open preprocessor menu
   Main Menu> Preprocessor
4. Define Keypoints
   Preprocessor> Modeling> Create> Keypoints> In Active CS...
   Define 3 keypoints as given in the following table:

Keypoints	Coordinates (x, y)
1	(0,50)
2	(50,50)
3	(50,0)

5. Create Area
   Preprocessor> Modeling> Create> Areas> Arbitrary> Through KPs
   Define an area through keypoints 1, 2, 3. Select keypoints 1, 2 and 3 and then select 'OK'.
6. Define the Type of Element
   Preprocessor> Element Type> Add/Edit/Delete...> Add...> Solid> Quad 4 node 182
7. Select Plane Stress with Thickness
   In the Element Types window, select 'Options' and in Element behavior select 'Plane strs w/thk'
8. Define Real Constants
   Preprocessor > Real Constants > Add/Edit/Delete > Add...
   In the 'Real Constants for PLANE182' window, enter the thickness: 10
9. Define Element Material Properties
   Preprocessor> Material Props> Material Models> Structural> Linear> Elastic> Isotropic
   In the window that appears, enter the following geometric properties for steel:
   (a) Young's modulus EX: 2e5_
   (b) Poisson's ratio PRXY: 0.3
10. Define Mesh Size
    Preprocessor> Meshing> Size Cntrls> ManualSize> Lines> All Lines...
    Set No. of element divisions to 2 to obtain the desired mesh.
11. Mesh the area
    Preprocessor> Meshing> Mesh> Areas> Pick All

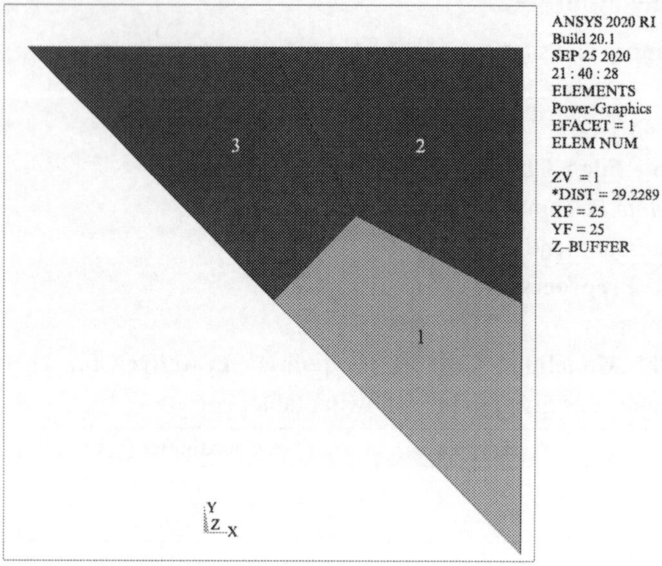

12. Mirror the geometry

 Create local coord system to mirror geom.

 Utility Menu> WorkPlane> Local Coordinate Systems> Create Local CS> At specified Loc

 First mirror the geometry about the diagonal from node 2 to 4. Click on the bottom right node (bottom corner) and select 'OK'

 As shown below, create a coordinate system moved 50 mm along X

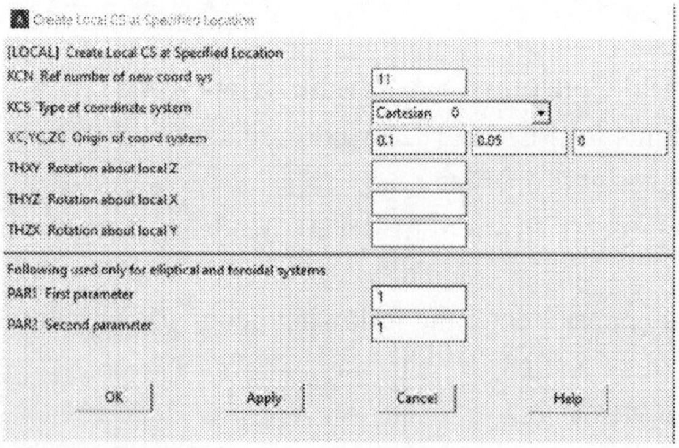

Create Local CS at Specified Location

[LOCAL] Create Local CS at Specified Location

KCN Ref number of new coord sys	11
KCS Type of coordinate system	Cartesian 0
XC,YC,ZC Origin of coord system	0.1 0.05 0
THXY Rotation about local Z	
THYZ Rotation about local X	
THZX Rotation about local Y	
Following used only for elliptical and toroidal systems	
PAR1 First parameter	1
PAR2 Second parameter	1

OK   Apply   Cancel   Help

Next, mirror the geometry. Select:

Preprocessor> Modeling> Reflect> Areas> Pick All

In the window that appears select Y-Z plane X and click 'OK'. This will mirror the geometry about the Y-Z plane.

Use the same technique to obtain the full geometry.

13. Re-activate the global coordinate system
    Utility Menu> WorkPlane> Change Active CS to> Global Cartesian
14. Plot Elements
    Utility Menu> Plot> Elements
    The mesh now appears as follows:

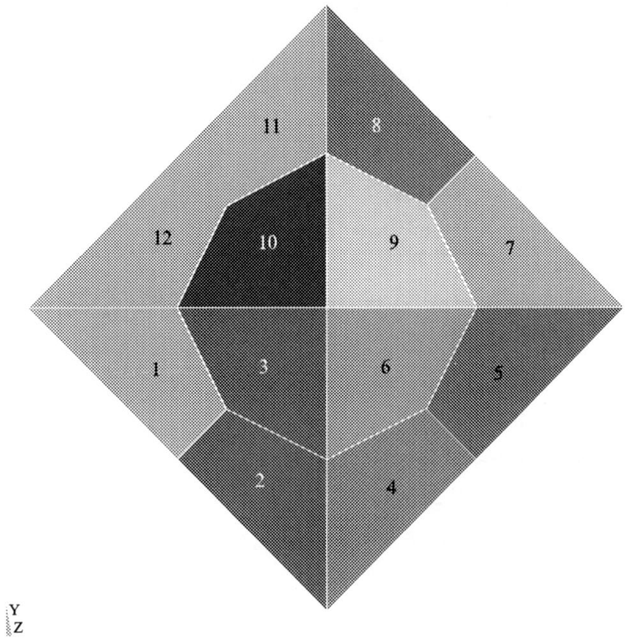

15. Merge duplicate nodes/elements
    Preprocessor> Numbering Ctrls> Merge Items> All

III. Gluing Areas Method

1. Clear the memory and start a new model
   Utility Menu> File> Clear & Start New ...
2. Give example a Title
   Utility Menu> File> Change Title ...
   'Meshing a plate by copying areas'
3. Open preprocessor menu
   Main Menu> Preprocessor
4. Define Keypoints
   Preprocessor> Modeling> Create> Keypoints> In Active CS...
   Define 7 keypoints as given in the following table:

Keypoints	Coordinates (x, y)
1	(0, 50)
2	(50, 50)
3	(50, 0)
4	(25, 50)
5	(25, 25)
6	(50, 25)
7	(33.33, 33.33)

5. Create Area

   Preprocessor> Modeling> Create> Areas> Arbitrary> Through KPs

   Now define 3 areas; (1, 2, 7, 6), (2, 3, 4, 7), (4, 5, 6, 7)

6. Mirror the geometry

   As shown in the previous section, create a local coordinate system and mirror the geometry

   Utility Menu> WorkPlane> Local Coordinate Systems> Create Local CS> At specified Loc

   Then, mirror the geometry, select:

   Preprocessor > Modeling> Reflect> Areas

   Do this twice to obtain the full geometry

7. Re-activate the global coordinate system

   Utility Menu> WorkPlane> Change Active CS to> Global Cartesian

8. Glue the areas together

   Preprocessor> Modeling> Operate> Booleans> Glue> Areas

   Glue the areas together so that the areas are attached but that the subdivided areas remain to give the elements

9. Define the Type of Element

   Preprocessor> Element Type> Add/Edit/Delete...> Add... > Solid> Quad 4 node 182

   As in the previous mesh, use the PLANE182

10. Select Plane Stress with Thickness

    In the Element Types window, select 'Options' and in 'Element behavior' select 'Plane strs w/thk'

11. Define Real Constants

    Preprocessor> Real Constants> Add/Edit/Delete> Add...

    In the 'Real Constants for PLANE182 window, enter thickness: 10

12. Define Element Material Properties

    Preprocessor> Material Props> Material Models> Structural> Linear> Elastic> Isotropic

In the window that appears, enter the following geometric properties for steel:

(a) Young's modulus EX: 2e5

(b) Poisson's ratio PRXY: 0.3

13. Define Mesh Size

Preprocessor> Meshing> Size Cntrls> ManualSize> Areas> All Areas...

Set Element edge length = 1 to obtain the desired mesh

14. Mesh the area

Preprocessor> Meshing> Mesh> Areas> Pick All

And again the desired mesh is obtained:

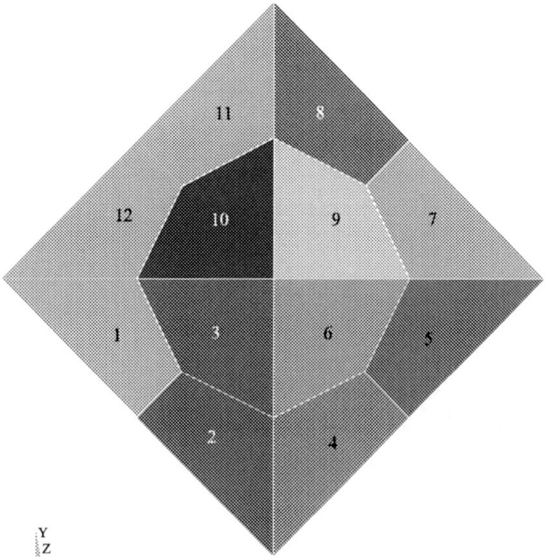

W6: Create pattern shown below and mesh the same.

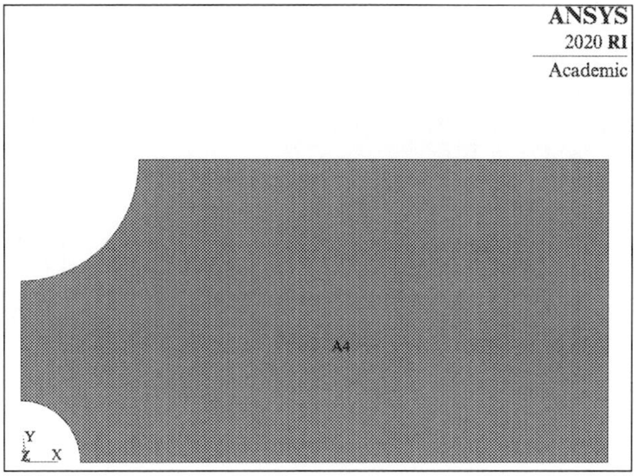

Solution:

1. Turn ON area numbers
   PlotCntls> Numbering> Areas> ON
2. Create the base rectangle
   Preprocessor> Modeling> Create> Areas> Rectangle> By 2Corners
   Enter WP X = 0, WP Y = 0, Width = 10 and Height = 5

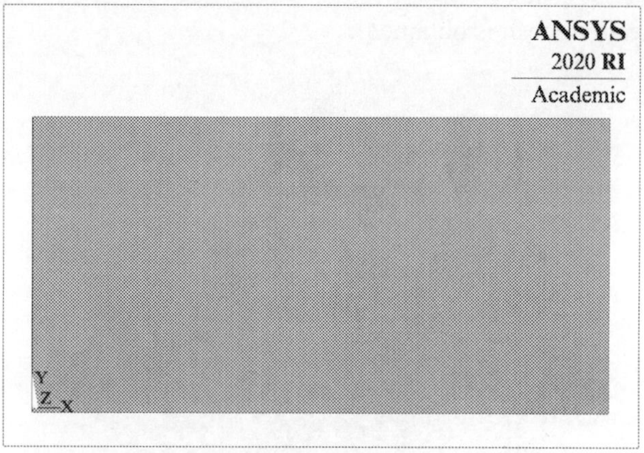

3. Create a circle on lower left corner of the rectangle.
   Preprocessor> Modeling> Create> Areas> Circle> Solid Circle
   Enter WP X = –5, WP Y = – 2.5) and Radius = 1
4. Repeat circle creation
   Enter center WP X = – 5, WP Y = 2.5 and Radius = 2

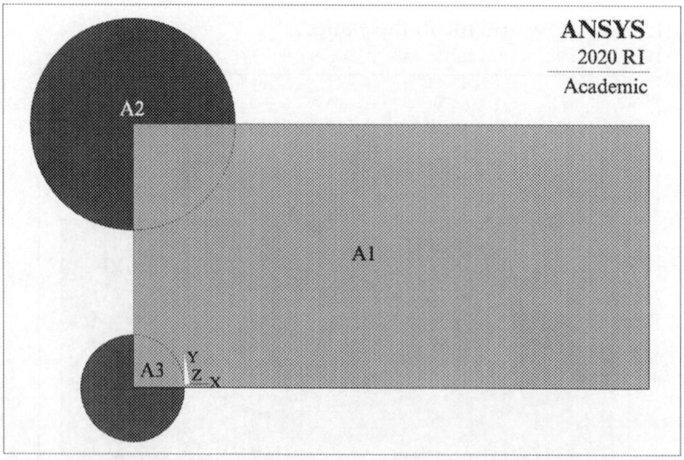

5. Subtract circles from rectangle
   Preprocessor> Modeling> Operate>Subtract> Areas
   Pick base area, pick rectangle, click 'OK'
   Pick areas to be subtracted, pick the two circles, click 'OK'

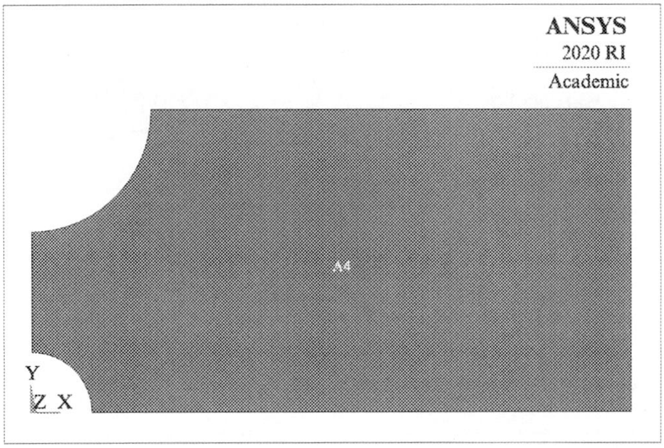

6. Mesh the pattern
   Element Type> Add/Edit/Delete> Add> Solid> Quad 4 node 182
   Meshing> Mesh> Areas> Free
   Pick the area.

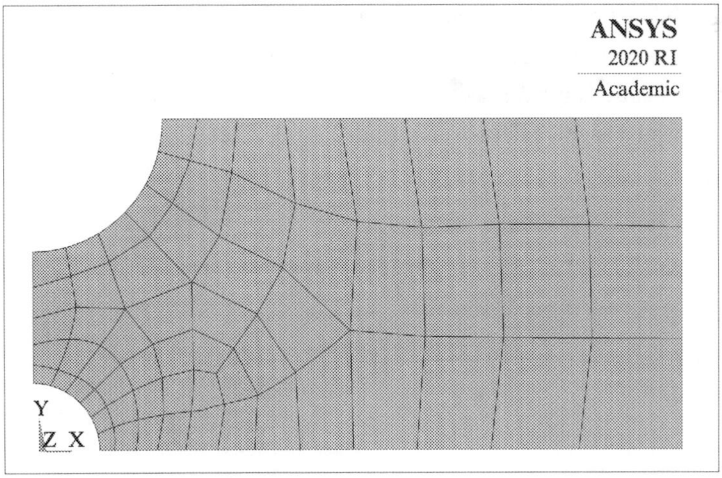

W7: Model a chain link shown below and mesh the same.

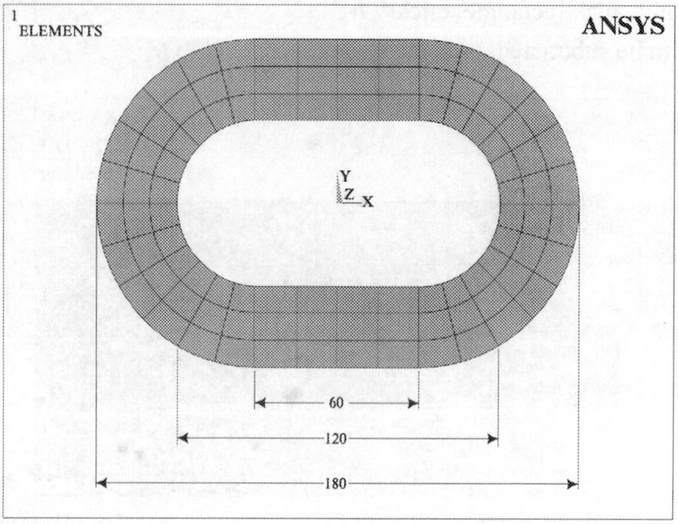

Solution: Since the object is symmetric, first draw one quadrant of the chain and reflect twice to get the full object.

1. Draw the rectangle.

   Main Menu> Modeling> Areas> Rectangle> By Dimensions

   X1 = 0, X2 = 30, Y1 = 30, Y2 = 60
2. Now draw the partial annulus

   Main Menu> Modeling> Areas> Circle> Partial Annulus

   X = 0, Y = 0, Rad-1 = 30, Theta-1 = 0, Rad-2 = 60, Theta-2 = 90

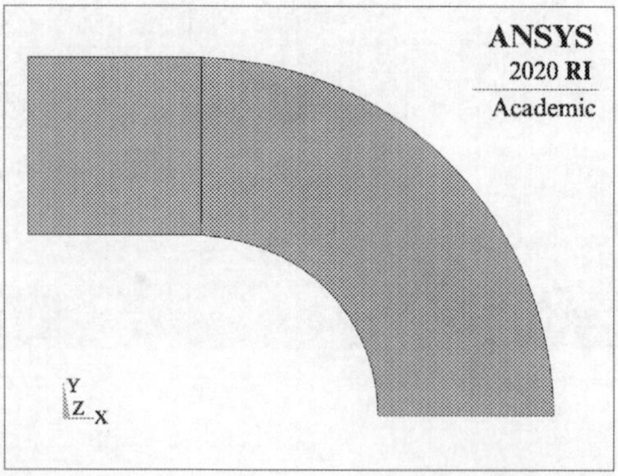

3. Mesh the rectangle

   Main Menu> Meshing> Size Contrls> ManualSize> Picked Lines

   Select vertical lines of the rectangle. Click OK

Set No. of element divisions = 3
Now select horizontal lines.
Set No. of element divisions = 2
Main menu> Preprocessor> Meshing> SizeCntrls> Manual Size> Picked Lines
Select 2 arcs. Set No. of element divisions = 6. Now select 2 radial lines.
Set No. of element divisions = 3.
Main Menu> Preprocessor> Meshing> MeshTool> Mesh
Click partial annulus, OK

4. Draw the left hand portion of the part.
   Main Menu> Preprocessor> Modeling> Reflect> Areas
   Select the two areas. Under Reflect Areas Select YZ plane X under Reflect Areas
5. Repeat reflect operation. Select areas, OK. Select XZ plane y under Reflect Areas.

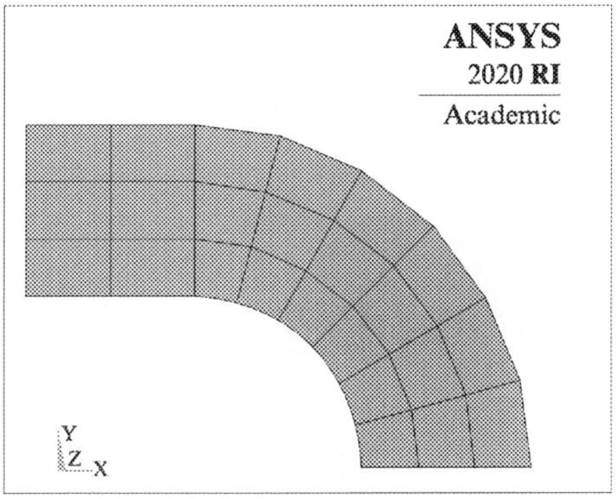

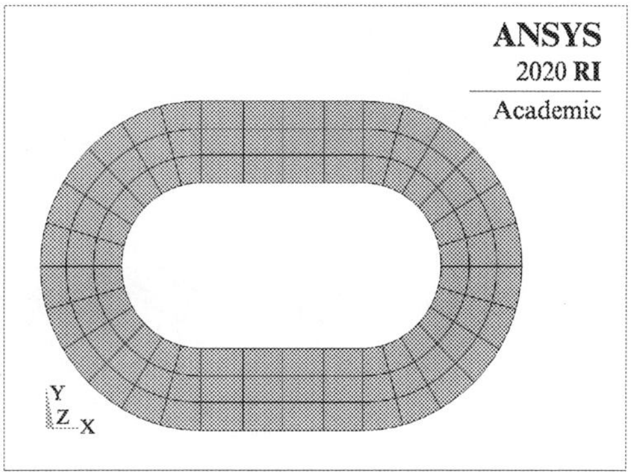

W8: Model a spanner shown below and mesh the same.

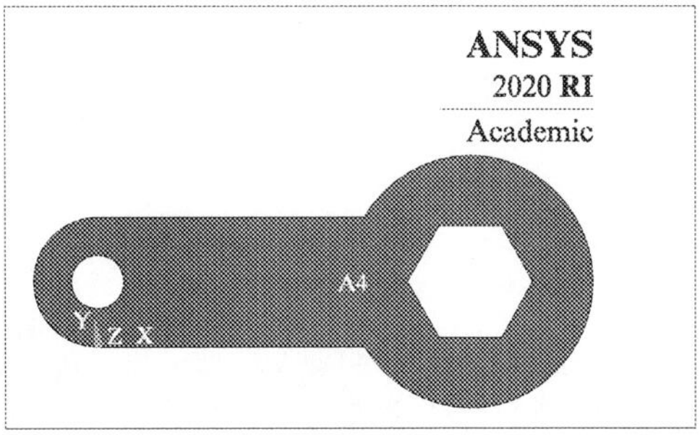

Solution:

1. Turn ON area numbers
   PlotCntls> Numbering> Areas> ON

2. Create the base rectangle
   Preprocessor> Modeling> Create> Areas> Rectangle> By 2 Corners
   Enter WP X = 0, WP Y = 0, Width = 150 and Height = 50

3. Create a circle with center at the mid-point of the left side of the rectangle.
   Preprocessor> Modeling> Create> Areas> Circle> Solid Circle
   Enter WP X = 0, WP Y = 25, and Radius = 25
4. Draw a circle with center at mid-point of the right side.
   Enter WP X = 150, WP Y = 25 and Radius = 50

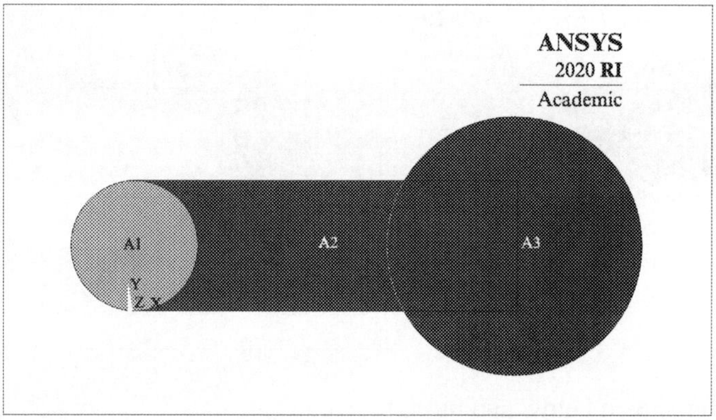

5. Add circles to rectangle
   Preprocessor> Modeling> Operate> Add> Areas
   Pick all areas, then 'OK'.

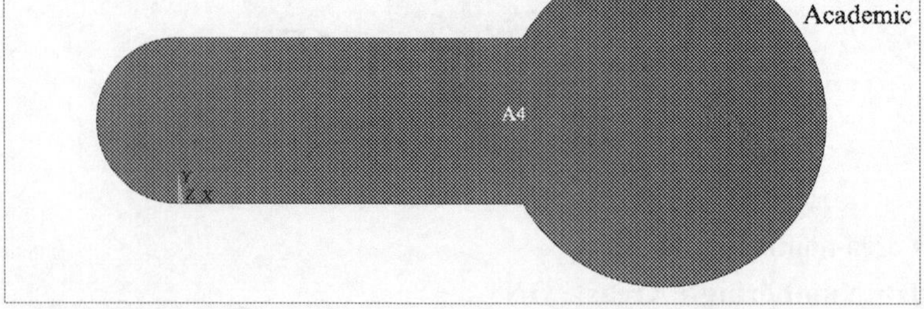

6. Create a circle at the mid-point of the left side of the rectangle.
   Preprocessor> Modeling> Create> Areas> Circle> Solid Circle
   Enter WP X = 0, WP Y – 25) and Radius = 10
7. Create a hexagon at the mid-point of the bigger circle.
   Preprocessor> Modeling> Create> Areas> Polygon> Hexagon
   Enter WP X = 150, WP Y = 25 and Radius = 25
8. Subtract circle and hexagon from rectangle
   Preprocessor> Modeling> Operate>Add> Areas> Pick all
9. Define element type
   Element Type> Add/Edit/Delete> Add> Solid> Quad 4 node 182
10. Mesh the model
   Meshing> Mesh> Areas> Mapped> 3 or 4 sided
   Pick the area

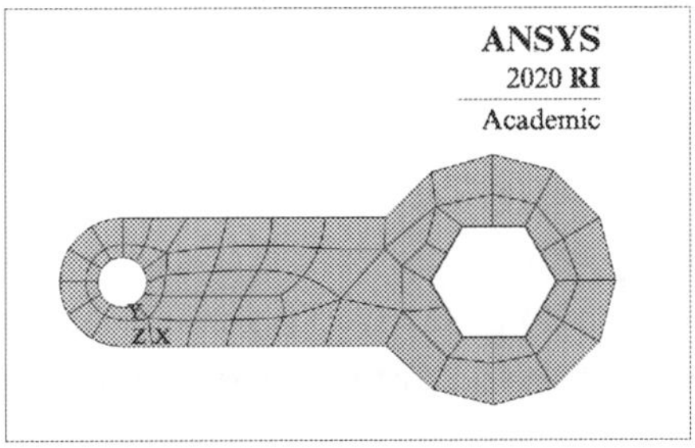

W9: Build a solid 2-D model of an automobile connecting rod using bottom-up modeling technique. Mesh the 2-D connecting rod. Then extrude the area with the mesh to create a 3-D mesh.

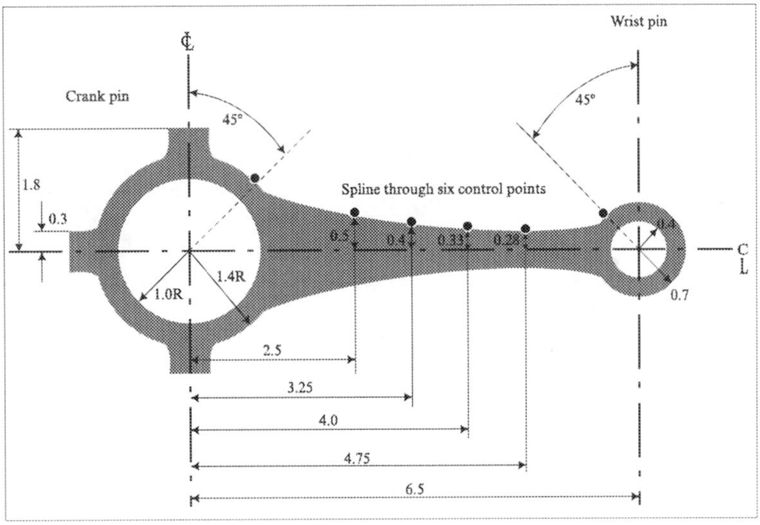

Solution:

1. Enter ANSYS in the working directory using '*C-Rod-2D*' as the jobname.
2. Create two circular areas
   Main Menu> Preprocessor> Modeling> Create> Areas> Circle> By Dimensions ...
   Enter RAD1=1.4, RAD2 = 1, THETA1 = 0, THETA2 = 180
   Next, set THETA1 = 45
3. Turn area numbering on
   Utility Menu> PlotCtrls> Numbering ...
   Set Area numbers ON
4. Create two rectangular areas
   Main Menu> Preprocessor> Modeling Create> Areas> Rectangle> By Dimensions ...
   X1 = – 0.3, X2 = 0.3, Y1 = 1.2, Y2 = 1.8
   X1 = – 1.8, X2 = –1.2, Y1 = 0, Y2 = 0.3
5. Offset working plane to XYZ location (X = 6.5)
   Utility Menu> WorkPlane> Offset WP to> XYZ Locations +
   Enter 6.5 in the ANSYS Input window
6. Set the active coordinate system to be the working plane coordinate system
   Utility Menu> WorkPlane> Change Active CS to> Working Plane
7. Create two more circular areas
   Main Menu> Preprocessor> Modeling> Create> Areas> Circle> By Dimensions ...
   RAD1 = 0.7, RAD2 = 0.4, THETA1 = 0, THETA2 = 180
   Next, set THETA2 = 135

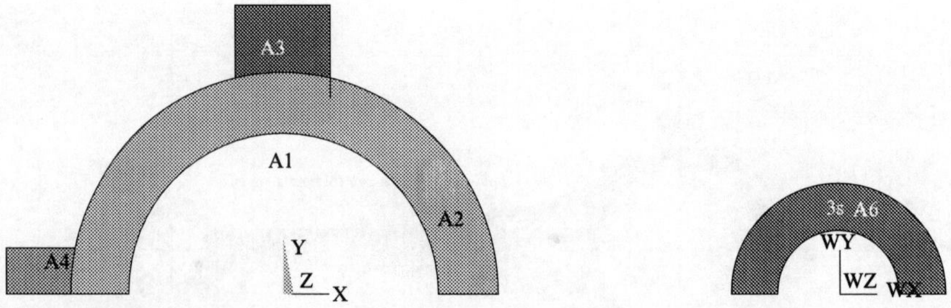

8. Issue separate Boolean overlaps on each area group
   Main Menu> Preprocessor> Modeling> Operate> Booleans> Overlap> Areas +
   First select the left group of areas
   Next, select the right group of areas

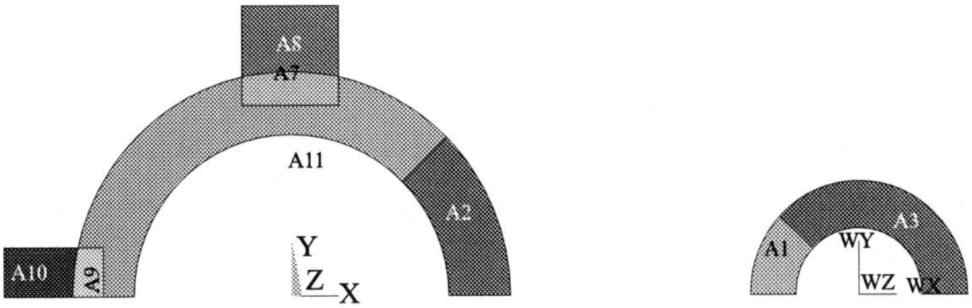

9. Set the active coordinate system to be Global Cartesian
   Utility Menu> WorkPlane> Change Active CS to> Global Cartesian
10. Define four new keypoints
   Main Menu> Preprocessor> Modeling> Create> Keypoints> In Active CS ...
   1st KP, X = 2.5, Y = 0.5, then [Apply], 2nd KP, X = 3.25, Y = 0.4
   3rd KP, X = 4, Y = 0.33, then [Apply], 4th KP, X = 4.75, Y = 0.28
11. Set the active coordinate system to be global cylindrical
   Utility Menu> WorkPlane> Change Active CS to> Global Cylindrical
12. Create a single line from a spline fit to a series of keypoints
   Main Menu> Preprocessor> Modeling> Create> Lines> Splines> With Options> Spline thru KPs +
   Pick, the six keypoints as shown in the graphics window below.

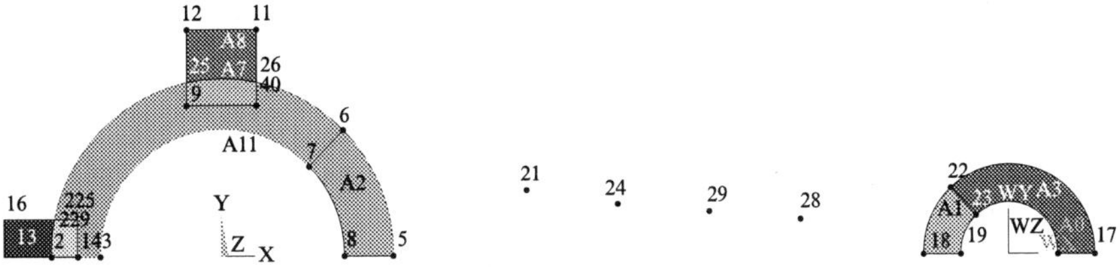

XV1 = 1 (radius at keypoint 1 in Global Cylindrical coordinate system)
YV1 = 135 (theta at keypoint 1 in Global Cylindrical coordinate system)
XV6 = 1 (radius at keypoint 6 in Global Cylindrical coordinate system)
YV6 = 45 (theta at keypoint 6 in Global Cylindrical coordinate system)
Click 'OK'.

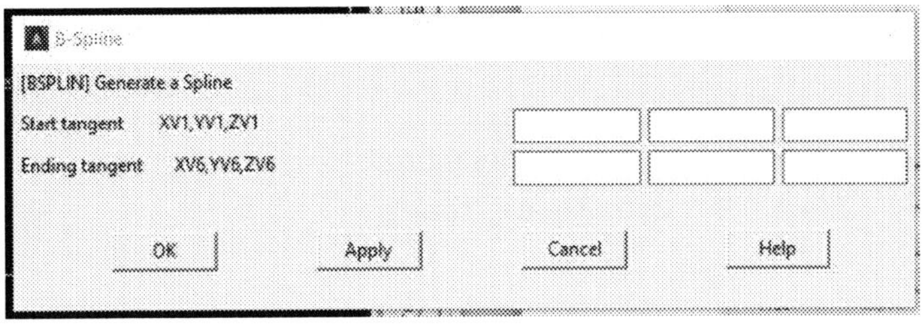

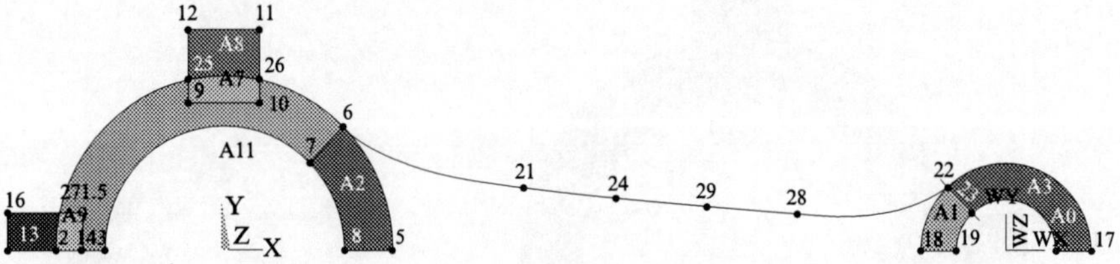

13. Create a straight line between keypoints 5 and 18
    Main Menu> Preprocessor> Modeling> Create> Lines> Lines> Straight Line +
    Pick the two keypoints as shown in the graphics window below.

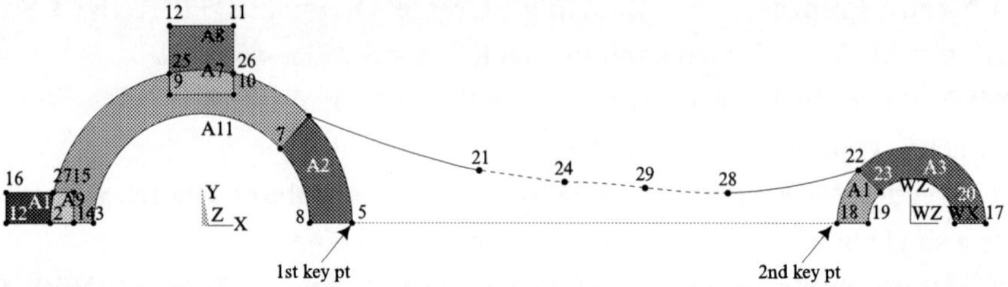

14. Turn line numbering on and plot lines
    Utility Menu> PlotCtrls> Numbering ...
    Set Line numbers ON
    Utility Menu> Plot> Lines
15. Create a new area bounded by previously defined lines 6, 1, 7, 25
    Main Menu> Preprocessor> Modeling> Create> Areas> Arbitrary> By Lines +
    Pick the four lines (6, 1, 7, and 25)

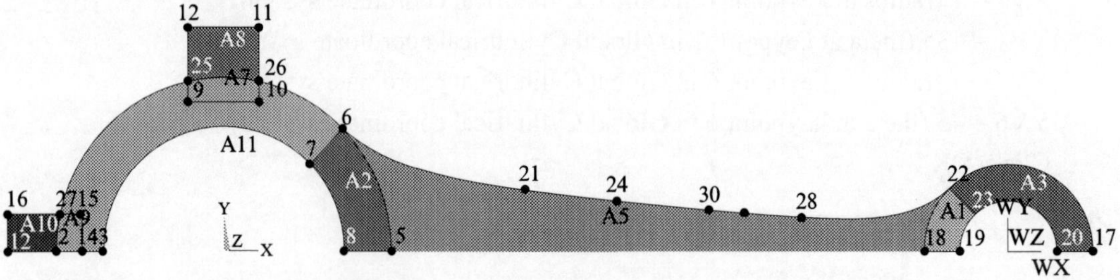

16. Zoom in on the left portion of the connecting rod
    Utility Menu> PlotCtrls> Pan, Zoom, Rotate ...
    Box Zoom

17. Create three line fillets
    Main Menu> Preprocessor> Modeling> Create> Lines> Line Fillet+
    Pick lines 36 and 40, RAD = 0.25
    Pick lines 40 and 31, pick lines 30 and 39
    Utility Menu> Plot> Lines

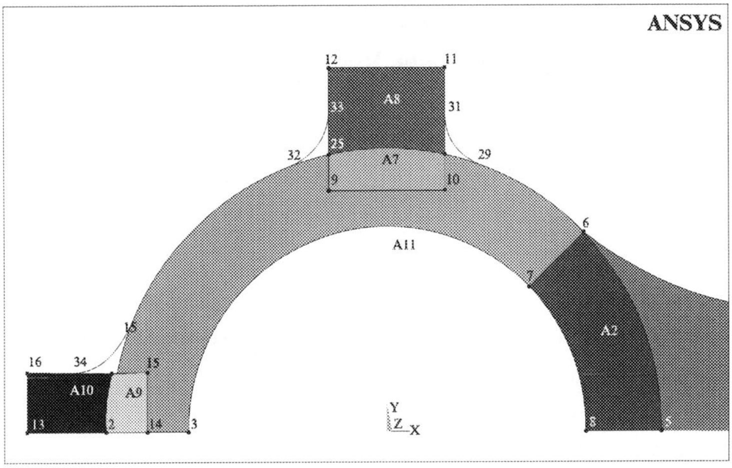

18. Create three new areas bounded by the previously defined fillet lines
    Main Menu> Preprocessor> Modeling> Create> Areas> Arbitrary> By Lines+
    Pick lines 12, 10, and 13
    Pick lines 17, 15, and 19

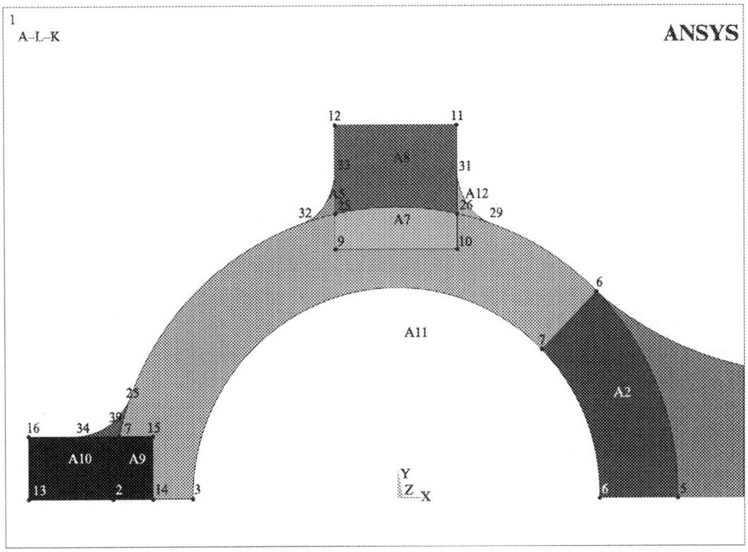

Pick lines 23, 21, and 24

Utility Menu> Plot> Areas

19. Add all areas together to form one single area

Main Menu> Preprocessor> Modeling> Operate> Add> Areas +

Pick All

20. Fit the entire model within the graphics window

Utility Menu> PlotCtrls> Pan, Zoom, Rotate …

Fit

21. Turn off line and area numbering

Utility Menu> PlotCtrls> Numbering ...

Set Line and Area numbers off

Utility Menu> Plot> Areas

22. Set the active coordinate system to be global cartesian

Utility Menu> WorkPlane> Change Active CS to> Global Cartesian

23. Reflect the area about the X-Z plane (in Y direction)

Main Menu> Preprocessor> Modeling> Reflect> Areas +

Pick All. Select 'X-Z plane'.

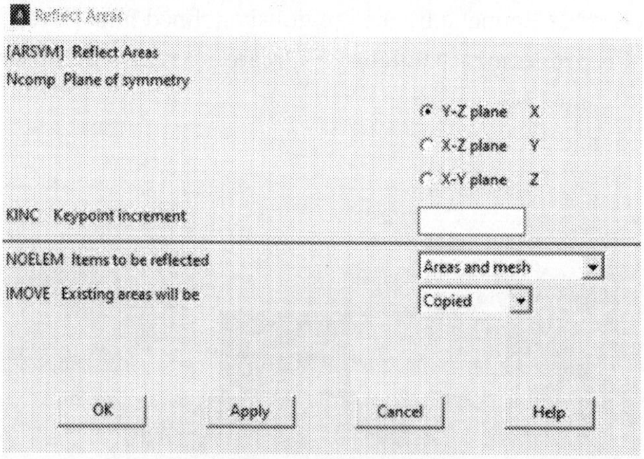

24. Add all areas together to form one single area

Main Menu> Preprocessor> Modeling> Operate> Add> Areas +

Pick All

25. Turn working plane off

Utility Menu> WorkPlane> Display Working Plane

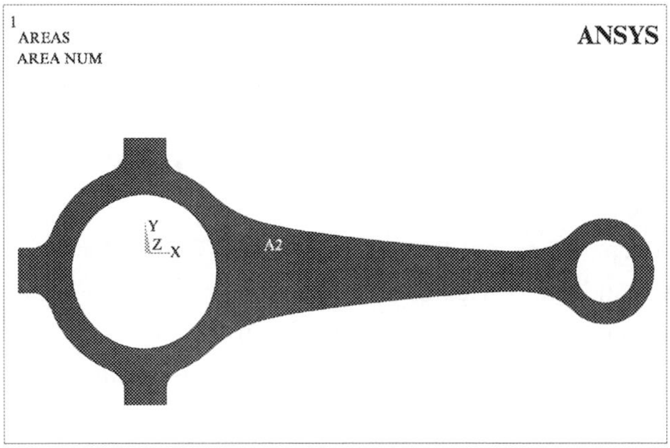

26. Save

    Pick the 'SAVE_DB' button in the Toolbar

Meshing

1. Change the jobname to be '*C-Rod-2D-Mesh*'

   Utility Menu> File> Change Jobname ...
2. Enter the preprocessor and specify the element type to be Solid and 'Quad 8 node 183'

   Main Menu> Preprocessor> Element Type> Add/Edit/Delete ...

   Add ...
3. Set element size to 0.2 and free mesh the model with quad elements:

   Main Menu> Preprocessor> MeshTool ...

   Pick 'Set' under Size Controls: Global

   SIZE = 0.2, OK, Mesh, Pick All

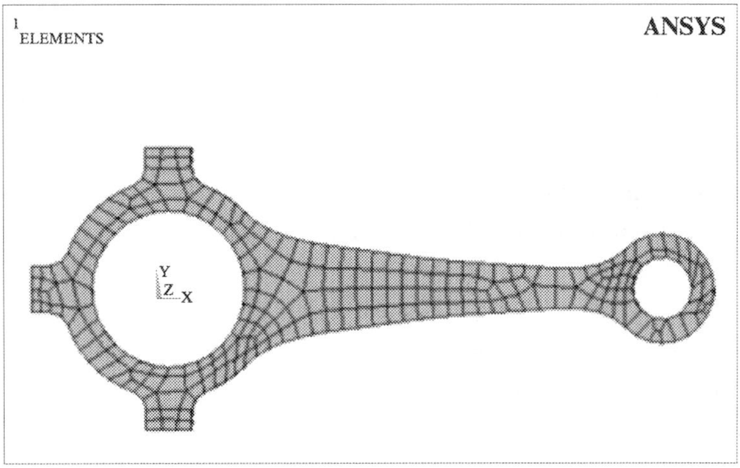

4. Save the meshed model
   Utility Menu> File> Save as ...
   Enter '*C-Rod-2D-Mesh-Quad.db*' as the database name
5. Extrude the meshed area along its normal to create a 3D brick mesh model
(a) Add a 3-D brick element
   Main Menu> Preprocessor> Element Type> Add/Edit/Delete ...
   Add ..., choose 'Solid' and 'Brick 20node 186'
(b) Set element divisions along extrusion and then extrude the area
   Main Menu> Preprocessor> Modeling> Operate> Extrude> Elem Ext Opts ...
   Enter VAL1 = 3, then 'OK'
   Main Menu> Preprocessor> Modeling> Operate> Extrude> Areas> Along Normal +
   Pick area number 2, then OK, DIST = 0.5, then OK
6. Change to isometric view
   Utility Menu> PlotCtrls> Pan, Zoom, Rotate ...
   ISO

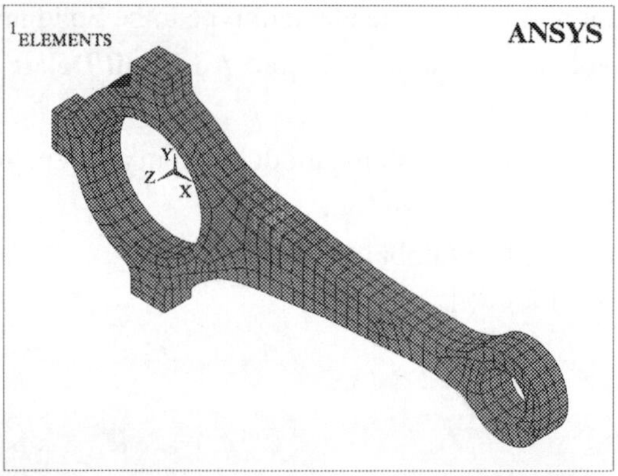

7. Save the meshed model
   Utility Menu> File> Save as ...
   Enter '*C-Rod-3D-Mesh-Brick.db*' as the database name
   Clear the mesh and save the model as '*C-Rod-3d-Model*'.
8. Use VSWEEP to create a 3-D brick element mesh
(a) Resume the 'C-Rod-3D-Model.db' database
   Utility Menu> File> Resume from ...
   Select the 'C-Rod-3D-Model.db'
(b) Enter the preprocessor and specify the element type to be SOLID186
   Main Menu> Preprocessor> Element Type> Add/Edit/Delete ...
   Pick 'Add', choose 'Solid' and 'Brick 20node 186'

(c) Mesh the volume using VSWEEP

Main Menu> Preprocessor> MeshTool ...

Pick [Set] under Size Controls: Global, SIZE = 0.15

Select 'Hex' and 'Sweep', and leave the default 'Auto Src/Trg' setting

Select 'Sweep', then 'Pick All'

(d) Save the meshed model

Utility Menu> File> Save as ...

Enter 'C-Rod-Mesh-Sweep.db' as the database name

3.11.3 3-D Problems

W10: Model and mesh the segment of a cylinder shown below. Draw a rectangular block at the bottom of the partial cylinder. Glue the two objects and mesh the objects.

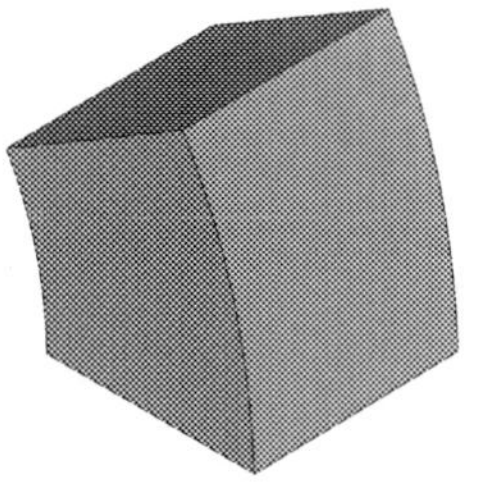

Solution:

1. Create the partial cylinder

   MainMenu> Preprocessor> Modeling> Create> Volumes> Cylinder> Partial Cylinder

   X = 0,Y = 0, Rad 1 = 1.5, Theta – 1 = 0, Rad 2 = 2.5, Theata – 2 = 30, Depth = 1, OK

2. Create a brick mapped mesh,

   First select the brick elements from the element library.

   Element Type> Add/Edit/Delete> Add> Solid> Brick 8 node 185 Meshing> Mesh> Volumes> Mapped> 4 to 6 sided

   Pick the volume shown

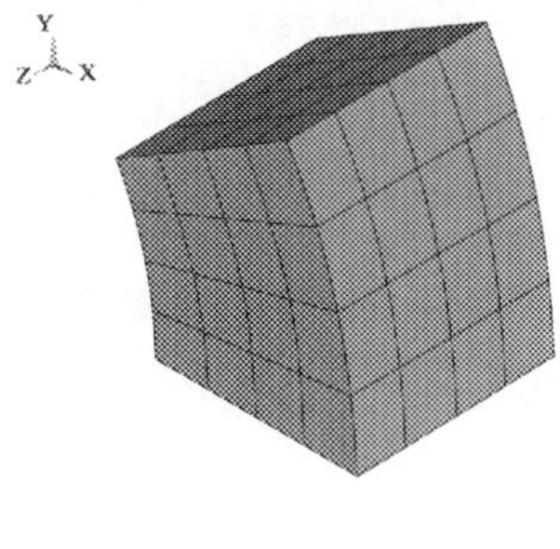

3. If a specific number of divisions is required along an edge of one of the 2D or 3D regions described above, go first

   Meshing> Size Cntrls> Manual Size> Picked Lines (or All Lines)

   Set the element edge length or number of divisions and then pick the lines to be divided.
4. Create the cylinder segment and the brick separately and then **glue** together. Gluing two volumes creates a common boundary. Adding two volumes creates a single volume and the single entity must be meshed.

   MainMenu> Preprocessor> Modeling> Create> Volumes> Block> 2 Corners & Z

   X = 1.5, Y = 0, Width = 1, Height = – 1, Depth = 1

   MainMenu> Preprocessor> Modeling> Operate> Booleans> Glue> Volumes> PickAll

   Meshing> Mesh> Volumes> Mapped> 4 to 6 sided

   Pick the two volumes shown

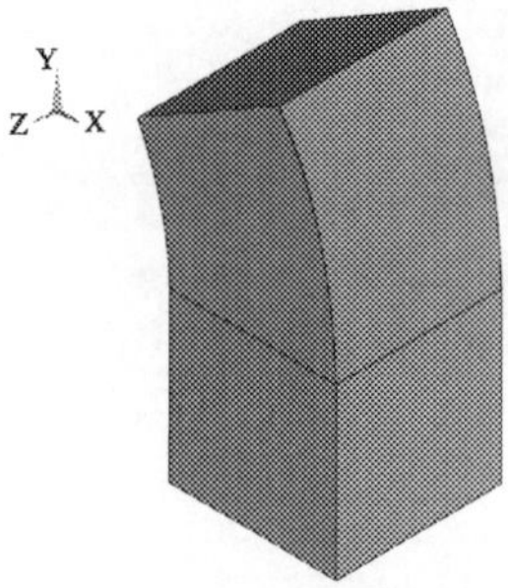

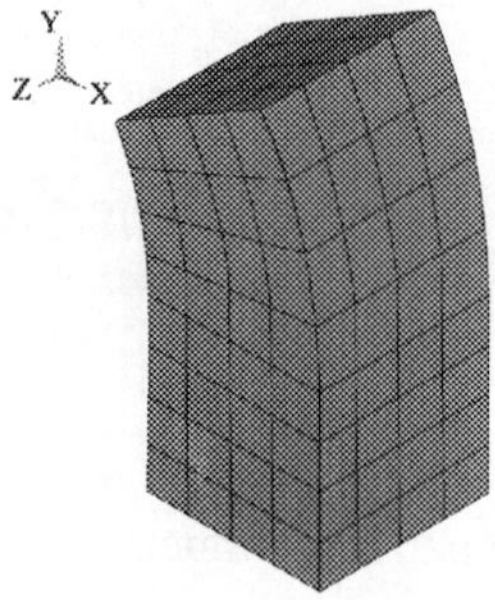

W11: Build and mesh a half symmetry solid model of the pillow block shown below. Use both free meshing and sweep meshing techniques.

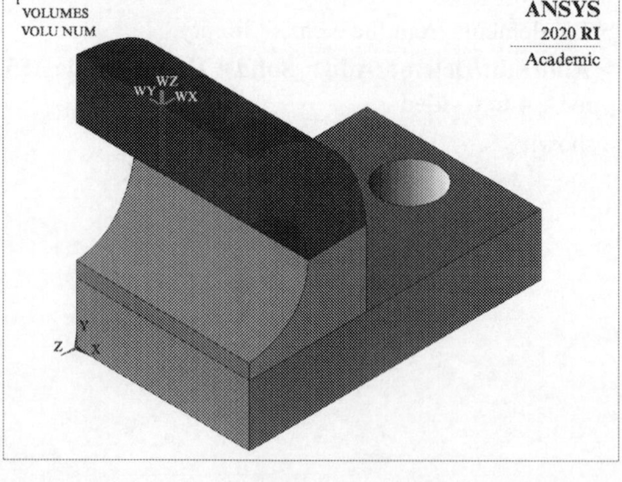

Solution:

1. Enter ANSYS. Enter 'Solid Bearing' as the jobname.
2. Switch to isometric view
   Utility Menu> PlotCtrls> Pan, Zoom, Rotate ...
3. Create the sole of the Solid Bearing

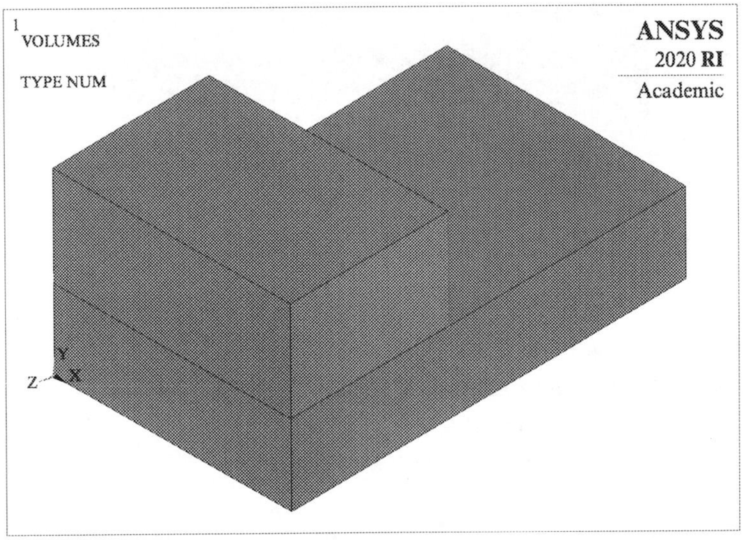

   Main Menu> Preprocessor> Modeling> Create> Volumes> Block> By Dimensions ...
   X1 = 0, X2 = 60, Y1 = 0, Y2 = 20, Z1 = 0, Z2 = – 100, OK
   X1 = 0, X2 = 60, Y1 = 20, Y2 = 45, Z1 = 0, Z2 = – 40, OK
4. Offset working plane to location X = 30, Y = 20, Z = – 85
   Utility Menu> WorkPlane> Offset WP by Increments ...
   Set X, Y, Z Offsets = 30, 20, – 85
   Set XY, YZ, ZX Angles = 0, – 90, 0, then [OK]
5. Create a solid cylinder having a diameter of 20 mm and a depth of –20 mm:
   Main Menu> Preprocessor> Modeling> Create> Volumes> Cylinder> Solid Cylinder +
   Radius = 10, Depth = – 20, then 'OK'
6. Subtract the solid cylinder from the base
   Main Menu> Preprocessor> Modeling> Operate> Booleans> Subtract> Volumes +
   Pick the base volume (Vol. 1), then 'OK'
   Pick the cylinder volume (Vol. 3), then 'OK'

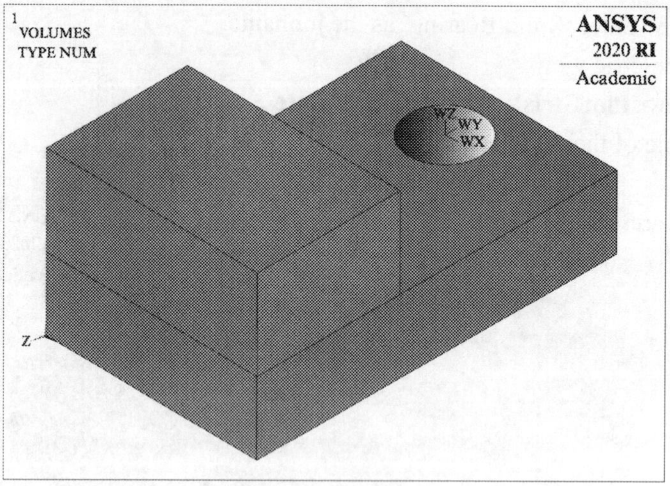

7. Align working plane with the Global Cartesian origin

   Utility Menu> WorkPlane> Align WP with> Global Cartesian

8. Offset working plane to location X = 0, Y = 45, Z = 0

   Utility Menu> WorkPlane> Offset WP by Increments ...

   Set X, Y, Z Offsets = 0, 45, 0

   Set XY, YZ, ZX Angles = 0, – 90, 0, click 'OK'

9. Create the arch of the solid bearing

   Main Menu> Preprocessor> Modeling> Create> Volumes> Cylinder> Partial Cylinder +

   WP X = 0, WP Y = 0, Rad – 1 = 0, Theta – 1 = 90, Rad – 2 = 40, Theta – 2 = 180, Depth = – 60, click 'OK'

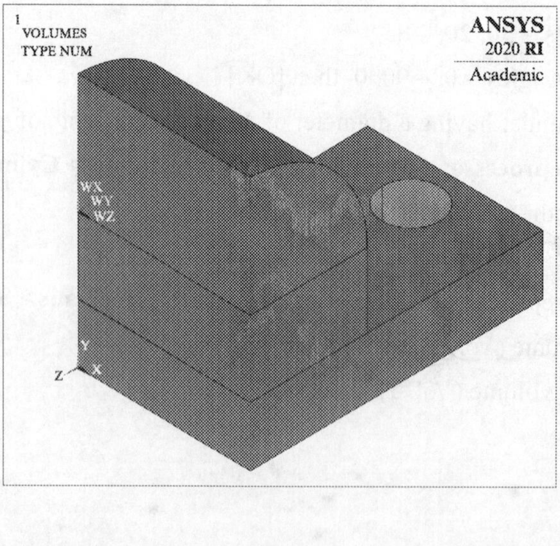

10. Create cylinder for the through hole in the solid bearing:
Main Menu> Preprocessor> Modeling> Create> Volumes> Cylinder> Solid Cylinder +
WP X = 0, WP Y = 0, Radius = 20, Depth = – 60, then 'Apply'

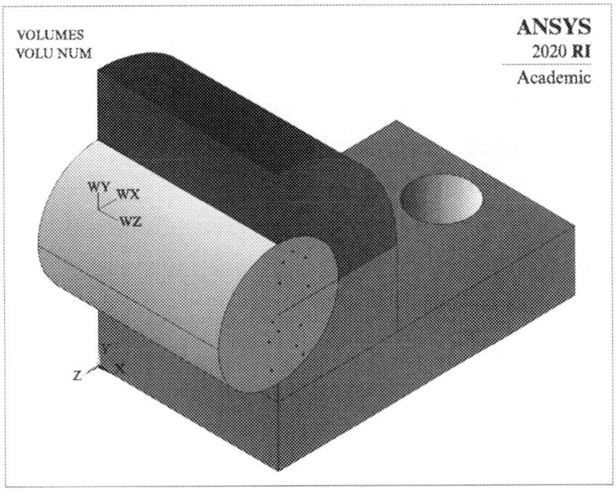

11. Subtract the solid cylinder to form the arch and base
Main Menu> Preprocessor> Modeling> Operate> Booleans-Subtract> Volumes +
Pick the two volumes that form the base and the arch of the solid bearing, 'Apply'
Pick the counterbore cylinder, 'Apply'
Pick the same two base volumes, 'Apply'
Pick the through-hole cylinder, 'OK'

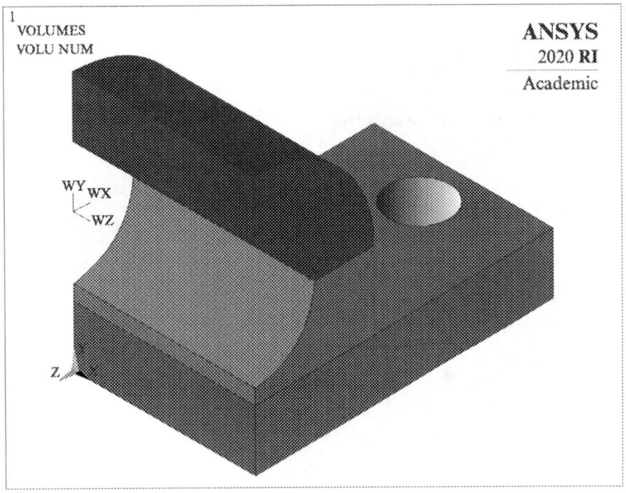

12. Offset working plane to location X = 0, Y = 45, Z = 0
    Utility Menu> WorkPlane> Offset WP by Increments ...
    Set X, Y, Z Offsets = 0, 40, – 30
    Set XY, YZ, ZX Angles = 0, –90, 0, then 'OK'
13. Create cylinder for the through hole in the solid bearing:
    Main Menu> Preprocessor> Modeling> Create> Volumes> Cylinder> Solid Cylinder +
    WP X = 0, WP Y = 0, Radius = 1.5, Depth = – 20, then 'Apply'
14. Subtract the solid cylinder to form the arch and base
    Main Menu> Preprocessor> Modeling> Operate> Booleans> Subtract> Volumes +
    Pick the volume that forms the arch of the solid bearing, Apply
    Pick the oil hole cylinder, 'Apply'
15. Merge coincident keypoints
    Main Menu> Preprocessor> Numbering Ctrls> Merge Items ...
    Set Label to 'Keypoints', then 'OK'
16. Glue the volumes together
    Main Menu> Preprocessor> Modeling > Operate> Booleans> Glue> Volumes +
    Pick All
17. Turn volume numbers on and then plot volumes
    Utility Menu> PlotCtrls> Numbering ...
    Set Volume numbers on, then OK
18. Save
    Pick the 'SAVE_DB' button in the Toolbar, OK

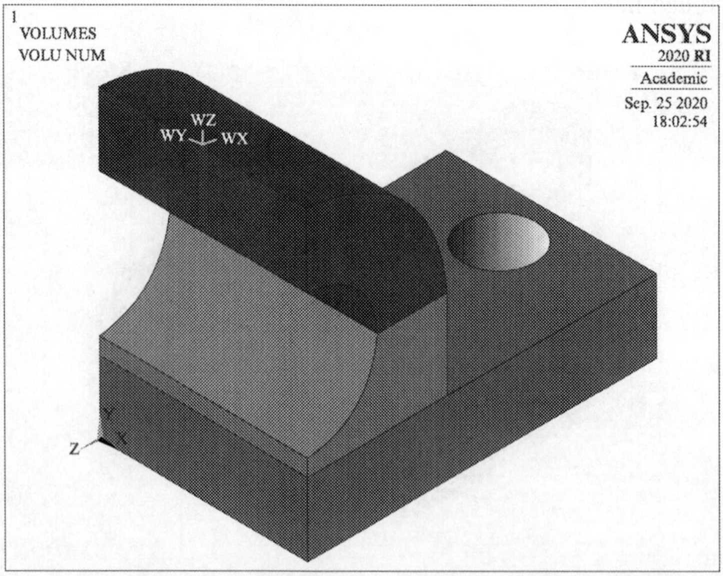

19. Specify the element type to be SOLID186

    Main Menu> Preprocessor> Element Type> Add/Edit/Delete ...

    Add ...

    Choose 'Solid' and 'Brick 20node 186', then 'OK', 'Close'

20. Activate Smartsize meshing and free mesh the model with tetrahedral elements:

    Main Menu> Preprocessor> MeshTool ...

    Activate 'Smart Size'. Set Smart Size level to 4, 'Mesh', 'Pick All', 'Close'

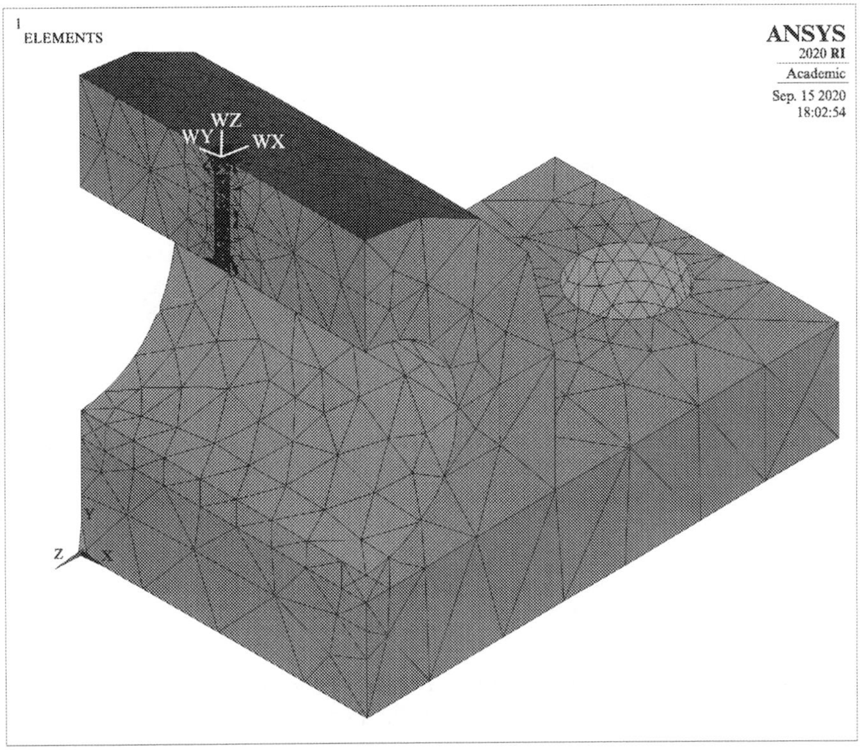

21. Save the meshed model

    Utility Menu > File > Save as ...

    Enter '*P-Block-Mesh-Free.db*' as the database name, then 'OK'

22. Sweep mesh the model

(a) Clear the free mesh

Main Menu> Preprocessor> MeshTool ...

Clear, Pick All

Utility Menu> Plot> Volumes

(b) Divide the base volume in two in order to make it topologically consistent for sweep meshing

Utility Menu> WorkPlane> Align WP with> Keypoints +
Pick the three keypoints displayed in the graphics window below, then OK
Main Menu> Preprocessor> Modeling> Operate> Booleans> Divide> Volu by WrkPlane +
Pick the base volume, then OK
Utility Menu> WorkPlane> Display Working Plane
Utility Menu> Plot> Volumes

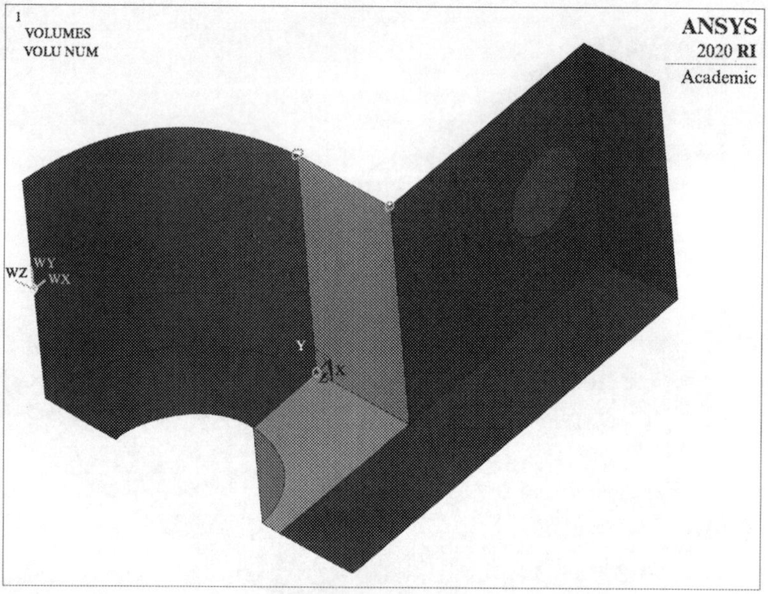

(c) Activate 'tet-mesh' for volumes that cannot be swept
Main Menu> Preprocessor> Meshing> Mesh> Volume Sweep> Sweep Opts ...
Select 'Tet mesh in nonsweepable volumes', then 'OK'

(d) Leave Smartsize level at 4 and set global element size to 0.125, then sweep mesh the model with brick/tet elements
Main Menu> Preprocessor> MeshTool ...
Pick [Set] under Size Controls: Global
Set SIZE = 5, OK
Select 'Hex' and 'Sweep', and leave the default 'Auto Src/Trg' setting active
Select 'Sweep', 'Pick All'
'Yes' - to mesh volume 6 with tetrahedral elements

(e) Save the meshed model
Utility Menu> File> Save as ...
Enter '*P-Block-Mesh-Sweep.db*' as the database name, 'OK'

W12: Build a half symmetry solid model of the pillow block shown below.

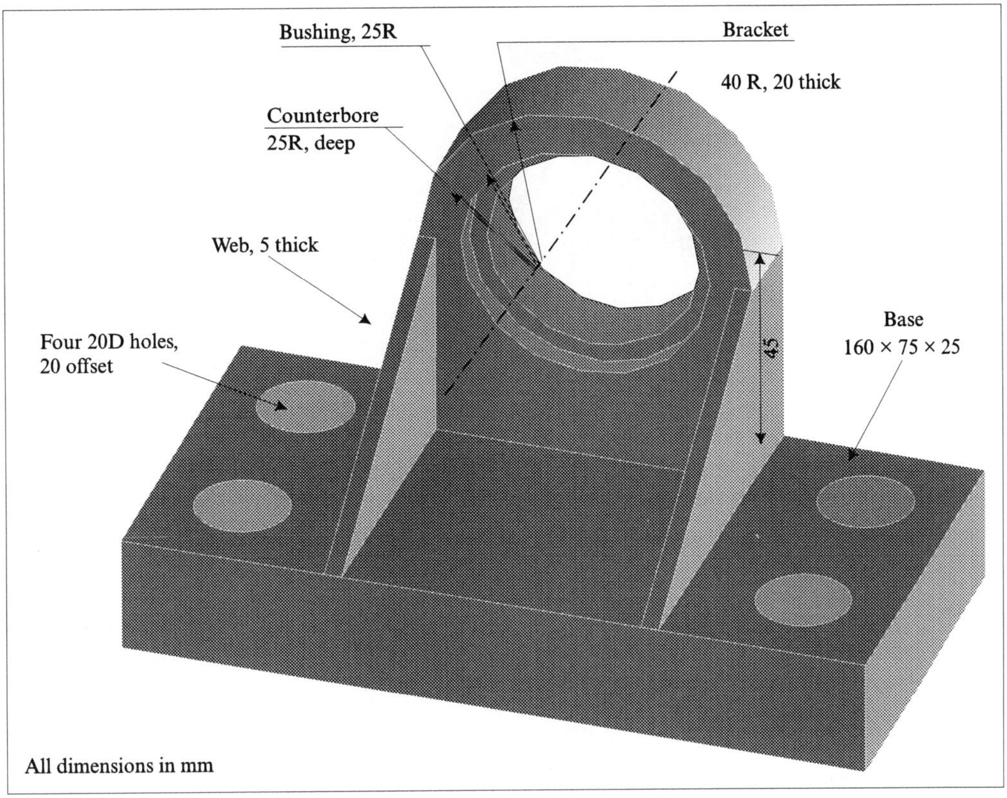

Solution:

1. Enter ANSYS. Enter '*P-Block*' as the jobname.
2. Switch to isometric view

   Utility Menu> PlotCtrls> Pan, Zoom, Rotate ...
3. Create the base of the pillow block

   Main Menu> Preprocessor> Modeling> Create> Volumes> Block> By Dimensions ...

   X1 = 0, X2 = 80, Y1 = 0, Y2 = 30, Z1 = 0, Z2 = 75, then 'OK'
4. Offset working plane to location X = 60, Y = 30, Z = 20

   Utility Menu> WorkPlane> Offset WP by Increments ...

   Set X, Y, Z Offsets = 60, 30, 20

   Set XY, YZ, ZX Angles = 0, – 90, 0, then 'OK'

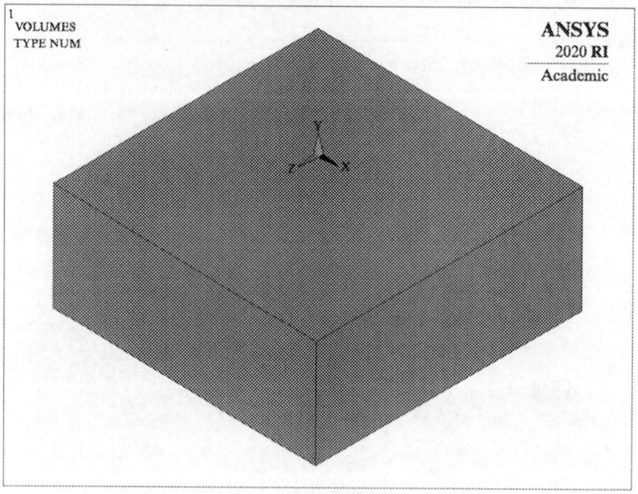

5. Create a solid cylinder having a diameter of 20 mm and a depth of – 30 mm
   Main Menu> Preprocessor> -Modeling> Create> Volumes> Cylinder> Solid Cylinder +
   Radius = 10 mm
   Depth = – 30, then OK
6. Copy the solid cylinder to new location with DZ = 35
   Main Menu> Preprocessor> Copy> Volumes +
   Pick the cylinder volume (Vol. Number 2), then 'OK'
   DZ = 35, then 'OK'
7. Subtract the two solid cylinders from the base
   Main Menu> Preprocessor> Modeling> Operate> Booleans> Subtract> Volumes+
   Pick the base volume (Vol. 1), then 'OK'
   Pick the two cylinder volumes (Vols. 2 and 3), then 'OK'

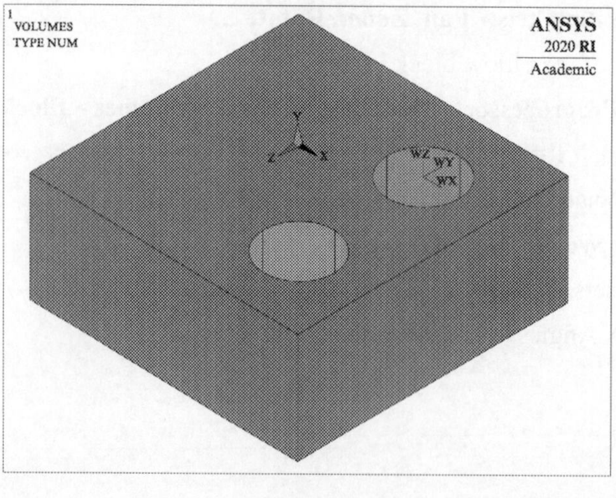

8. Align working plane with the Global Cartesian origin
   Utility Menu> WorkPlane> Align WP with> Global Cartesian
9. Create the base of the bushing bracket
   Main Menu> Preprocessor> Modeling> Create> Volumes> Block> By 2 Corners & Z+
   WP X = 0, WP Y = 30, Width = 40, Height = 45, Depth = 20, then 'OK'

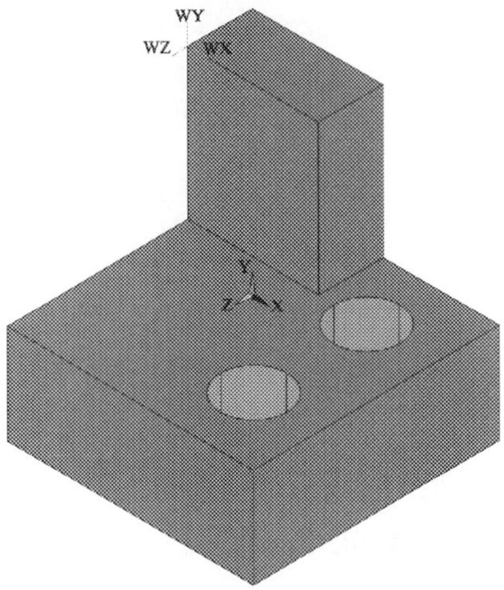

10. Offset working plane to the front face of the bushing bracket
    Utility Menu> WorkPlane> Offset WP to> Keypoints +
    Pick keypoint at the top left corner of the front face, then 'OK'
11. Create the arch of the bushing bracket
    Main Menu> Preprocessor > Modeling> Create> Volumes> Cylinder > Partial Cylinder+
    WP X = 0, WP Y = 0, Rad – 1 = 0, Theta – 1 = 0, Rad – 2 = 40, Theta – 2 = 90, Depth = – 20

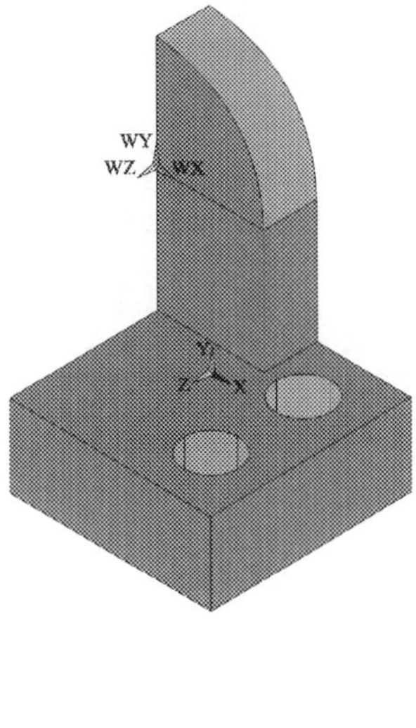

12. Create cylinders for the counter bore and the through hole in the bushing bracket
    Main Menu> Preprocessor> Modeling> Create> Volumes> Cylinder> Solid Cylinder+
    WP X = 0, WP Y = 0, Radius = 25, Depth = – 5, then 'Apply', WP X = 0
    WP Y = 0, Radius = 20, Depth = – 50, then 'OK'

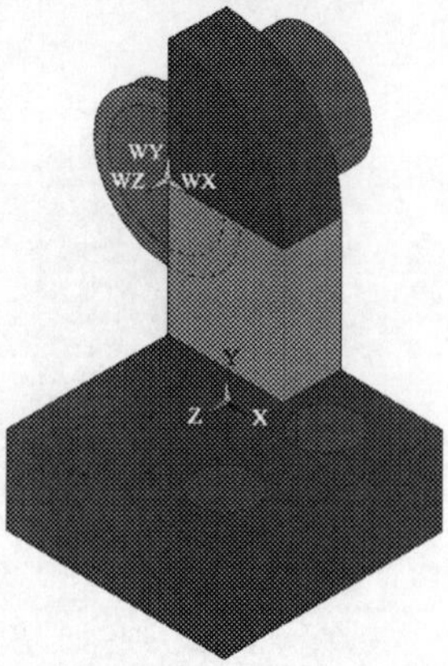

13. Subtract the two solid cylinders to form the counterbore and bushing through-hole
    Main Menu> Preprocessor> Modeling> Operate> Booleans> Subtract> Volumes+
    Pick the two volumes that form the base and the arch of the bushing bracket, Apply
    Pick the counterbore cylinder, 'Apply'
    Pick the same two base volumes, 'Apply'
    Pick the through-hole cylinder, 'OK'
14. Merge coincident keypoints
    Main Menu> Preprocessor> Numbering Ctrls> Merge Items …
    Set Label to 'Keypoints', then 'OK'

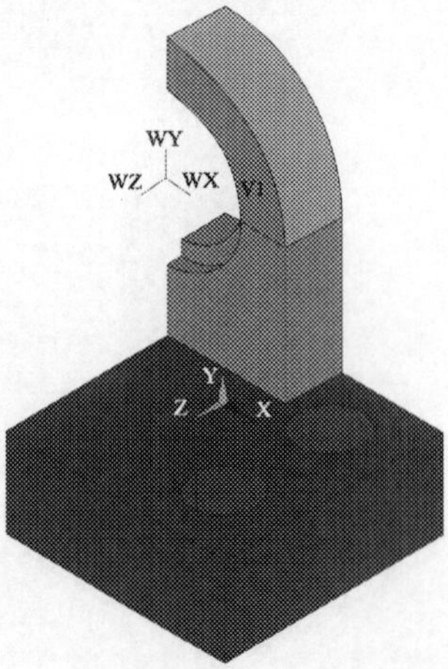

15. Create the web

(a) Create a keypoint in the middle of the front top edge of the base

Main Menu> Preprocessor> Modeling> Create> Keypoints> KP between KPs +

Pick the two upper front corner keypoints of the base, then 'OK'

RATI = 0.5, then 'OK'

(b) Create a triangular area

Main Menu> Preprocessor> Modeling> Create> Areas> Arbitrary> Through KPs+

Pick the 1st keypoint where the base of the bushing bracket intersects the base of the pillow block at X = 40

Pick the 2nd keypoint where the base of the bushing bracket intersects the bottom arch surface at X = 40

Pick the 3rd keypoint that was created in step 14a at X = 40, Y = 30, Z = 75, 'OK'

(c) Extrude area along area normal

Main Menu> Preprocessor> Modeling> Operate> Extrude> Areas> Along Normal

Pick the triangular area created in step 14b, then 'OK'

DIST = – 5, then 'OK'

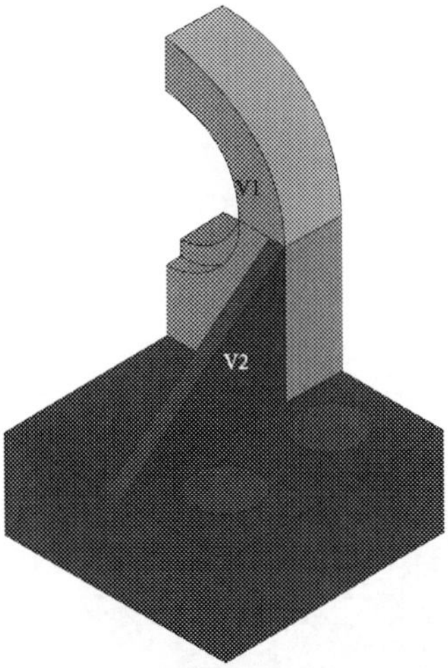

16. Glue the volumes together

Main Menu > Preprocessor> Modeling> Operate> Booleans> Glue> Volumes + Pick All

17. Turn volume numbers on and then plot volumes:

Utility Menu> PlotCtrls> Numbering ...

Set Volume numbers on, then 'OK'

18. Save and exit ANSYS
    Pick the '**SAVE_DB**' button in the Toolbar
    Pick the 'QUIT' button in the Toolbar
    Select '**Quit – No Save!**', 'OK'

W13: Mesh the pillow block that was created in previous example. Use both free meshing and sweep meshing techniques.

Solution:

1. Enter ANSYS using '*P-Block-Mesh*' as the jobname.
2. Resume the '*P-Block.db*' database file created in previous problem:
   Utility Menu> File> Resume from ...
   Select the 'P-Block.db' database file, then 'OK'
3. Enter the preprocessor and specify the element type to be SOLID186
   Main Menu> Preprocessor> Element Type> Add/Edit/Delete ...
   Add ...
   Choose 'Solid' and 'Brick 20node 186', then 'OK'
4. Activate Smartsize meshing and free mesh the model with tetrahedral elements:

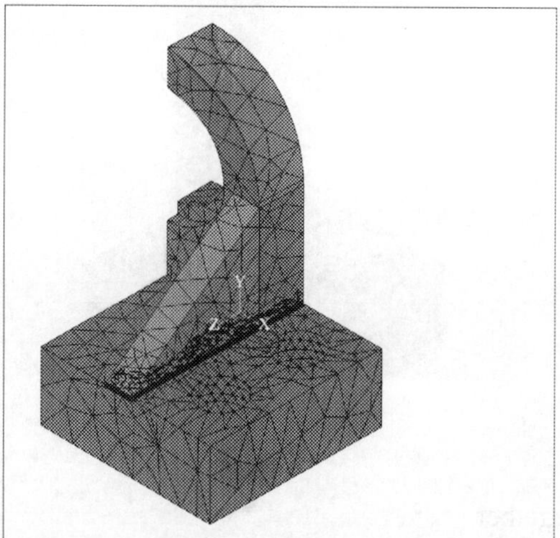

   Main Menu> Preprocessor> MeshTool ...
   Activate 'Smart Size'. Set Smart Size level to 4, Mesh, Pick All, Close
5. Save the meshed model:
   Utility Menu> File> Save as ...
   Enter 'P-Block-Mesh-Free.db' as the database name, then 'OK'

6. Sweep mesh the model

   (a) Clear the free mesh

   Main Menu> Preprocessor> MeshTool ...

   Clear, Pick All

   Utility Menu> Plot> Volumes

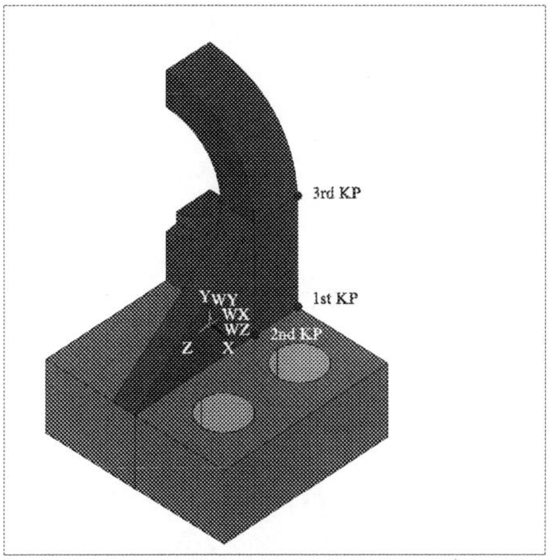

   (b) Divide the base volume in two in order to make it topologically consistent for sweep meshing

   Utility Menu> WorkPlane> Align WP with> Keypoints +

   Pick the three keypoints displayed in the graphics window below, then 'OK'

   Main Menu> Preprocessor> Modeling> Operate> Booleans> Divide> Volu by WrkPlane +

   Pick the base volume, then 'OK'

   Utility Menu> WorkPlane> Display Working Plane

   Utility Menu> Plot> Volumes

   (c) Activate 'tet-mesh' for volumes that cannot be swept

   Main Menu> Preprocessor> Meshing> Mesh> Volume Sweep> Sweep Opts ...

   Select 'Tet mesh in nonsweepable volumes', then 'OK'

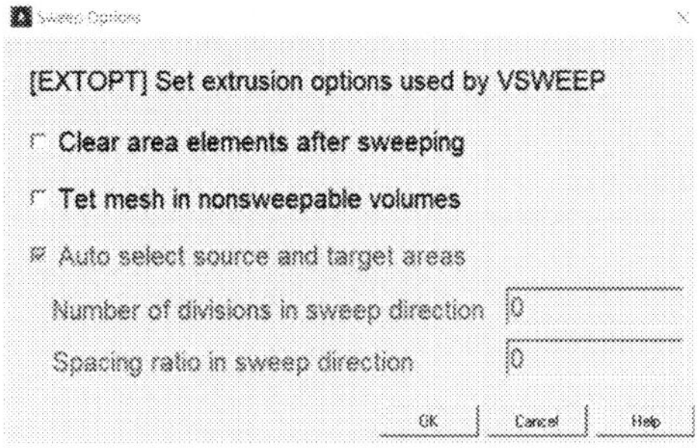

(d) Leave Smartsize level at 4 and set global element size to 3, then sweep mesh the model with brick/tet elements:

Main Menu> Preprocessor> MeshTool ...

Pick [Set] under Size Controls: Global, Set SIZE = 3, then 'OK'

Select 'Hex' and 'Sweep', and leave the default 'Auto Src/Trg' setting active

Sweep, Pick All, Yes - to mesh volume 6 with tetrahedral elements

(e) Save the meshed model

Utility Menu> File> Save as ...

Enter 'P-Block-Mesh-Sweep.db' as the database name, then 'OK'

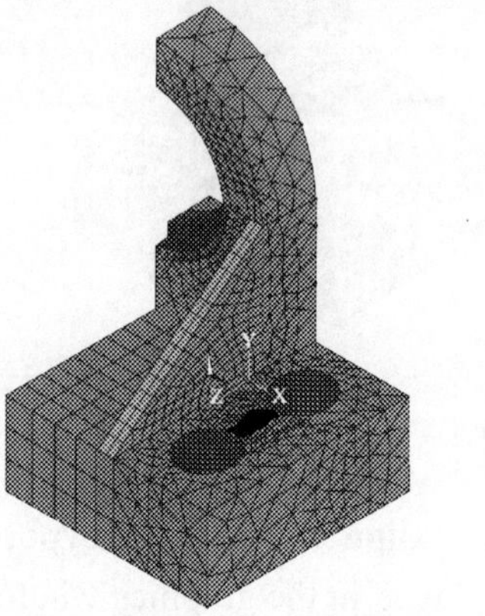

W14: Create a solid model of the Spindle Base shown in the following figure.

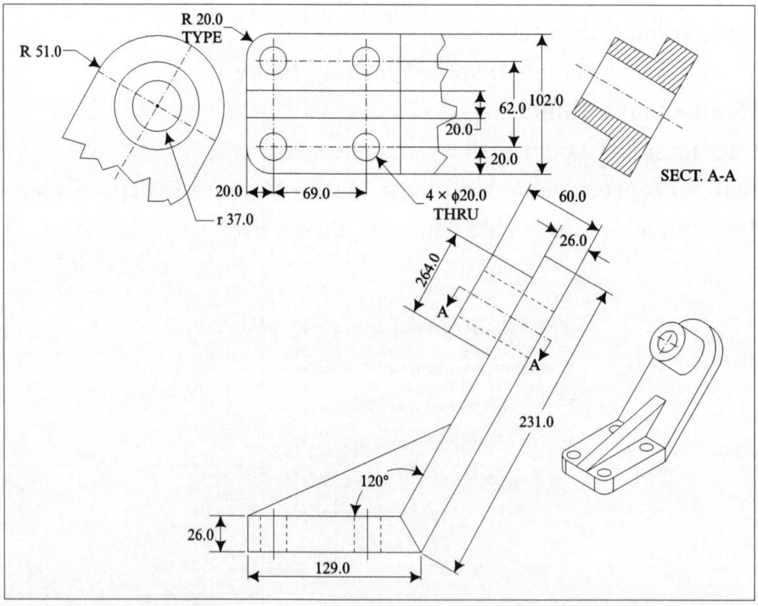

Solution:

Preprocessing

1. Create the base rectangle

WP X (X Corner)	WP Y (Y Corner)	Width	Height
0	0	109	102

2. Create the curved edge (using keypoints and lines to create an area)

Create the following keypoints

	X	Y	Z
Keypoint 5	– 20	82	0
Keypoint 6	– 20	20	0
Keypoint 7	0	82	0
Keypoint 8	0	20	0

The graphic obtained should look similar to the following image:

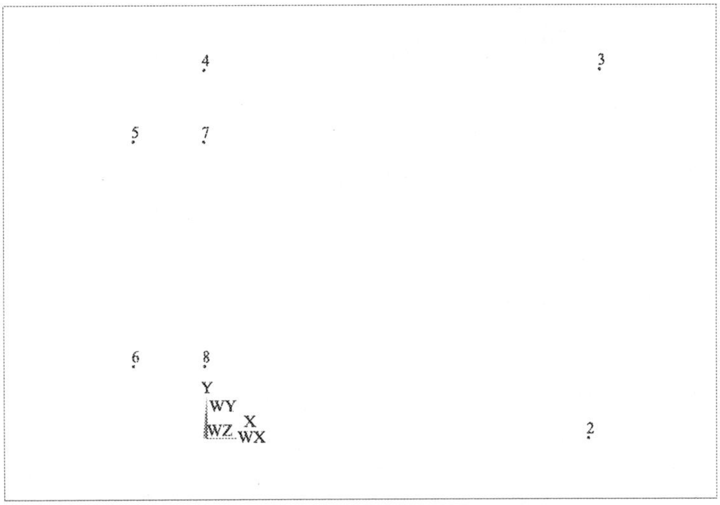

Create arcs joining the keypoints

Main Menu > **Preprocessor** > **Modeling** > **Create** > **Lines Arcs** > **By End KPs & Rad**

- Select keypoints 4 and 5 (either click on them or type 4, 5 into the command line) when prompted.
- Select keypoint 7 as the center-of-curvature when prompted.
- Enter the radius of the arc (20) in the 'Arc by End KPs & Radius' window
- Repeat to create an arc from keypoints 1 and 6

Create a line from keypoints 5 to 6

Main Menu> Preprocessor> Modeling> Create> Lines> Lines> Straight Line

Create an arbitrary area within the bounds of the lines

Main Menu> Preprocessor> Modeling> Create> Areas> Arbitrary> By Lines

Combine the 2 areas into 1 (to form Area 3)

Main Menu> Preprocessor> Modeling> Operate > Booleans> Add> Areas

The graphic obtained should look similar to the following image:

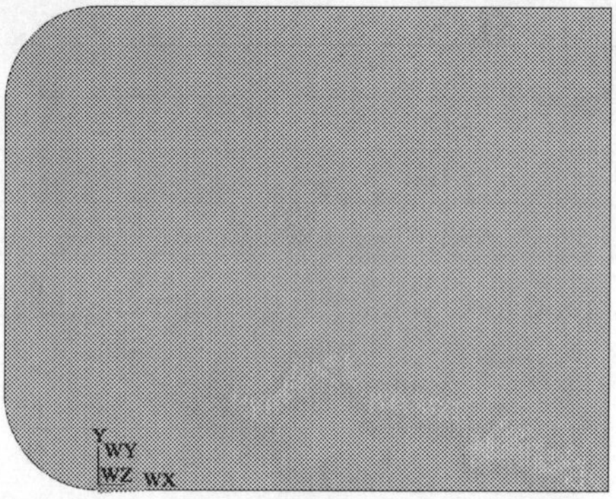

3. Create the 4 holes in the base

   Use the 'copy' feature in ANSYS to create all 4 holes

   Create the bottom left circle (XCENTER = 0, YCENTER = 20, RADIUS = 10)

   Copy the area to create the bottom right circle (DX = 69)

   Copy both circles to create the upper circles (DY = 62)

   Subtract the four circles from the main base

   The graphic obtained should look similar to the following image:

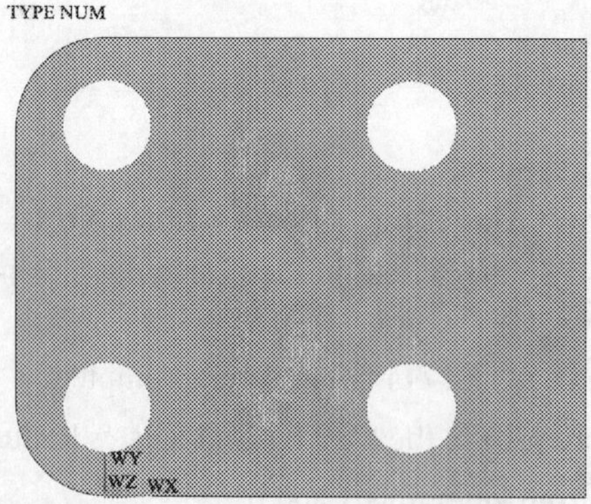

4. Extrude the base

   Preprocessor> Modeling> Operate> Extrude> Areas> Along Normal

The following window will appear once the area is selected

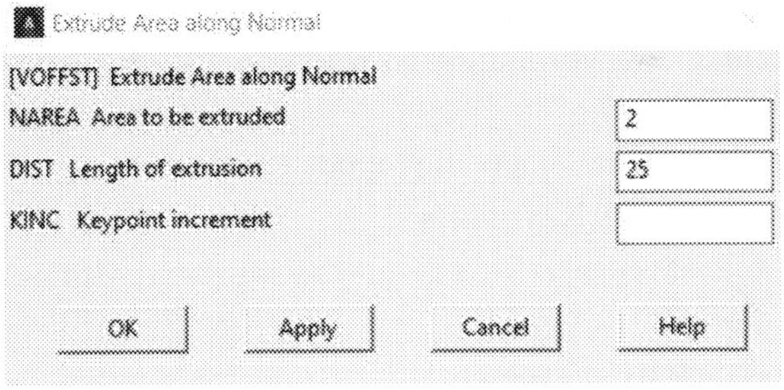

Fill in the window as shown (length of extrusion = 26 mm). Note, to extrude the area in the negative z direction simply enter – 26.

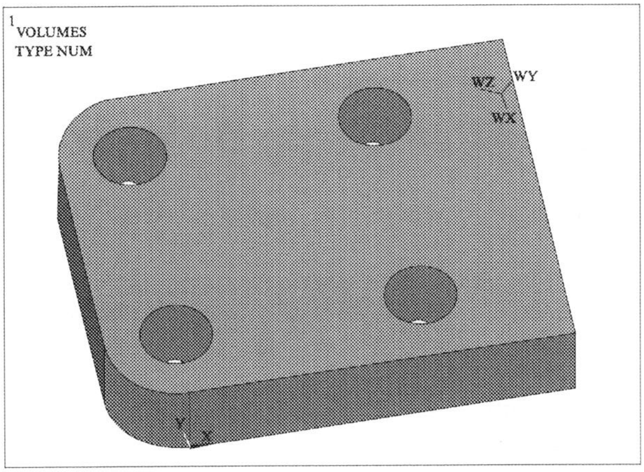

5. Align the working plane with the back of the spindle base

Change the working plane. Geometry can only be created in the X - Y plane. Therefore, in order to create the back of the Spindle Base, it is necessary to create a new working plane where the X – Y plane is parallel to the back.

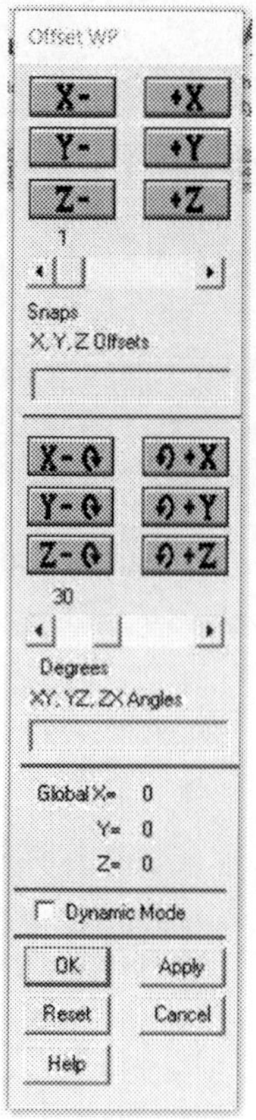

Move the working plane to the diagonally opposite corner.

Workplane> Offset WP to> Keypoints

Select diagonally opposite corner. Now align working plane with base of the back.

Workplane> Offset WP by Increments...

Click 'OK'.

6. Create the back area

   Create the base rectangle (XCORNER = 0, YCORNER = 0, WIDTH = 102, HEIGHT = 180)

   Create a circle to obtain the curved top (XCENTER = 51, YCENTER = 180, RADIUS = 51)

   Add the 2 areas together

7. Extrude the area (length of extrusion = 26 mm)

   Preprocessor> Modeling> Operate> Extrude> Areas> Along Normal

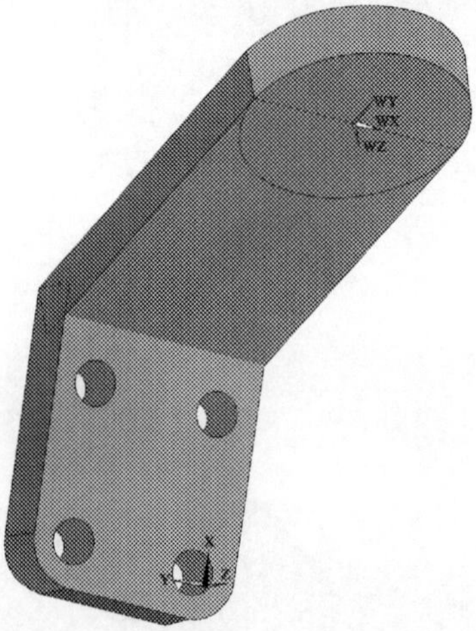

8. Add the base and the back together

   Add the two volumes together

   Preprocessor> Modeling> Operate> Booleans> Add> Volumes

   Note that the planar areas between the two volumes were not added together.

   Add the planar areas together

Preprocessor> Modeling> Operate> Booleans> Add> Areas

9. Create the Upper Cylinder

   Create the outer cylinder (XCENTER = 51, YCENTER = 180, RADIUS = 30, DEPTH = 50)

   Preprocessor> Modeling> Create > Volumes> Cylinder> Solid Cylinder

   Add the volumes together

   Create the inner cylinder (XCENTER = 51, YCENTER = 180, RADIUS = 20, DEPTH = 50)

   Subtract the volumes to obtain a hole

10. Change the working plane

    First change the active coordinate system back to the global coordinate system

    Utility Menu> WorkPlane> Align WP with> Global Cartesian

    Create keypoints at the vertices of triangular rib.

    Align the working plane to the 3 keypoints

    Recall when defining the working plane; the first keypoint defines the origin, the second keypoint defines the x-axis orientation, while the third defines the orientation of the working plane.

11. Change active coordinate system

    Now update the coordinate system to follow the working plane changes (i.e., make the new Work Plane origin the active coordinate).

    Utility Menu> WorkPlane> Change Active CS to> Working Plane

    Create the area

    Create first keypoint at the centre of top left base edge, second at centre of intersection line of base and back and third at bottommost point of outer circle. Create the rib area through these three keypoints.

    Preprocessor> Modeling> Create> Areas> Arbitrary> Through KPs

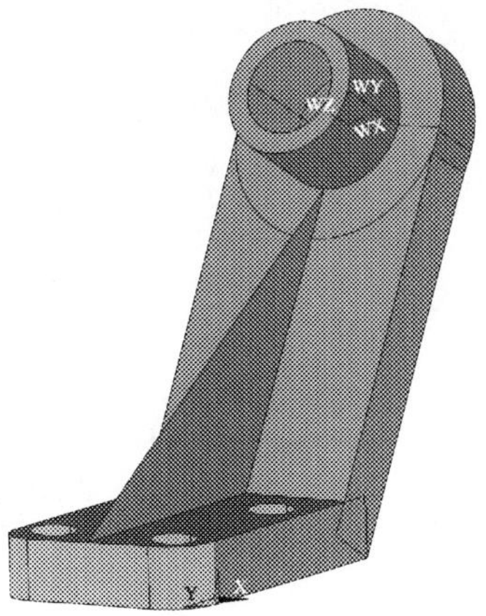

Extrude the area on either side (length of extrusion = 3 mm)

12. Add the volumes together

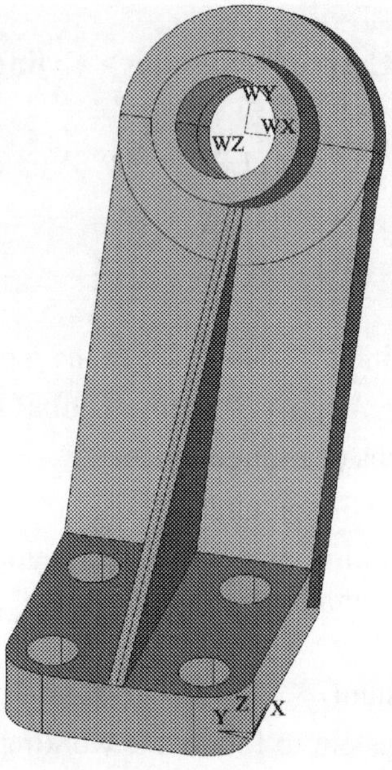

13. Quit ANSYS

Select 'QUIT' from the ANSYS Toolbar or select 'Utility Menu'/'File'/'Exit...'. In the dialog box that appears, click on 'Save Everything' and then click on 'OK'.

W15: Model and mesh an impeller shown below.

Solution: First create 2D model of an impeller. Mesh it with quadratic elements and then extrude 2-D quad meshed areas to form a 3-D brick meshed volume.

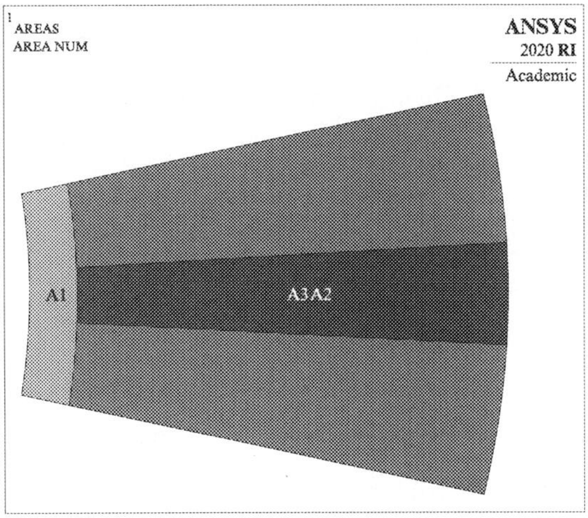

1. Enter ANSYS using '*Impeller-2D*' as the jobname.
2. Draw inner ring segment.

   Main Menu> Preprocessor> Modeling> Create> Areas> Circle> Partial Annulus

   WPX = 0, WPY = 0, Rad – 1 = 30, Rad – 2 = 33, Theta1 = –12, Theta 2 = 12, click 'OK'
3. Draw the fin segment

   Main Menu> Preprocessor> Modeling> Create> Areas> Circle> Partial Annulus

   WPX = 0, WPY = 0, Rad – 1 = 33, Theta – 1 = – 3, Rad – 2 = 60, Theta – 2 = 3, Click 'OK'
4. Draw the rear disc segment

   Main Menu> Preprocessor> Modeling> Create> Areas> Annulus

   WPX = 0, WPY = 0, Rad – 1 = 30, Theta1 = – 12, Rad – 2 = 60, Theta 2 = 12,Click 'OK'

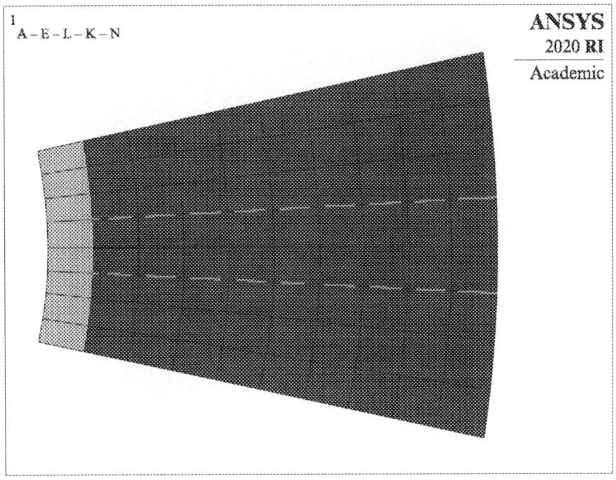

5. Mesh the 2-D model
   Add element type 1 as PLANE183 and set the default TYPE number to 1:
   Main Menu> Preprocessor> Element Type> Add/Edit/Delete ...
   'Add ...', Select 'Solid' and 'Quad 8node 183', then 'OK', 'Close'
   Select 'Mesh Attributes'
   Main Menu> Preprocessor> MeshTool> Mesh attributes...
   Pick 'Set' under Default Attributes: Global
   Set TYPE = 'PLANE 183', then 'OK'
   Mesh inner ring, fin and rear disc segments.
   Add element type '2' as SOLID185 and set the default TYPE number to '2'
   Main Menu> Preprocessor> Element Type> Add/Edit/Delete ...
   'Add ...', Select 'Solid' and 'Brick 8node 185', then 'OK', 'Close'
   Select mesh attributes
   Main Menu> Preprocessor> MeshTool> Mesh attributes...
   Pick [Set] under Default Attributes: Global
   Set TYPE = 'SOLID 185', then 'OK'
6. Define extrude options and extrude all areas
   (a) Set the number of element divisions for extrude operation
   Main Menu> Preprocessor> Modeling> Operate> Extrude> Elem Ext Opts ...
   (b) Extrude three areas and then plot elements
   Main Menu> Preprocessor> Modeling> Operate> Extrude> Areas> Along Normal +
   Utility Menu> Plot> Elements

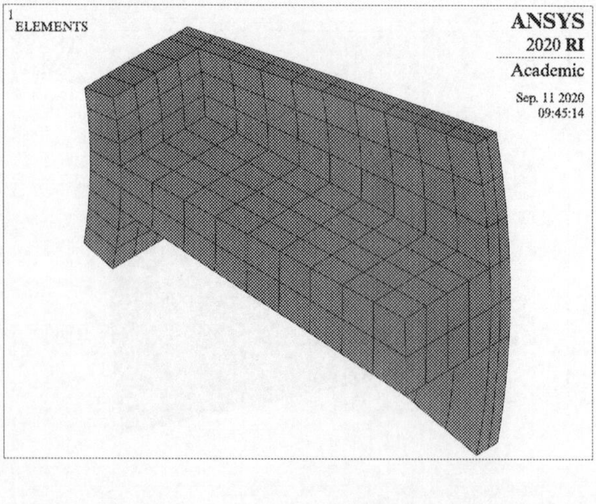

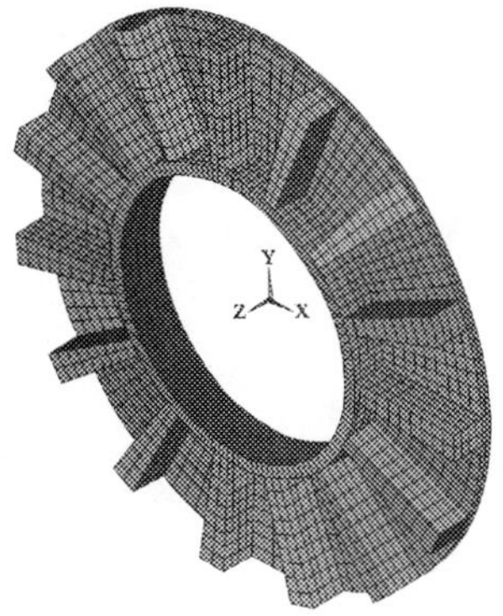

Utility Menu> Plot> Elements

7. Draw the rest of the model

   Change Coordinate System to Global Cylindrical

   Utility Menu> WorkPlane> Change Active CS to> Global Cylindrical

   Copy the three segments.

   Main Menu> Preprocessor> Modeling> Copy> Areas

   Pick the first Fin and Click 'OK'

   Enter Number of copies = 15, Y-offset in active CS = 24

8. Save the meshed model

   Utility Menu> File> Save as ...

   Enter 'Impeller-Mesh.db' as the database name, then 'OK'

   Pick the **'QUIT'** button in the Toolbar

   Select 'Quit - No Save!', 'OK'

W16: A silo model is shown in the figure below. The model consists of both shells and beams. The lines represent beams. Concrete is used to make cylinder areas and steel for cone areas and stiffeners. The shell thickness for cylinder, upper part of cone and lower part of cone is 25, 10 and 6 respectively. Construct the model and then mesh it. Use SHELL181 for the cylinder and BEAM188 for the stiffener. Material properties of concrete are: EX = 45 GPa, NUXY = 0.29. Material properties of steel are: EX = 200 MPa, NUXY = 0.3.

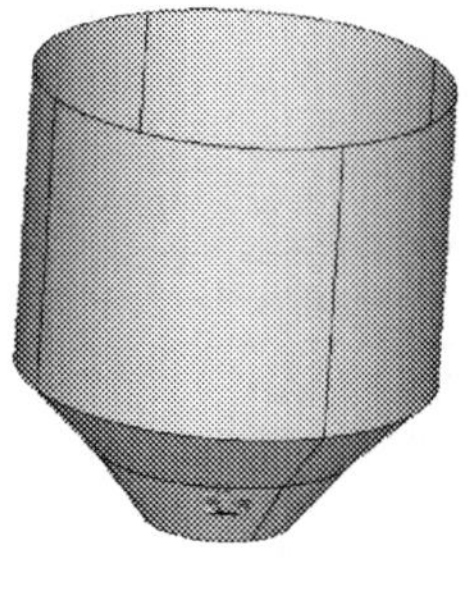

Solution:

1. Give example a Title
   Utility Menu> File> Change Title ...
   '*Silo*'
2. Open preprocessor menu
   Main Menu> Preprocessor
3. Create Circles
   Preprocessor> Modeling> Create> Lines> Arcs> By Cent & Radius ...

Draw 4 circles as given in the following table.

Circle	Centre (x, y)	Radius
1	(0, 0, 0)	9
2	(0, 0, 9)	12
3	(0, 0, 18)	18
4	(0, 0, 54)	18

Create Cylinder and cone walls
Preprocessor> Modeling> Create> Areas> Arbitrary> By Skinning
Pick any corresponding arcs cylinder, click apply. Repeat the procedure.

4. Create Line
   Preprocessor> Modeling> Create> Line> Lines> Straight Line
   Define lines through extreme left keypoints.
   Select extreme left keypoints and then select 'OK'.
   Change Coordinate System to Global Cylindrical
   UtilityMenu> WorkPlane> Change CS to> Global Cylindrical
   Enter 12 for No of copies and 30 for Y as shown below and select 'OK'
5. Copy the line to the other radial positions.
   Preprocessor> Modeling> Copy> Lines..
   Select extreme left lines and then select 'OK'.

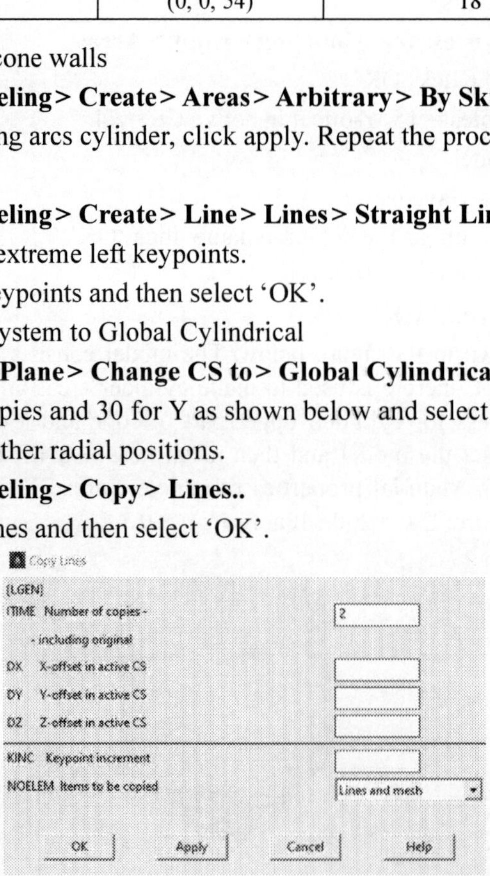

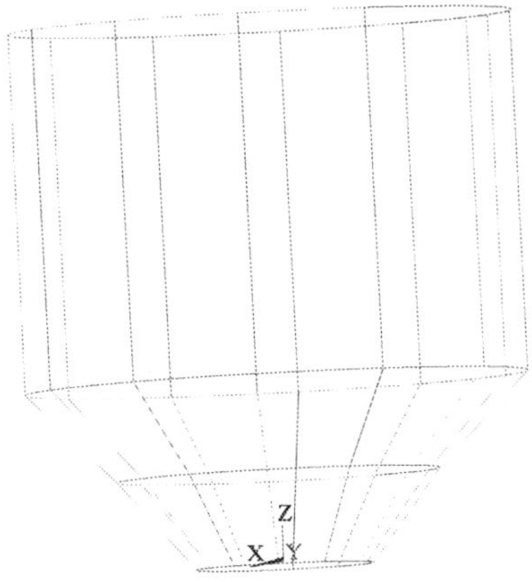

6. Define the Type of Element

   Preprocessor> Element Type> Add/Edit/Delete...> Add...> Solid> 2 node 188

7. Define section properties

   Preprocessor> Sections> Shell> Layup> Add/Edit

   In the Sections for 'SHELL181' window, enter the thickness: 0.025, 0.01, 0.006

   In the Sections for 'BEAM188' window, enter B = 0.006, H = 0.006

8. Define Element Material Properties

   Preprocessor> Material Props> Material Models> Structural> Linear> Elastic> Isotropic

   In the window that appears, enter the following geometric properties for concrete:

   (a) Young's modulus EX: 4.5e4

   (b) Poisson's ratio PRXY: 0. 3.

   In the window that appears, enter the following geometric properties for steel.

   (a) Young's modulus EX = 2e5

   (b) Poisson's ratio NUXY = 0.3

9. Define Mesh Size

   Preprocessor> Meshing> Size Cntrls> ManualSize> Lines> All Lines...

   To obtain the desired mesh set NDIV to 9 on all lines corresponding to cylinder. Divide diagonal lines of the conical parts into 3 divisions. Divide the arcs of cone into 9 divisions each.

10. Bring up the 'Select Entities' menu and select the cylinder areas:

    Utility Menu> Select> Entities ...

    Select 'Areas', 'By Location', 'Z coordinates'

    Min,Max = 18,54, Apply, Plot

11. Assign solid model attributes to the cylinder areas:

    Main Menu> Preprocessor> MeshTool ...

    Select 'Areas' under Element Attributes:, then 'Set'

    [Pick All], MAT = 1, Section ID = 1

    TYPE = '1 SHELL181', OK

12. Select the areas of the upper part of the cone and assign the solid model attributes.

    Utility Menu> Select> Entities ...

    Select 'Areas', SELE All, Plot

    Utility Menu> Select> Entities ...

    Select 'Areas', 'By Num / Pick', Apply, Pick areas shown in the picture to the right. OK, Plot

    Main Menu> Preprocessor> MeshTool ...

    Select 'Areas' under Element Attributes: then 'Set'

    Pick All, MAT = 2, Section ID = 2, TYPE = '1 SHELL181', OK

13. Select the areas of the lower part of the cone and assign the solid model attributes

    Utility Menu> Select> Entities ...

    Select 'Areas', 'By Location', 'Z coordinates'

    Min,Max = 0,9, Appl], Replot

    Main Menu> Preprocessor> MeshTool ...

    Select 'Areas' under Element Attributes:, then 'Set', Pick All, MAT = 2, Section ID = 3

    TYPE = '1 SHELL181', OK

14. Select the lines in the cone and assign the solid model attributes:

    Utility Menu> Select> Entities...

    Select 'Lines', 'By Location', 'Z coordinates'

    Min,Max = 0,18, Apply, Plot

    Main Menu> Preprocessor> MeshTool...

    Select 'Lines' under Element Attributes:, then 'Set'

    Pick All, MAT = 1, ID = 1, TYPE = '2 BEAM188', OK

15. Mesh the lines with BEAM188 elements:

    Main Menu> Preprocessor> MeshTool ...

    Select 'Lines' under Mesh:, then 'Mesh', Pick All

16. Select everything and mesh the areas with SHELL181 elements:

    Utility Menu> Select> Everything

    Utility Menu> Plot> Areas

    Main Menu> Preprocessor> MeshTool ...

    Select 'Areas' under Mesh:, then Mesh, Pick All

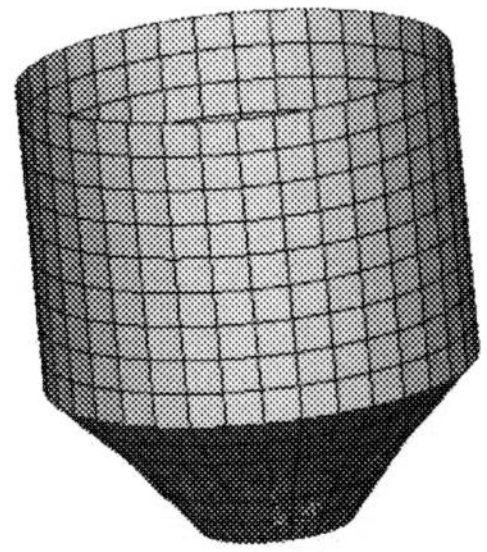

17. Turn on material numbers for 'Elem / Attrib numbering' and plot the elements:
    Utility Menu > PlotCtrls > Numbering ...
    Elem / Attrib numbering = 'Material numbers', OK

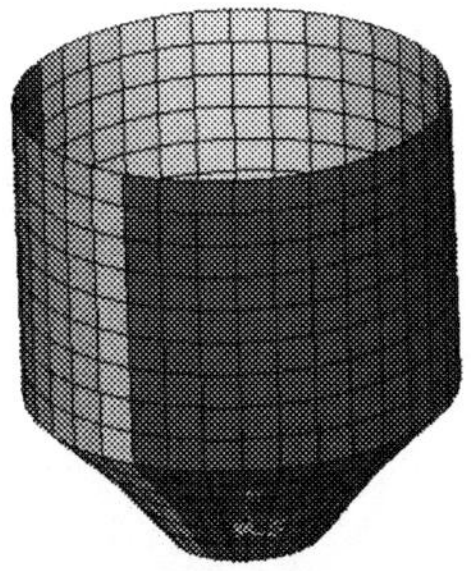

18. Turn on real constant numbers for 'Elem / Attrib numbering' and plot the elements
    Utility Menu > PlotCtrls > Numbering ...
    Elem / Attrib numbering = 'Section num', OK

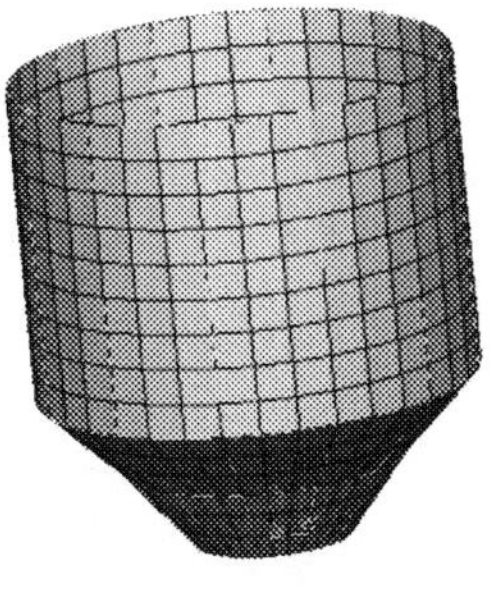

19. Save the meshed model and exit ANSYS
    Utility Menu > File > Save as...
    Enter 'Silo-Mesh.db' as the database name, then 'OK'
    Pick the 'QUIT' button in the Toolbar, Select 'Quit - No Save!', OK

4 CHAPTER

STATIC STRUCTURAL ANALYSIS

Structural analysis is probably the most common application of the finite element method. The term *structural* (or *structure*) implies not only civil engineering structures such as bridges and buildings, but also naval, aeronautical and mechanical structures such as ship hulls, aircraft bodies, and machine housings, as well as mechanical components such as pistons, machine parts and tools.

Types of Structural Analysis

The following types of structural analyses can be performed using ANSYS:

- *Static Analysis* – Used to determine displacements, stresses, etc., under static loading conditions, both linear and nonlinear static analyses. Nonlinearities can include plasticity, stress stiffening, large deflection, large strain, hyperelasticity, contact surfaces and creep.
- *Modal Analysis* – Used to calculate the natural frequencies and mode shapes of a structure. Several mode- extraction methods are available.
- *Harmonic Analysis* – Used to determine the response of a structure to harmonically time-varying loads.
- *Transient Dynamic Analysis* – Used to determine the response of a structure to arbitrarily time-varying loads. All nonlinearities mentioned under static analysis above are allowed.
- *Spectrum Analysis* – An extension of the modal analysis, used to calculate stresses and strains due to a response spectrum or a PSD input (random vibrations).
- *Buckling Analysis* – Used to calculate the buckling loads and determine the buckling mode shape. Both linear (eigenvalue) buckling and nonlinear buckling analyses are possible.
- *Explicit Dynamic Analysis* – This type of structural analysis is available via the ANSYS LS-DYNA product, which provides an interface to the LS-DYNA explicit finite element program. Explicit dynamic analysis calculates fast solutions for large deformation dynamics and complex contact problems.

The primary unknowns (nodal degrees of freedom) calculated in a structural analysis are called *displacements*. Other quantities such as strains, stresses and reaction forces are then derived from the nodal displacements.

Selecting Elements

Most element types are structural elements, ranging from simple spars and beams to more complex layered shells and large strain solids. Most types of structural analyses can use any of these elements.

Selecting Material Models

For analyses described in this book, the analyst must specify the material that s/he intends to simulate. Using the GUI, the analyst can specify many materials that s/he intends to simulate via an intuitive material model interface. The interface uses a hierarchical tree structure of material categories, intended to assist in choosing the appropriate model for the analysis.

Solution Method

The ANSYS family of products uses the h-method for solving structural problems. The h-method can be used for any type of analysis.

STATIC ANALYSIS

A static analysis calculates the effects of *steady* loading conditions on a structure, while ignoring inertia and damping effects, such as those caused by time-varying loads. A static analysis can, however, include steady inertia loads and time-varying loads that can be approximated as static equivalent loads.

Static analysis determines the displacements, stresses, strains, and forces in structures or components caused by loads that do not induce significant inertia and damping effects. Steady loading and response conditions are assumed; that is, the loads and the structure's response are assumed to vary slowly with respect to time. The types of loading that can be applied in a static analysis include:

- Externally applied forces and pressures
- Steady-state inertial forces
- Imposed displacements
- Temperatures
- Fluences.

Linear vs Nonlinear Static Analyses

Static analysis can be either linear or nonlinear. All types of nonlinearities are allowed—large deformations, plasticity, creep, stress stiffening, contact elements, hyperelastic elements, and so on. This section focuses on linear static analyses, with brief references to nonlinearities.

PERFORMING STATIC ANALYSIS

Build the Model

Keep the following points in mind while doing static analysis:

- Both linear and nonlinear structural elements can be used.
- Material properties can be linear or nonlinear, isotropic or orthotropic, and constant or temperature-dependent.
- Define stiffness in some form.
- For inertia loads, define the data required for mass calculations, such as density.
- For thermal loads, define the coefficient of thermal expansion.

Note the following information about mesh density:

- Regions where stresses or strains vary rapidly require a relatively finer mesh than regions where stresses or strains are nearly constant.
- While considering the influence of nonlinearities, remember that the mesh should be able to capture the effects of the nonlinearities. For example, plasticity requires a reasonable integration point density in areas with high plastic deformation gradients.

Set Solution Controls

Setting solution controls involves defining the analysis type and common analysis options for an analysis, as well as specifying load step options for it. Take advantage of a streamlined solution interface when performing a structural static analysis. The Solution Controls dialog box (**Main Menu> Solution> Analysis Type> Sol'n Controls**) provides default settings that will work well for many structural static analyses. It is necessary to set only a few of the options. Because the streamlined solution interface is the *recommended* tool for setting solution controls in structural static analysis, it is the method that is presented in this section. Solution controls for the analysis can also be set using the menu paths (**Main Menu> Solution> Unabridged Menu>** ***option***).

Apply the Loads

After the desired solution options are set, apply loads to the model. The load types applicable in a static analysis are displacements, velocities, forces, moments, pressures, temperatures, fluences, gravity, spinning, etc.

Except for inertia loads and velocity loads, the analyst can define loads either on the solid model or on the finite element model. S/he can also apply boundary conditions via TABLE type array parameters or as function boundary conditions.

Loads can also be applied using TABLE type array parameters. In a structural analysis, valid primary variables are TIME, TEMP, and location (X, Y, Z). When defining the table, TIME must be in ascending order in the table index (as in any table array).

Solve the Analysis

ANSYS is now ready to solve the problem.

1. Save a backup copy of the database to a named file (**Utility Menu> File> Save as**). Then retrieve the model by reentering the ANSYS program and issuing **RESUME**.
2. Start solution calculations (**Main Menu> Solution> Solve> Current LS**).
3. If it is necessary to include additional loading conditions (that is, multiple load steps), repeat the process of applying loads, specifying load step options, saving, and solving for each load step.
4. Leave SOLUTION. Close the Solution menu.

Review the Results

Results from a static analysis are written to the structural results file, `Jobname.RST`. They consist of the following data:

- Primary data: Nodal displacements (UX, UY, UZ, ROTX, ROTY, ROTZ)
- Derived data: Nodal and element stresses, nodal and element strains, element forces, nodal reaction forces and so on

Postprocessors

These results can be viewed using POST1, the general postprocessor, and POST26, the time-history processor.

- POST1 is used to review results over the entire model at specific sub-steps (time-points). Some typical POST1 operations are explained below.
- POST26 is used in nonlinear static analyses to track specific result items over the applied load history.

Reviewing Results Data

1. Read in the database from the database file using **Utility Menu> File> Resume from**
2. Read in the desired set of results. Identify the data set by load step and substep numbers or by time. If a time value is specified for which no results are available, the ANSYS program will perform linear interpolation on all the data to calculate the results at that time.
   Main Menu> General Postproc> Read Results> By Load Step
3. Perform the necessary POST1 operations. Typical static analysis POST1 operations are explained below.

Postprocessing Operations

- Use **Main Menu> General Postproc> Plot Results> Deformed Shape** to display deformed shape.

- Use **Main Menu> General Postproc> List Results> Reaction Solu** to list reaction forces and moments.
- Use **Main Menu> General Postproc> List Results> Element Solution** to list nodal forces and moments.

The sum of all nodal forces and moments for a selected set of nodes can be listed. Select a set of nodes and use this feature to find out the total force acting on those nodes using **Main Menu> General Postproc> Nodal Calcs> Total Force Sum.**

The total force and total moment at each selected node can also be checked. For a body in equilibrium, the total load is zero at all nodes except where an applied load or reaction load exists: **Main Menu> General Postproc> Nodal Calcs> Sum @ Each Node**

For line elements, such as beams, spars, and pipes, use **Main Menu> General Postproc> Element Table> Define Table** to gain access to derived data. Results data are identified by a combination of a label and a sequence number or component name on the **ETABLE** command.

Use **Main Menu> General Postproc> Plot Results> Contour Plot> Nodal Solu** or **Element Solu** to display contours.

Use **Main Menu> General Postproc> Element Table> Plot Element Table** and **Main Menu> General Postproc> Plot Results> Contour Plot> Line Elem Res** to contour element table data and line element data.

Use **Main Menu> General Postproc> Plot Results> Vector Plot> Predefined** to display vectors and **Main Menu> General Postproc> List Results> Vector Data** to view vector listings.

Use **Main Menu> General Postproc> List Results>** ***solution option*** to produce tabular listings.

4.1 1-D PROBLEMS

W1: Consider a 2 m long steel bar of 50 mm^2 cross-sectional area as shown below. Use two element mesh to model this problem. Find nodal displacements, element stresses and reaction. Take Young's modulus, $E = 2 \times 10^5$ N/mm^2, load = 100 N.

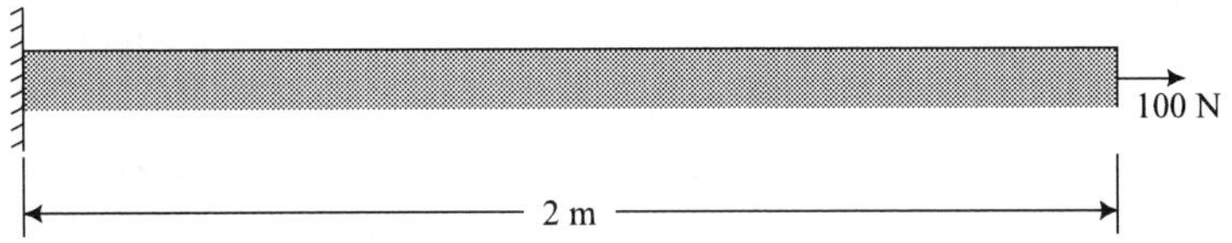

Data: Length of bar = 2000 mm, cross-sectional area = 50 mm^2, $E = 2 \times 10^5$ N/mm^2, load = 100 N

Solution:

Preprocessing

1. Change jobname

   File> Change Jobname

   Enter *'Bar with end load*'' and click on 'OK'.

2. Define element types

   Preprocessor> Element Type> Add/Edit/Delete

   Click on 'Add..', highlight 'Link', then '3D finit stn 180', click on 'OK', then 'Close'.
3. Define the cross-sectional area for the Link180

   Preprocessor> Sections> Link> Add

   Click 'OK' for 'Type 1 LINK180'

   Type '1' for'Add Link Section with ID'. Type 'Bar' for Section Name'. In this problem, the area is 50. Fill in the Link area value. Choose 'Tension' for 'Tension Key and click on 'OK', then 'Close'.
4. Define Material Properties

   Preprocessor> Material Properties> Material Models> Structural> Linear> Elastic> Isotropic

   'OK' for material set number 1, then enter 2e5 for EX, then 'OK'.
5. Create nodes at bar nodes

   Preprocessor> Modeling> Create> Nodes> In Active CS

Enter 1 for node number (ANSYS would automatically number nodes if this column is left blank). Enter the location as (X,Y,Z) = (0,0,0). Click on 'Apply'. Continue defining nodes 2&3 using the locations based on the sketch of the bar, but after entering the node 3 location, click on 'OK' instead of 'Apply'. So, node 2 is at (X,Y,z)=(1000,0,0) and node 3 is at (X,Y,Z) = (2000,0,0).

   As a check to ensure all nodes were entered correctly, list the nodes.

   Utility Menu> List> Nodes

   Turn on node numbering.

   Utility Menu> PlotCtrls> Numbering
6. Create link elements between nodes

   Preprocessor> Modelling> Create> Elements> Auto Numbered> Thru Nodes

   A picking menu appears. Pick node 1, then node 2, and click on 'Apply' in the Picking Menu. Pick node 2, then node 3, and click on 'OK' in the Picking Menu.
7. Apply constraints and forces on the model.

   Solution> Loads> Apply> Structural> Displacement> On Nodes

Pick node 1, then click 'OK' in the picking menu that has appeared. Choose 'ALL DOF', and use the default displacement value of zero.

   Apply the forces.

   Solution> Loads> Apply> Structural> Force/Moment> On Nodes

   Pick node 3, then 'OK' in the picking menu, choose 'FY' as the direction of the force, and enter 100 for the force value. Click on 'OK'.

Solution:

1. Solve the problem

   Solution> Solve> Current LS

   Click 'OK' in the 'Solve Current Load Step' Box

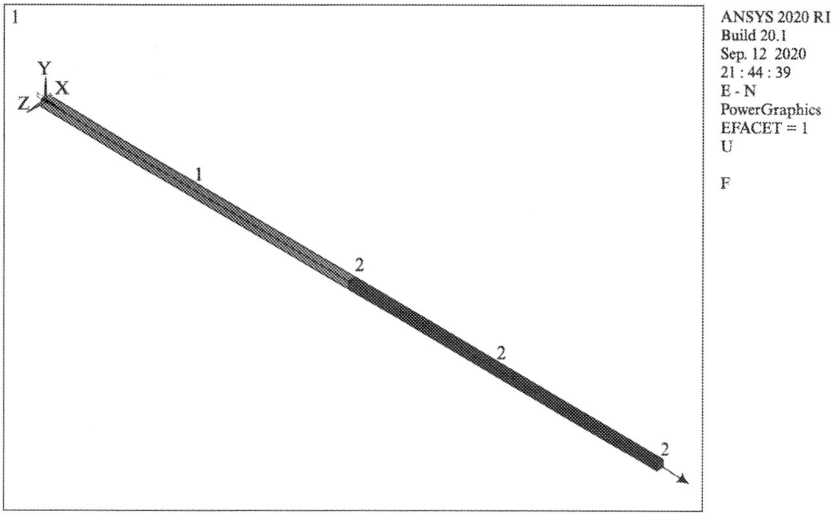

Postprocessing

1. Plot the deformed shape

   General Postproc> Plot Results> Deformed Shape

   Choose 'Def + undeformed', then 'OK'

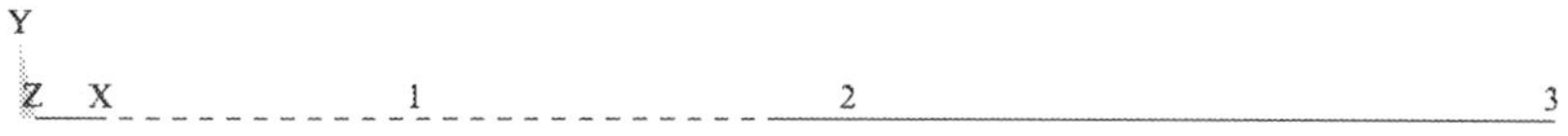

2. List reaction forces

   General Postproc> List Results> Reaction Solution

   Click on 'All struc forc F', and 'OK'

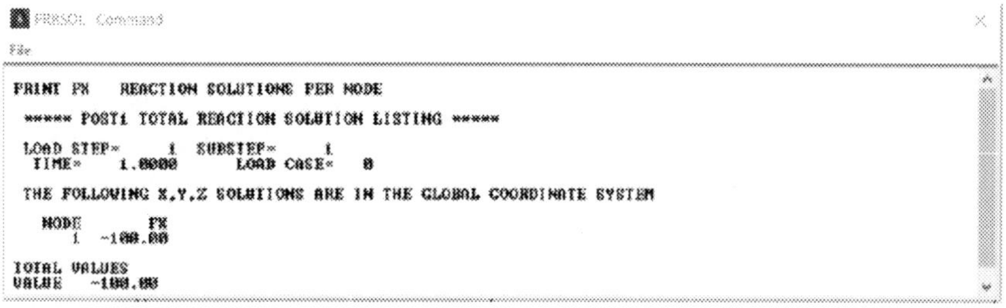

```
PRINT FX    REACTION SOLUTIONS PER NODE

 ***** POST1 TOTAL REACTION SOLUTION LISTING *****

 LOAD STEP=      1  SUBSTEP=      1
  TIME=    1.0000      LOAD CASE=   0

 THE FOLLOWING X,Y,Z SOLUTIONS ARE IN THE GLOBAL COORDINATE SYSTEM

   NODE       FX
      1   -100.00

TOTAL VALUES
VALUE   -100.00
```

Print the information in the Listing Box, by clicking, in that box, on **File> Print**. Or else, just write the information down.

3. List a summary, which includes the axial force and stress in each element

   General Postproc> List Results> Element Solution> LineElem results> Structural ELEM

Click 'OK', and a summary for each element is printed, including the node numbers, from which the corresponding bar element can be identified, and also the axial force (MFORX), and the axial stress (SAXL). Record or print this information.

4. List the x and y direction deflections for each node.

   General Postproc> List Results> Nodal Solution> DOF Solution> ALL DOFs

   Click 'OK'. Again, record or print, this information.

5. Exit ANSYS

   Quit> Save Everything> OK

4.2 2-D PROBLEMS

4.2.1 Truss

W2: A simple truss is shown below. Modulus for material is 200 GPa, A = 500 mm^2, force at node 2 in the + y direction of 1,000 N. Find the displacements and stresses.

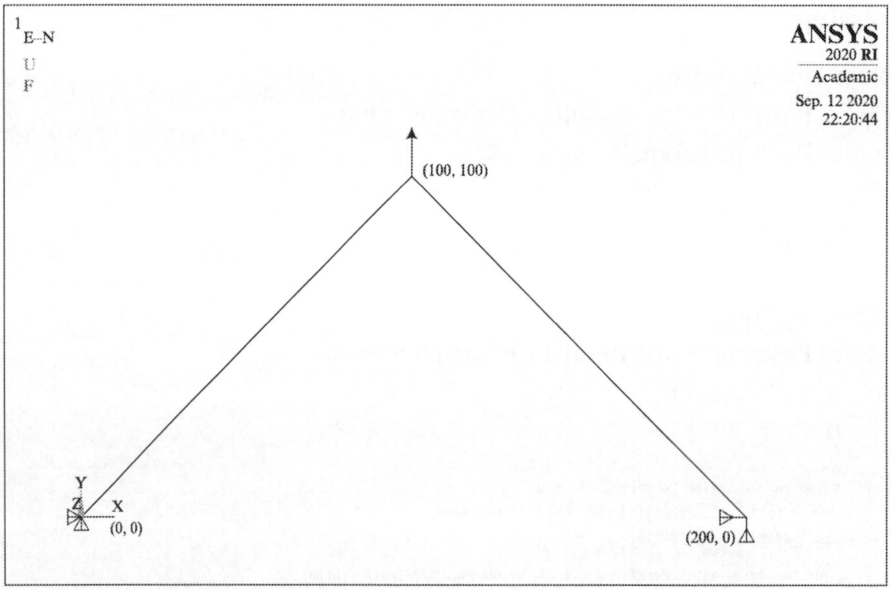

Data: E = 2e5 N/mm^2, A = 500 mm^2 F = 1000 N

Solution:

Preprocessing

1. Change jobname

   File> Change Jobname

   Enter 'Truss-1', and click on 'OK'.

2. Define element types

   Preprocessor> Element Type> Add/Edit/Delete

   Click on 'Add..', highlight 'Link', then '3D finit stn 180', click on 'OK', then 'Close'.
3. Define the cross-sectional area for the Link180

   Preprocessor> Sections> Link> Add

   Type '1' for'Add Link Section with ID'. Type 'Truss 1' for Section Name'. In this problem, the area is 500. Fill in the Link area value. Choose 'Tension' for 'Tension Key and click on 'OK', then 'Close'.
4. Define Material Properties

   Preprocessor> Material Properties> Constant> Isotropic

   'OK' for material set number 1, then enter 2e5 for EX, then 'OK'.
5. Create nodes at truss joints

   Preprocessor> Modeling> Create> Nodes> In Active CS

   Enter 1 for node number. Enter the location as (X,Y,Z) = (0,0,0). Click on 'Apply'. Continue defining nodes 2&3 using the locations based on the sketch of the truss, but after entering the node 3 location, click on 'OK' instead of 'Apply'. So, node 2 is at (X,Y) = (100,100) and node 3 is at (X,Y) = (200,0).

   As a check to ensure all nodes were entered correctly, list the nodes.

   Utility Menu> List> Nodes

   Turn on node numbering.

   Utility Menu> PlotCtrls> Numbering
6. Create link elements between nodes

   Preprocessor> Create> Elements> Auto Numbered> Thru Nodes

   A picking menu appears. Pick node 1, then node 2, and click on 'Apply' in the Picking Menu. Pick node 2, then node 3, and click on 'OK' in the Picking Menu.
7. Apply constraints and forces on the model.

   Solution> Loads> Apply> Structural> Displacement> On Nodes

   Pick nodes 1 & 3, then click 'OK' in the picking menu that has appeared. Choose ALL DOF, and use the default displacement value of zero.

   Apply the forces.

   Solution> Loads> Apply> Structural> Force/Moment> On Nodes

   Pick node 2, then 'OK' in the picking menu, choose 'FY' as the direction of the force, and enter 1,000 for the force value. Click on 'OK'.

Solution:

1. Solve the problem

   Solution> Solve> Current LS

   Click 'OK' in the 'Solve Current Load Step' Box.

Postprocessing

1. Plot the deformed shape

   General Postproc> Plot Results> Deformed Shape

   Choose 'Def + undeformed', then 'OK'.

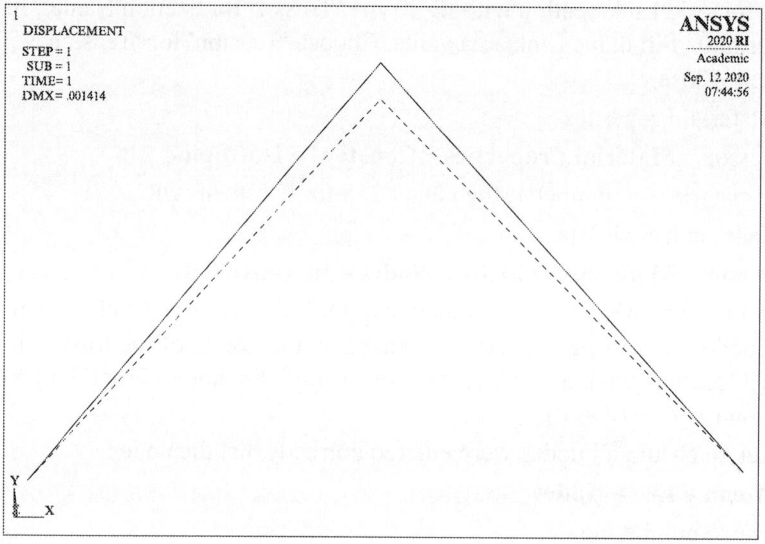

2. List reaction forces

   General Postproc> List Results> Reaction Solution

   Click on 'All struc forc F', and 'OK'.

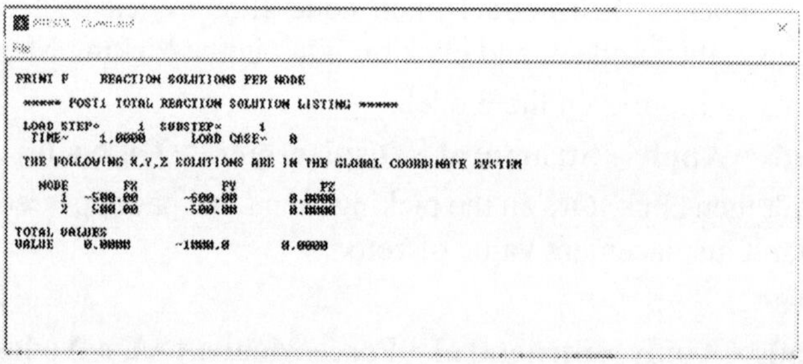

```
PRINT F    REACTION SOLUTIONS PER NODE

 ***** POST1 TOTAL REACTION SOLUTION LISTING *****

 LOAD STEP=     1  SUBSTEP=     1
  TIME=    1.0000      LOAD CASE=   0

 THE FOLLOWING X,Y,Z SOLUTIONS ARE IN THE GLOBAL COORDINATE SYSTEM

   NODE       FX            FY            FZ
      1   -500.00       -500.00        0.0000
      2    500.00       -500.00        0.0000

TOTAL VALUES
VALUE     0.0000       -1000.0        0.0000
```

   Print the information in the Listing Box, by clicking, in that box, on **File> Print**. Or else, just write the information down.

3. List a summary, which includes the axial force and stress in each element.

   General Postproc> List Results> Element Solution> LineElem results> Structural ELEM

   Click 'OK', and a summary for each element is printed, including the node numbers, from which the corresponding truss member can be identified, and also the axial force (MFORX), and the axial stress (SAXL). Record or print, this information.

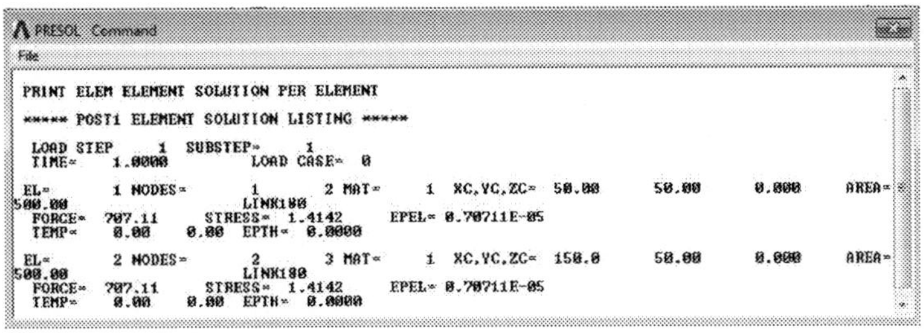

```
PRESOL Command
File
PRINT ELEM ELEMENT SOLUTION PER ELEMENT

***** POST1 ELEMENT SOLUTION LISTING *****

 LOAD STEP     1  SUBSTEP=     1
 TIME=    1.0000          LOAD CASE=  0

 EL=       1 NODES=         1       2 MAT=     1  XC,YC,ZC=  50.00      50.00      0.000      AREA=
500.00                   LINK180
 FORCE=  707.11      STRESS=  1.4142     EPEL= 0.70711E-05
 TEMP=    0.00     0.00  EPTH=  0.0000

 EL=       2 NODES=         2       3 MAT=     1  XC,YC,ZC=  150.0      50.00      0.000      AREA=
500.00                   LINK180
 FORCE=  707.11      STRESS=  1.4142     EPEL= 0.70711E-05
 TEMP=    0.00     0.00  EPTH=  0.0000
```

4. List the x and y direction deflections for each node.

General Postproc> List Results> Nodal Solution> DOF Solution> ALL DOFs

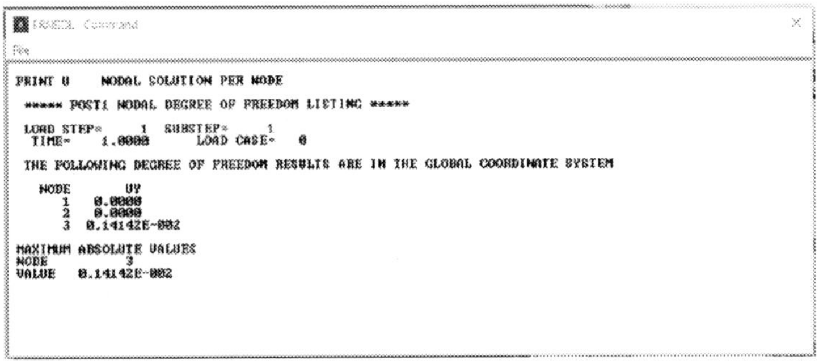

```
PRESOL Command
File
PRINT U    NODAL SOLUTION PER NODE

 ***** POST1 NODAL DEGREE OF FREEDOM LISTING *****

 LOAD STEP=     1  SUBSTEP=     1
  TIME=    1.0000        LOAD CASE=   0

 THE FOLLOWING DEGREE OF FREEDOM RESULTS ARE IN THE GLOBAL COORDINATE SYSTEM

   NODE       UY
     1   0.0000
     2   0.0000
     3  0.14142E-002

MAXIMUM ABSOLUTE VALUES
NODE          3
VALUE   0.14142E-002
```

Click 'OK'. Again, record or print this information.

5. Exit ANSYS

   Quit> Save Everything> OK

W3: Find the displacements and stresses in the truss shown below. Consider area = 300 mm^2, E = 200 GPa.

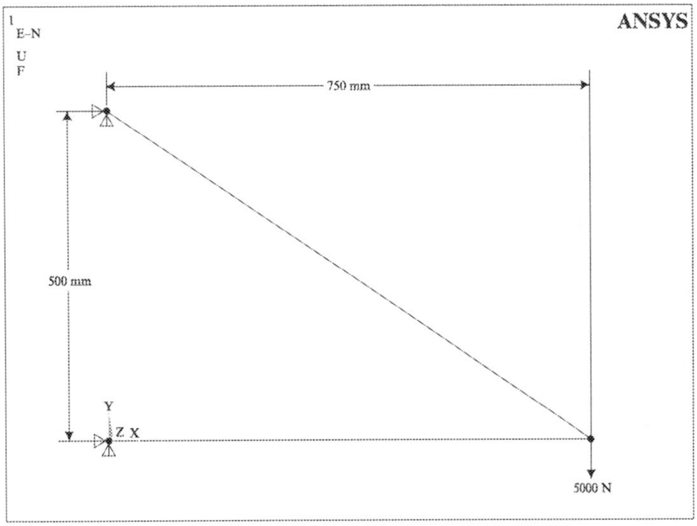

Data: A = 300 mm^2, E = 2e5 N/mm^2, F = 4000 N

Solution:

Preprocessing

1. Change jobname

   File> Change Jobname

   Enter 'Truss-2', and click on 'OK'.

2. Define element types

   Preprocessor> Element Type> Add/Edit/Delete

   Click on 'Add..', highlight 'Link', then '3D finit stn 180', click on 'OK', then 'Close'.

3. Define the real constants for the Link180, which are cross-sectional area and initial strain

   Preprocessor> Real Constants> Add

   Click 'OK' for 'Type 1 LINK180'

In this problem, there is no initial strain (leave blank), and the area is 300. After filling in the area value, click on 'OK', then 'Close'.

4. Define Material Properties

   Preprocessor> Material Properties> Material Models> Structural Linear> Elastic> Isotropic>

   'OK' for material set number 1, then enter 2e5for EX and '0.3' for 'PRXY' , then 'OK'.

5. Create nodes at truss joints

   Preprocessor> Modeling> Create> Nodes> In Active CS

   Enter 1 for node number. Enter the location as (X, Y, Z) = (0, 0, 0). Click on 'Apply'. Continue defining nodes 2 & 3 using the locations based on the sketch of the truss, but after entering the node 3 location, click on 'OK' instead of 'Apply'. So, node 2 is at (X, Y) = (750, 0) and node 3 is at (X, Y) = (0, 500).

   As a check to ensure all nodes were entered correctly, list the nodes

   Utility Menu> List> Nodes

   Turn on node numbering.

   Utility Menu> PlotCtrls> Numbering

6. Create link elements between nodes

   Preprocessor> Create> Elements> Auto Numbered> Thru Nodes

   A picking menu appears. Pick node 1, then node 2, and click on 'Apply' in the Picking Menu. Pick node 2, then node 3, and click on 'OK' in the Picking Menu.

7. Apply constraints and forces on the model

   Solution> Loads> Apply> Structural> Displacement> On Nodes

   Pick nodes 1 & 3, then click 'OK' in the picking menu that has appeared.

   Choose ALL DOF.

   Apply the forces.

Solution> Loads> Apply> Structural> Force/Moment> On Nodes

Pick node 2, then 'OK' in the picking menu, choose 'FY' as the direction of the force, and enter –5000 for the force value. Click on 'OK'.

Solution:

1. Solve the problem

   Solution> Solve> Current LS

   Click 'OK' in the 'Solve Current Load Step' Box.

   Postprocessing

2. Plot the deformed shape

   General Postproc> Plot Results> Deformed Shape

   Choose 'Def + undeformed', then 'OK'.

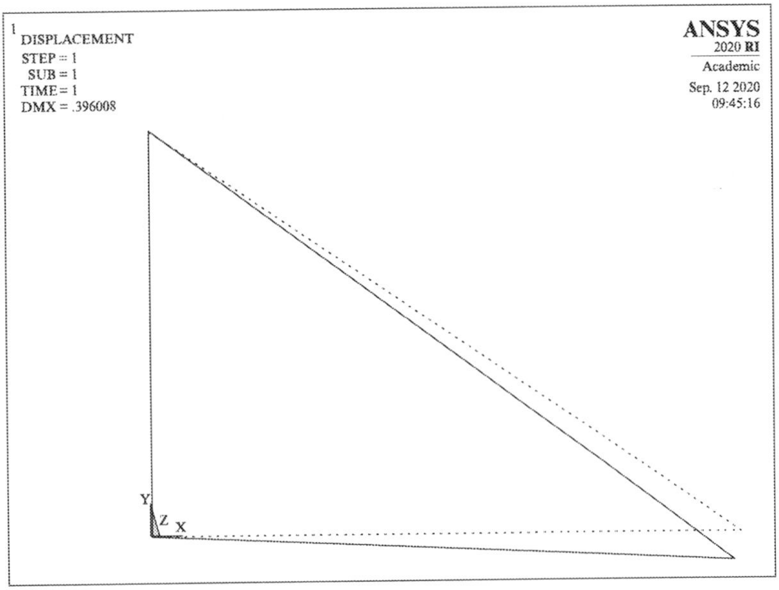

3. List reaction forces

   General Postproc> List Results> Reaction Solution

   Click on 'All items', and 'OK'.

```
PRINT REACTION SOLUTIONS PER NODE

 ***** POST1 TOTAL REACTION SOLUTION LISTING *****

 LOAD STEP=     1  SUBSTEP=     1
  TIME=    1.0000      LOAD CASE=   0

 THE FOLLOWING X,Y,Z SOLUTIONS ARE IN THE GLOBAL COORDINATE SYSTEM

   NODE       FX           FY          FZ
      1    7500.0       0.0000      0.0000
      3   -7500.0       5000.0      0.0000

TOTAL VALUES
VALUE   0.18190E-011  5000.0       0.0000
```

Print the information in the Listing Box, by clicking, in that box, on '**File> Print**'. Or else, just write the information down.

4. List a summary, which includes the axial force and stress in each element.

   General Postproc> List Results> Element Solution>Favourites> Element Solution>

```
PRNSOL Command
File

PRINT U    NODAL SOLUTION PER NODE

 ***** POST1 NODAL DEGREE OF FREEDOM LISTING *****

 LOAD STEP=     1  SUBSTEP=     1
  TIME=    1.0000      LOAD CASE=   0

 THE FOLLOWING DEGREE OF FREEDOM RESULTS ARE IN THE GLOBAL COORDINATE SYSTEM

    NODE       UX           UY           UZ           USUM
       1   0.0000       0.0000       0.0000       0.0000
       2   0.0000       0.0000       0.0000       0.0000
       3  -0.26786     -1.0993       0.0000       1.1315

 MAXIMUM ABSOLUTE VALUES
 NODE          3            3            0            3
 VALUE   -0.26786     -1.0993       0.0000       1.1315
```

Click 'OK', and a summary for each element is printed, including the node numbers, from which the corresponding truss member can be identified, Record or print, this information.

5. List the x and y direction deflections for each node:

   General Postproc> List Results> Nodal Solution> DOF Solution> Displacement vector sum

   Click 'OK'. Again, record or print, this information.

6. Exit ANSYS

   Quit> Save Everything> OK

W4: Model and analyze the following simple truss structure geometry, loading, and boundary conditions. The model consists of three elements connected by three nodes. F = 5,000N. Use ANSYS Link180 elements.

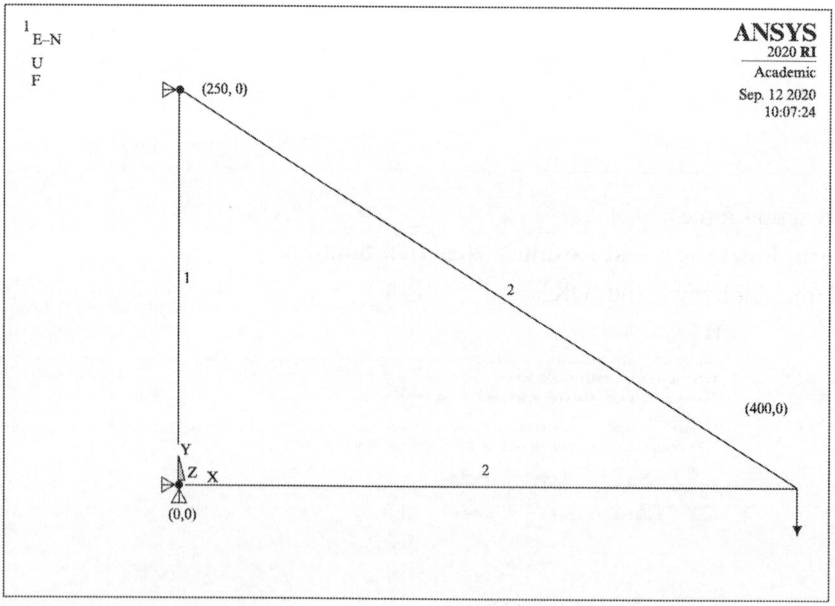

Data: A = 300 mm², E = 7e⁴, F = 5,000N

Solution:

Preprocessing

1. Change jobname

   File> Change Jobname

   Enter 'Truss-3', and click on 'OK'.

2. Define element types

   Preprocessor> Element Type> Add/Edit/Delete

   Click on 'Add..', highlight 'Link', then '3D finit stn 180', click on 'OK', then 'Close'.

3. Define the sections for the Link180

   Preprocessor> Sections > Link> Add

   Enter unique section ID, say '1' and Click 'OK' for 'Add Link Selection'. Enter unique Section Name, say 'A'

In this problem, and the area is 300. Fill in the area value, click on 'OK'.

4. Define Material Properties

   Preprocessor> Material Properties> Constant> Isotropic

   'OK' for material set number 1, then enter 7e4 for EX, then 'OK'.

5. Create nodes at truss joints

   Preprocessor> Modeling> Create> Nodes> In Active CS

   Enter 1 for node number (ANSYS would automatically number nodes if this column is left blank). Enter the location as (X,Y,Z) = (0,0,0). Click on 'Apply'. Continue defining nodes 2 & 3 using the locations based on the sketch of the truss, but after entering the node 3 location, click on 'OK' instead of 'Apply'. So, node 2 is at (X,Y) = (400,0) and node 3 is at (X,Y) = (0,250).

   As a check to ensure all nodes were entered correctly, list the nodes:

   Utility Menu> List> Nodes

   Turn on node numbering.

   Utility Menu> PlotCtrls> Numbering

6. Create link elements between nodes

   Preprocessor> Create> Elements> Auto Numbered> Thru Nodes

   A picking menu appears. Pick node 1, then node 2, and click on 'Apply' in the Picking Menu. Next pick node 2, then node 3, and click on 'Apply' in the Picking Menu. Lastly pick node 3, then node 1, and click on 'OK' in the Picking Menu.

7. Apply constraints and forces on the model

   Solution> Loads> Apply> Structural> Displacement> On Nodes

   Pick node 1, then click 'OK' in the picking menu that has appeared. Choose ALL DOF, and use the default displacement value of zero. Pick node 2, then click 'OK'. Choose UX, and use the default displacement value of zero. Apply the forces.

Solution > Loads > Apply > Structural > Force/Moment > On Nodes
Pick node 2, then 'OK' in the picking menu, choose 'FY' as the direction of the force, and enter -5000 for the force value. Click on 'OK'.

Solution:

1. Solve the problem
   Solution> Solve> Current LS
   Click 'OK' in the 'Solve Current Load Step' Box.

Postprocessing

1. Plot the deformed shape
   General Postproc> Plot Results> Deformed Shape
   Choose 'Def + undeformed', then 'OK'.

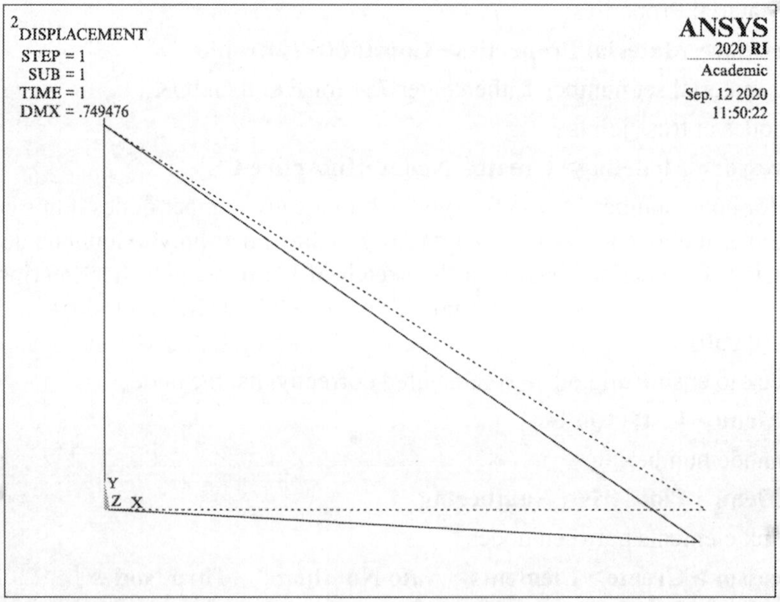

2. List reaction forces
   General Postproc> List Results> Reaction Solution
   Click on 'All struc forc F', and 'OK'.
   Print the information in the Listing Box, by clicking, in that box, on '**File> Print**'. Or else, just write the information down.
3. List a summary, which includes the axial force and stress in each element
   General Postproc> List Results> Element Solution>Favourites> Element Solution>

Click ‘OK’, and a summary for each element is printed, including the node numbers, from which the corresponding truss member can be identified,. Record or print, this information.

4. List the x and y direction deflections for each node:

   General Postproc> List Results> Nodal Solution> DOF Solution> ALL DOFs

   Click ‘OK’. Again, record or print, this information.

5. Exit ANSYS

   Quit> Save Everything> OK

W5: Model and analyze the truss shown below. Cross-sectional area of truss members = 3.0E-4m^2; modulus of elasticity = 207 GPa.

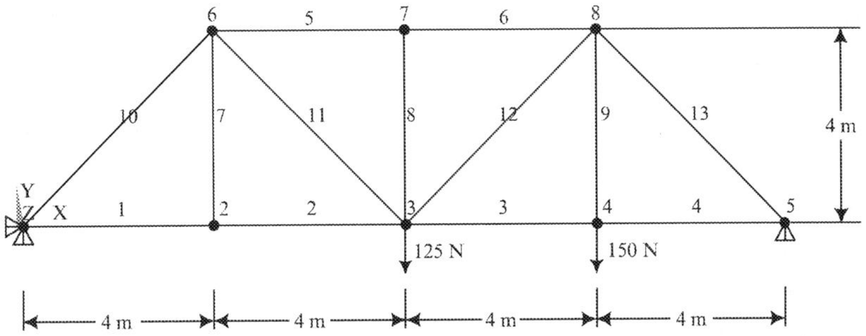

Data: E = 2e11 N/m^2, Area = 3e – 4 m^2

Solution:

Preprocessing

1. Change jobname

   File> Change Jobname

   Enter ‘Bridge Truss’, and click on ‘OK’.

2. Define element types

   Preprocessor> Element Type> Add/Edit/Delete

   Click on ‘Add..’, highlight ‘Link’, then ‘3D finit stn 180’, click on ‘OK’, then ‘Close’. Note that in ANSYS this element is sometimes referred to as ‘LINK 180’, because it is element type 180 in the ANSYS element library.

3. Define the section for the

   Preprocessor> Sections> Add

   Enter area = 3e-4.

4. Define Material Properties

   Preprocessor> Material Properties> Constant> Isotropic

   ‘OK’ for material set number 1, then enter 2E911 for EX, then ‘OK’.

5. Create nodes at truss joints

Preprocessor> Modeling> Create> Nodes> In Active CS

Enter 1 for node number. Enter the location as (X, Y, Z) = (0,0,0). Note that the locations are to be entered in meters, with node 1 located at the origin of the global x-y-z Cartesian coordinate system. Leave the entries for rotation angles blank. (Note: For this problem, all nodes will be in the X-Y plane, with Z = 0.) Click on 'Apply'. Continue defining nodes 2-8 using the locations based on the sketch of the truss, but after entering the node 8 location, click on 'OK' instead of 'Apply'. So, node 2 is at (X,Y) = (4,0), node 3 is at (X,Y) = (8,0), node 4 is at (X,Y) = (12,0), node 5 is at (X,Y) = (16,0), node 6 is at (X,Y) = (4,4), node 7 is at (X,Y) = (8,4), and node 8 is at (X,Y) = (12,4). The dimensions entered are in meters, and on the node definition menu, for this case, rotation angles are ignored, and for all nodes, Z = 0.

As a check to ensure all nodes were entered correctly, list the nodes:

Utility Menu> List> Nodes

If any errors were made in defining the nodes, redefine a node by repeating the procedure of step 5. Of course, it is not necessary to redefine all nodes simply to move one. Just repeat the node creation command for the incorrectly placed node.

Turn on node numbering.

Utility Menu> PlotCtrls> Numbering

Check 'Node Numbering', then click 'OK'. The node numbers may already be showing, but this will force the display of node numbers on subsequent plots.

6. Create link elements between nodes.

Preprocessor> Create> Elements> Auto Numbered >Thru Nodes

A picking menu appears. Pick node 1, then node 2, and click on 'Apply' in the Picking Menu. Continue creating elements using the definitions listed in the table below. After picking the nodes for the last element (element 13), choose 'OK' (instead of 'Apply') to define the element and close the picking menu.

Element	Node I	Node J
1	1	2
2	2	3
3	3	4
4	4	5
5	6	7
6	7	8
7	6	2
8	3	7
9	4	8
10	1	6
11	6	3
12	3	8
13	5	8

7. Apply constraints and forces on the model.

   Solution> Loads> Apply> Structural> Displacement> On Nodes

Pick node 1, then click 'OK' in the picking menu that has appeared. Choose ALL DOF, and use the default displacement value of zero. Click on 'Apply'. Pick node 5, then click 'OK' in the picking menu. To constrain this node in the Y-direction only, Click on 'UY'. Make sure to unselect the 'ALL DOF' label! If the 'ALL DOF' label is highlighted, unselect it by clicking on it. After confirming that only 'UY' is highlighted, click 'OK'.

Apply the forces.

Solution > Loads > Apply > Structural > Force/Moment > On Nodes

Pick node 4, then 'OK' in the picking menu, choose 'FY' as the direction of the force, and enter – 150 for the force value. Click on 'Apply'. Pick node 3, click 'OK' in the picking menu, and enter – 125 for the force value (still in the FY direction). Click on 'OK'.

Solution

1. Solve the problem

   Solution > Solve > Current LS

   Click 'OK' in the 'Solve Current Load Step' Box.

Postprocessing

1. Plot the deformed shape

   General Postproc > Plot Results > Deformed Shape

   Choose 'Def + undeformed', then 'OK'.

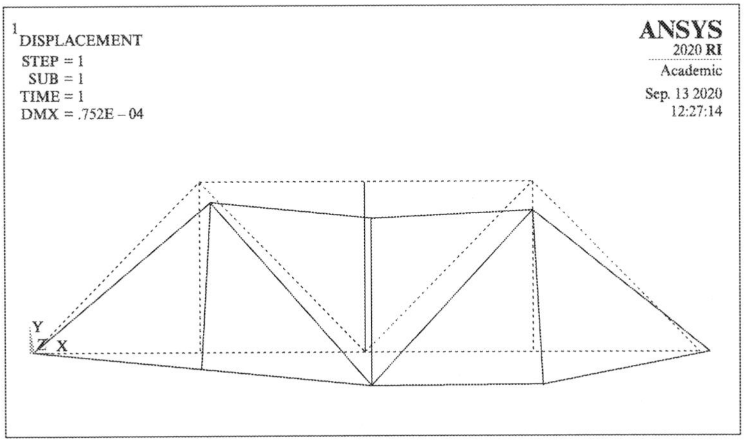

2. List reaction forces

   General Postproc > List Results > Reaction Solution

   Click on 'All struc forc F', and 'OK'.

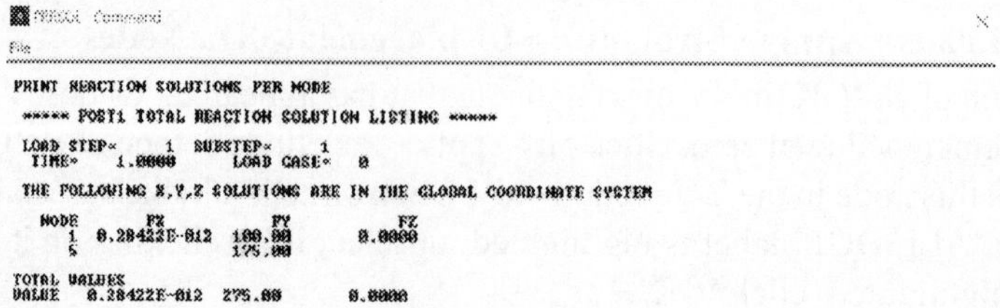

```
File
PRINT REACTION SOLUTIONS PER NODE

 ***** POST1 TOTAL REACTION SOLUTION LISTING *****

 LOAD STEP=     1  SUBSTEP=     1
  TIME=    1.0000      LOAD CASE=   0

 THE FOLLOWING X,Y,Z SOLUTIONS ARE IN THE GLOBAL COORDINATE SYSTEM

   NODE        FX           FY            FZ
     1   0.28422E-012   100.00         0.0000
     5                  175.00

TOTAL VALUES
VALUE    0.28422E-012  275.00        0.0000
```

Print the information in the Listing Box, by clicking, in that box, on '**File> Print**'. Or else, just write the information.

3. List a summary, which includes the axial force and stress in each element:
   General Postproc> List Results> Element Solution>Favourites> Element Solution>
   Click 'OK', and a summary for each element is printed, including the node numbers, from which the corresponding truss member can be identified. Record or print, this information.
4. List the x and y direction deflections for each node:
   General Postproc> List Results> Nodal Solution> DOF Solution>
   Click 'OK'. Again, record or print, this information.
5. Exit ANSYS
   Quit> Save Everything> OK

W6: A tower is made up of trusses. The four trusses at the top are each 2 m in length. The cross section of each of the truss members is 6.25e-3sq.m. Assume the structure is made of steel with modulus of elasticity $E = 200$GPa. The tower is constrained along X and Y directions at the bottom left corner, and along Y direction at the bottom right corner. The tower is loaded at the top. The load is in vertical direction only, and its magnitude is 4000N. Determine deflection at each joint. Determine stress in each member. Determine reaction forces at the base.

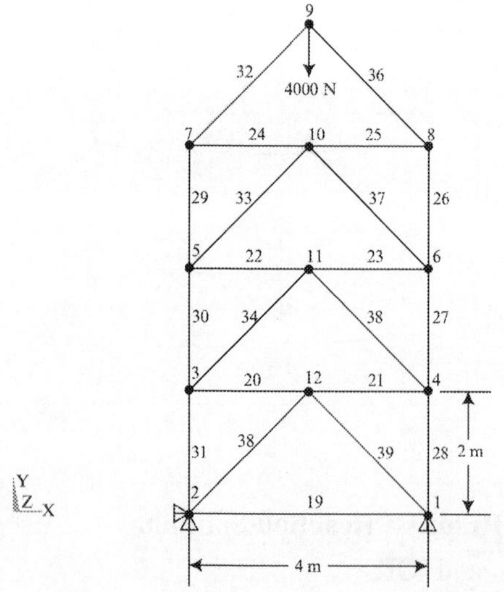

Data: Height of each truss = 2 m, cross-sectional area = 6.25e-3 m^2, E = 2e11N/m^2, Load = –4000N

Solution: A truss is a structural element that experiences loading only in the axial direction. In this example a 2-D truss element will be used. A power transmission tower is a common example of a structure that is made up of only truss members. These towers are actually 3-D structures, but for the sake of simplicity a cross-sectional face of the tower will be taken. The tower is mainly subjected to loading in the vertical direction due to the weight of the cables. Also it is subjected to forces due to wind. In this example only loading due to the weight of the cables, which is in the vertical direction, will be considered.

Preprocessing

1. Model the structure

   Utility Menu> Workplane> WP Settings

   The following window comes up

   Check the Cartesian and Grid Only buttons

   Enter Snap Incr = 0.5, Spacing = 0.5, Min = 0, Max = 10

   Main Menu> Preprocessor> Modeling> Create> Keypoints> On Working Plane

   The following window comes up. Now pick the end points of the truss.

   Select the keypoints on the workplane grid. The keypoints should look like this.

   If the complete workplane cannot be seen then go to **Utility Menu>**

   PlotCntrls> Pan Zoom Rotate and zoom out to see the entire workplane

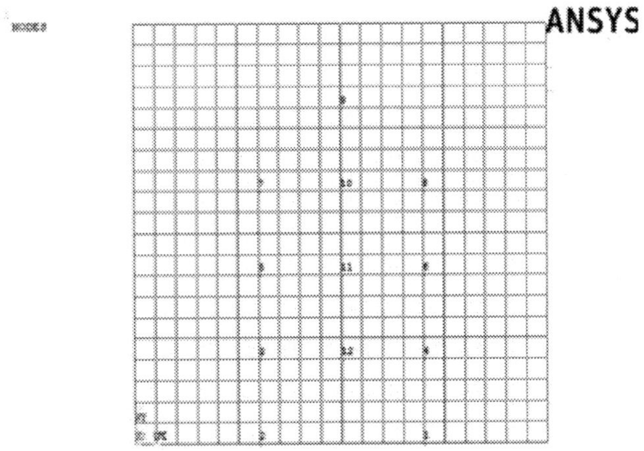

2. Now create lines connecting the keypoints

   Preprocessor> Modeling> Create> Lines> Lines> Straight Line

   Create lines by picking keypoints to make the figure shown below.

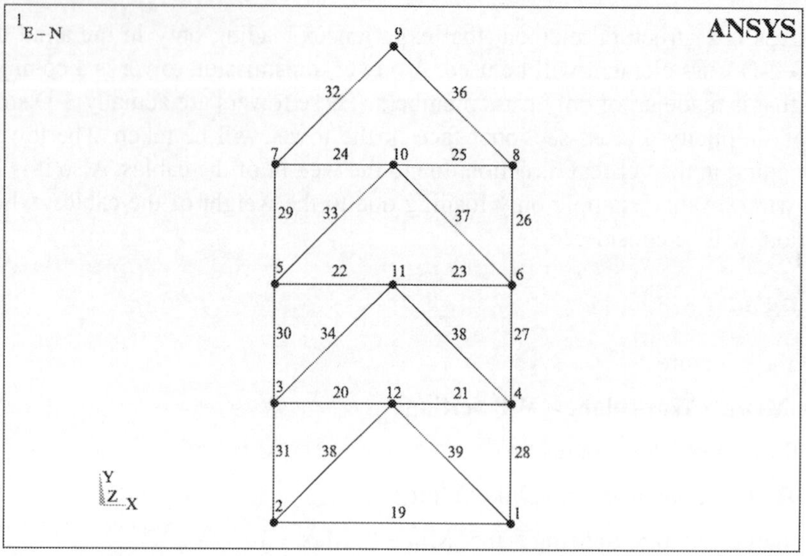

3. Define material properties

   Main Menu> Preprocessor> Material Props> Material Models

   In the window that comes up, for Material Model 1, choose **Structural> Linear> Elastic> Isotropic**

   Enter 1 for the Material Property Number and click 'OK'. The following window comes up.

   Fill in 2e11 for the Young's modulus and 0.3 for minor Poisson's Ratio. Click 'OK'.

4. Define element properties

   Select element type.

   Preprocessor> Element Type> Add/Edit/Delete...

   In the 'Element Types' window that opens click on Add...

   Type 1 in the Element type reference number.

   Click on Structural Link and select 3D finit stn. Click OK. Close the 'Element types' window.

   So now Element type 1 is selected to be a structural Link180 element. The trusses will be modeled as elements of Type 1, i.e., structural link element.

5. Now define the cross-sectional area for this element.

   Preprocessor> Real Constants

   In the 'Real Constants' dialog box that comes up click on Add

   In the 'Element Type for Real Constants' that comes up click 'OK'. The following window comes up.

   Type 6.25e-3 for cross-sectional area and click on 'OK'.

   Now the cross-sectional area of the link element is defined.

6. Mesh the model

   Divide the tower into elements

   Preprocessor> Meshing> Size Cntrls> ManualSize> Lines> All Lines

   In the menu that comes up type 1 in the field for 'Number of element divisions'.

   This divides each of the lines in the figure into 1 element.

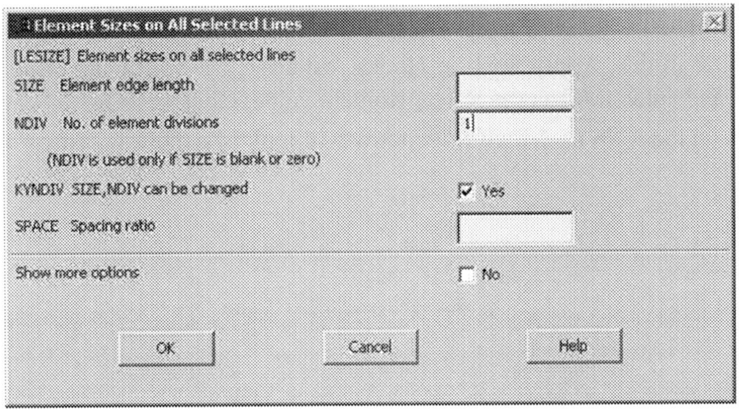

   Click on 'OK'. Now the figure is meshed, ANSYS will automatically divide each line into 1 element.

Preprocessor> Meshing> Mesh> lines

Select all the lines and click on OK in the 'Mesh Lines' dialog box.

Now each line is a truss element (Element 1).

7. Define boundary conditions and constraints

   Apply boundary conditions

   The tower is constrained in the X and Y directions at the bottom left corner and in the Y direction at the bottom right corner.

Main Menu> Preprocessor> Loads> Define

Loads> Apply> Structural> Displacement> On Keypoints

   Select the keypoint on displacement constraints are to be applied. The following window comes up.

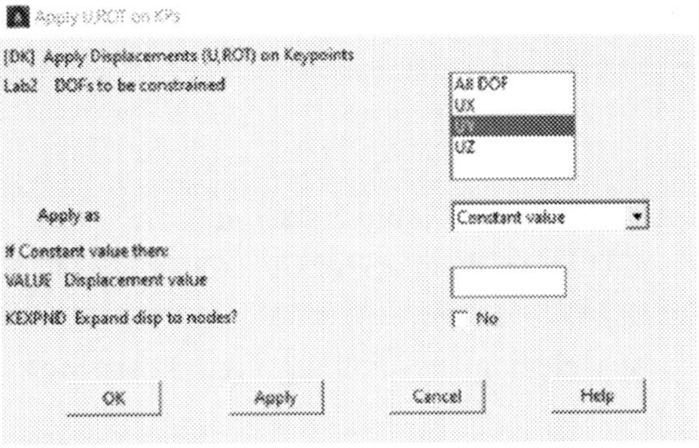

Select UX and UY for the bottom left corner and UY for the bottom right corner and click OK. The default displacement value is taken to be zero.

8. Apply forces

Main Menu> Preprocessor> Loads> Define Loads> Apply> Structural> Forces/ Moment> On Nodes

Select the top node.

Click on OK in the 'Apply F/M on Nodes' window. The following window will appear. Select FY and enter 4000 as the Force/Moment value. Click on 'OK'.

The figure on the ANSYS Graphics window will look like the following.

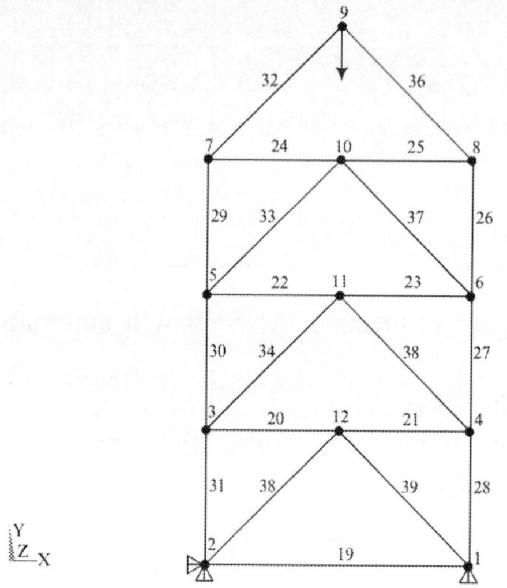

Solution

1. Solve the problem.

Main Menu> Solution> Analysis Type> New Analysis

Select static and click on 'OK'.

Solution> Solve> Current LS

Postprocessing

1. List the results.

General Postproc> List Results> Nodal Solution

Select DOF solution and Displacement vector sum. Click on 'OK'. The nodal displacements will be listed.

List the stresses for each element by clicking **Gen Postproc> List Results> Element Solution**.

Now select **LineElem Results.**

Select a stress to be plotted and click 'OK'.

2. Plot deformed shape

Main Menu> General Postproc> Plot Results> Deformed shape

Select 'Def + undeformed'

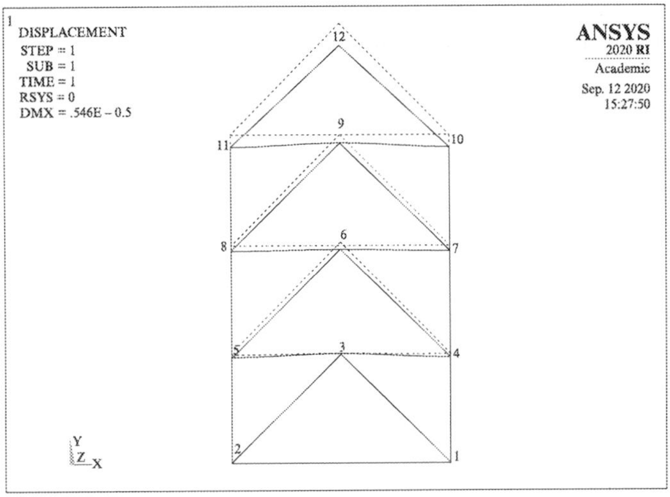

4.2.2 Frames

W7: The members have a annular cross section. The cross sections (A) of each of the truss members is 5.5e-3 m^2. The structure is made of steel with modulus of elasticity, E = 210 GPa. All the DOFs are constrained at the bottom end. The structure is loaded at the ends of the two arms. The load is in the negative Y direction. The load value is 4000N each. Determine deflections at the points of application of load. Determine the maximum stress in the structure.

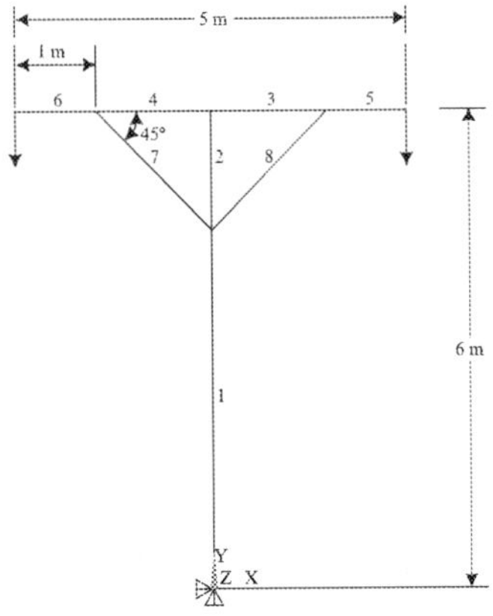

Data: E = 210 GPa, area = 5.5e-3 sqm, Load = 4000 N

Solution:

Preprocessing

1. Model the structure

   Utility Menu> Workplane> WP Settings

   Check the Cartesian and Grid Only buttons

   Enter Min = 0, Max = 8, Spacing = , Snap Increment = 0.5

   Go to the ANSYS Main Menu

   Preprocessor> Modeling> Create> Keypoints> On Working Plane

   Now create lines connecting the keypoints

   Preprocessor> Modeling> Create> Lines> Lines> Straight Line

   Create lines by picking keypoints

   Define material properties

   Main Menu> Preprocessor> Material Props> Material Models. choose **Structural> Linear> Elastic> Isotropic.**

   Fill in 2.1e11 for the Young's modulus and 0.3 for Poisson's ratio. Click OK.

2. Define element properties

   Select element type

   Preprocessor> Element Type> Add/Edit/Delete... Click on Add...

   Click on Structural Beam and select 2 node 188. Click OK.

3. Define the section properties for this element.

   Preprocessor> Sections> Beam> Common Sections

   Enter B = 20, H = 30

4. Mesh the model

   Divide the structure into elements

   Preprocessor> Meshing> Size Controls> Manual Size> Lines> All Lines.

   Type 1 in the field for 'Number of element divisions'. Click on OK.

   Preprocessor> Meshing> Mesh> Lines

   Select all the lines and click on OK in the 'Mesh Lines' dialog box.

5. Define boundary conditions and constraints

   Apply boundary conditions

   The tower is constrained in the DOFs at the bottom node.

   Go to Main Menu

   Preprocessor> Loads> Define Loads> Apply> Structural> Displacement> On Keypoints

   Select the keypoint on which displacement constraints are to be applied.

   Select All DOF and click 'OK'.

6. Apply Forces
   Preprocessor> Loads> Define Loads> Apply> Forces/Moment> On Nodes
   Select the top right node and the top left node.
   Click on OK in the 'Apply F/M on Nodes' window.
   Enter – 4000 for force.
   The figure looks like this now.

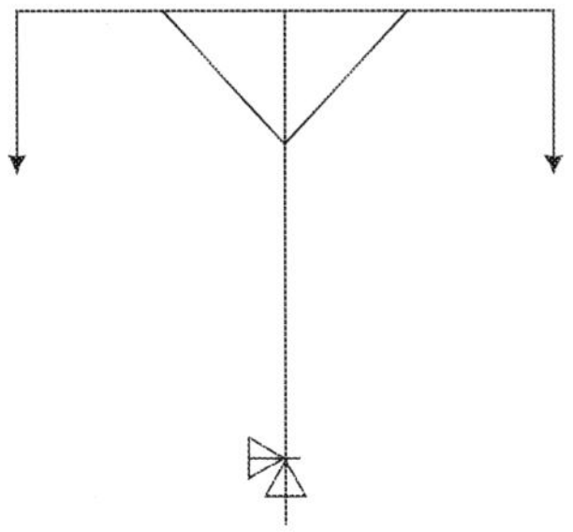

Solution

1. Solve the problem.
   Main Menu> Solution> Analysis Type> New Analysis
   Select static and click on 'OK'.
   Solution> Solve> Current LS

Postprocessing

1. List the results
   General Postprocessing> List Results> Nodal Solution.
   Select DOF solution and All Us. Click on 'OK'. The nodal displacements will be listed.
2. Similarly, list the stresses for each element by clicking **Gen Postprocessing> List Results> Element Solution.** Now select LineElem Results. The following table will be listed.
3. Plot the displacements and stress.
   General Postproc> Plot Results> Contour Plot> Element Solution. The following window will come up.

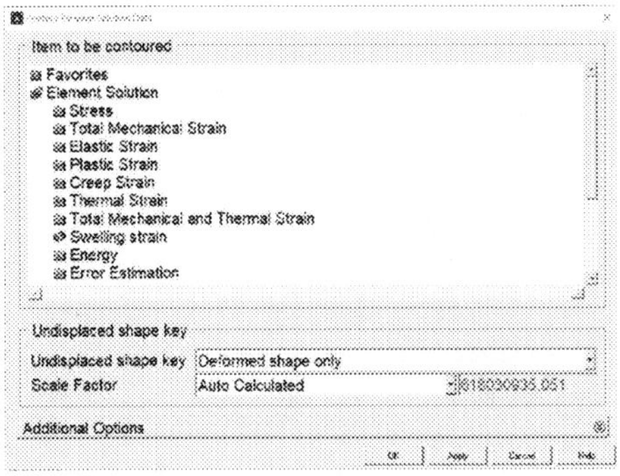

Select a stress to be plotted and click 'OK'. The output will be like this.

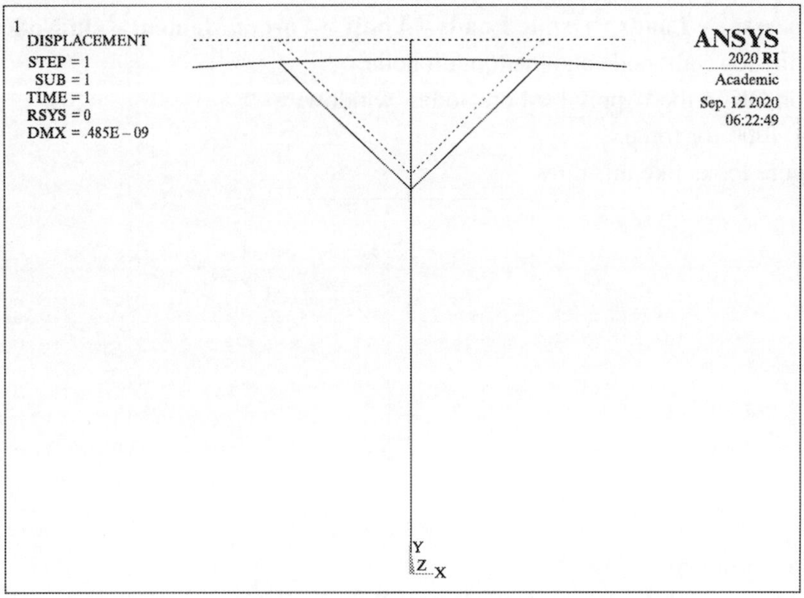

W8: A shelter truss is shown below. Truss members 1-5 are made with mild steel. Modulus for steel is 200 GPa, and cross-sectional area = 300 mm^2. Members 6-9 are made with aluminium. Modulus of aluminum is 70 GPa and cross-sectional area = 250 mm^2. A vertical downward force of 5000N acts at extreme right corner in the Y direction. Find the displacements and stresses.

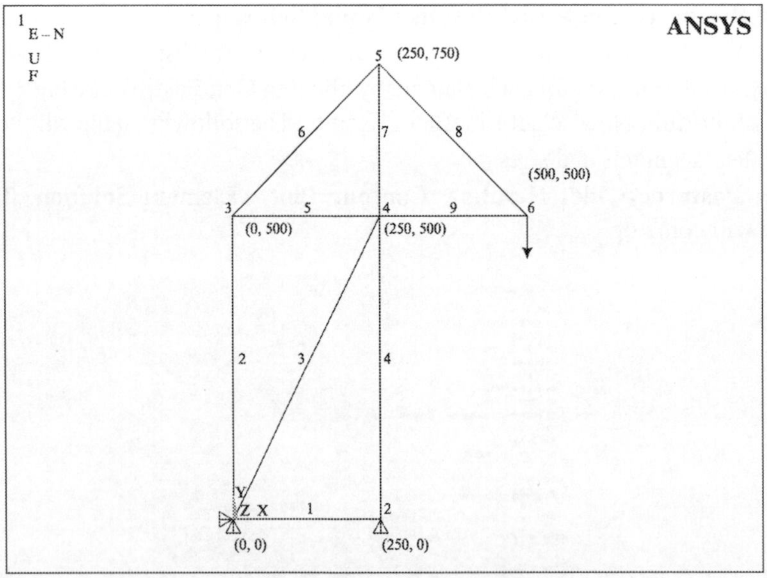

Data: E-1 = 2e5, R-1 = 300 mm^2, E-2 = 7e4, R-2 = 250 mm^2, F = – 5000 N

Solution:

Preprocessing

1. Change jobname

   File> Change Jobname

   Enter 'Shelter Truss', and click on 'OK'.
2. Define element types

   Preprocessor> Element Type> Add/Edit/Delete

   Click on 'Add..', highlight 'Link', then '3D finit stn 180', click on 'OK', then 'Close'.
3. Define first material properties

   Preprocessor> Material Properties> Constant- Isotropic

   'OK' for material set number 1, then enter *2E5* for EX, then 'OK'.

   Define second material properties

   Preprocessor> Material Properties> Material Models>

   Select 'New Model' under 'Material'

   Under 'Material Models Available' select **Structural> Linear> Elastic> Isotropic**

   'OK' for material set number 2, then enter 7E4 for EX, then 'OK'.
4. Define the real constants for first group of members:

   Preprocessor> Real Constants> Add/Edit/Delete

   Click 'OK' for 'Set 1 LINK180'

   In this problem, there is no initial strain (leave blank), and the area is 300. After filling in the area value, click on 'OK', then 'Close'.

   Define the real constants for second group of members:

   Preprocessor> Real Constants> Add/Edit/Delete

   Click Add. Real Constant Set Number 2, for LINK180 appears. Click 'OK' for 'Set 2 LINK180'. In this problem, there is no initial strain (leave blank), and the area is 250. After filling in the area value, click on 'OK', then 'Close'.
5. Create nodes at truss joints

   Preprocessor> Modeling> Create> Nodes> In Active CS

   Enter 1 for node number (ANSYS would automatically number nodes if this column is left blank). Enter the location as (X,Y,Z) = (0,0,0). Click on 'Apply'. Continue defining nodes 2-6 using the locations based on the sketch of the truss, but after entering the node 6 location, click on 'OK' instead of 'Apply'. So, node 2 is at (X,Y) = (250,0), node 3 is at (X,Y) = (0,500), node 4 at (X,Y) = 250,500, node 5 is at (X,Y) = (500,750) and node 6 at (X,Y) = (500,500).

   As a check to ensure all nodes were entered correctly, list the nodes.

   Utility Menu> List> Nodes

   Turn on node numbering.

Utility Menu> PlotCtrls> Numbering

6. Create first group of elements (1-5)
   Select first set of real constant & material properties
   Preprocessor> Create> Elements> Auto Numbered> Thru Nodes
   A picking menu appears. Pick node 3, then node 5, and click on 'Apply' in the Picking Menu. Similarly, create other elements.
   Now create second group of elements (6-9)
   Switch to second set of real & material properties. Pick node 3, then node 5 and click on 'Apply' in the Picking Menu. Similarly create other elements.
7. Apply constraints and forces on the model.
   Solution> Loads> Apply> Structural> Displacement> On Nodes
   Pick nodes 1, then click 'OK' in the picking menu that has appeared. Choose ALL DOF, and use the default displacement value of zero. Now pick nodes 2, then click 'OK' in the picking menu that has appeared. Choose UY, and use the default displacement value of zero.
   Apply the forces.
   Solution> Loads> Apply> Structural> Force/Moment> On Nodes
   Pick node 6, then 'OK' in the picking menu, choose 'FY' as the direction of the force, and enter – 5000 for the force value. Click on 'OK'.

Solution

1. Solve the problem
   Solution> Solve> Current LS
   Click 'OK' in the 'Solve Current Load Step' Box.

Postprocessing

1. Plot the deformed shape
   General Postproc> Plot Results> Deformed Shape
   Choose 'Def + undeformed', then 'OK'.

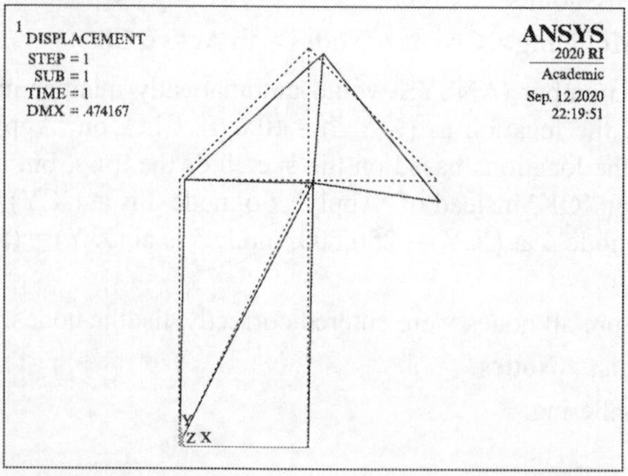

List reaction forces

General Postproc> List Results> Reaction Solution

Click on 'All struc forc F', and 'OK'.

Print the information in the Listing Box, by clicking in that box, on '**File> Print**'. Or else, just write the information.

2. List a summary, which includes the axial force and stress in each element:

   General Postproc> List Results> Element Solution> LineElem results> Structural ELEM

   Click 'OK', and a summary for each element is printed, including the node numbers, from which the corresponding truss member can be identified, and also the axial force (MFORX), and the axial stress (SAXL). Record or print, this information.

3. List the x and y direction deflections for each node:

   General Postproc> List Results> Nodal Solution> DOF Solution> ALL DOFs

   Click 'OK'. Again, record or print, this information.

4. Exit ANSYS

   Quit> Save Everything> OK

W9: Determine the end deflection and root bending stress of a steel cantilever beam of length 250 mm shown below.

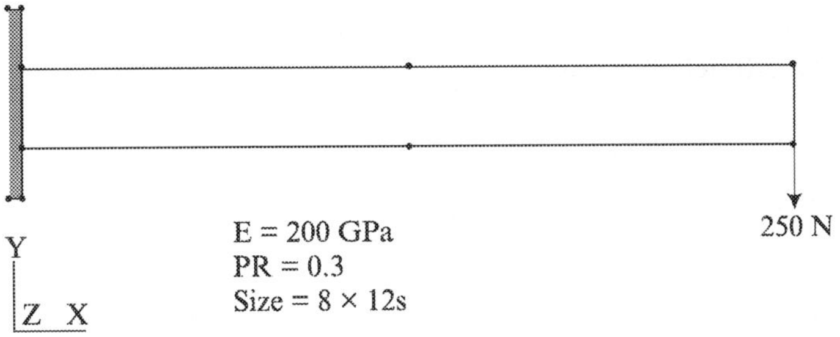

All diamensions in mm

Data: E = 2e5 N/mm², μ = 0.3, Length = 250 mm, Load = 250 N, Width = 8 mm, Height = 12 mm

Solution:

Preprocessing

1. Change jobname.

   File> Change Jobname

   Enter '*Cantilever with point load*', and click on 'OK'.

2. Define element type.

Preprocessor> Element Type> Add/Edit/Delete

Click on 'Add..', highlight 'Beam', then '2node 188', click on 'OK', click 'Options', select 'Quadratic Form' for Element behavior in BEAM188 element type options, then 'Close'.

3. Define the section properties for the BEAM188 elements.

   Preprocessor> Sections> Beam> Common Sections>

   The appropriate values for the given geometry are: B = 8; H = 12, Nb = 4, Nh = 6 and click 'OK'.

4. Define material properties.

   Preprocessor> Material Properties> Constant> Isotropic

   'OK' for material set number 1, then, enter 2e5 for EX, then 'OK'.

5. Create Keypoints.

   Preprocessor> Modeling> Create> Keypoints> In Active CS

   Enter 1 for node number (ANSYS would automatically number nodes if this is left blank). Enter the location as (x,y,z) = (0,0,0). Leave the entries for rotation angles blank. (For this problem, all nodes will be on the x-axis, with y = 0 and z = 0.) Click on 'Apply'. Define node 2 at (x,y,z) = (250,0,0), define node 3 at (x,y,z) = (0,30,0), then click on 'OK' instead of 'Apply'.

6. Create beam elements between keypoints.

   Draw line joining the keypoints.

   Preprocessor> Modeling Create> Lines> Line> Straight Line

   Pick key points 1 and 2, OK

   Preprocessor> Meshing> Mesh Tool

   Select 'Lines' under 'Element Attributes', click 'Set'

   Pick the line and click 'OK' in the Line Attributes.

   Preprocessor> Meshing> Mesh Attributes> Picked Lines

   Choose 'Yes' for Pick Orientation Key Point(s), OK. Select key point 3, OK.

   In the Mesh Tool window, click 'set' for lines under Size Controls. Click on line and OK. Assign 10 for No. of element divisions under Element on Picked Lines, OK.

   Click Mesh in Mesh Tool window. Click the line, and click OK in Mesh Lines window.

Solution:

1. Apply constraints and forces on the model.

   Solution> Loads> Apply> Structural> Displacement> On Nodes

   Pick Left extreme node, then click 'OK' in the picking menu that has appeared. Choose ALL DOF, and use the default value of zero. Click on 'OK'.

   Solution > Loads>Apply> Structural>Force/Moment> On Nodes

   Pick extreme right node , then 'OK' in the picking menu, choose 'FY' as the direction of the force, and enter '-250' for the force value. Click on 'OK'.

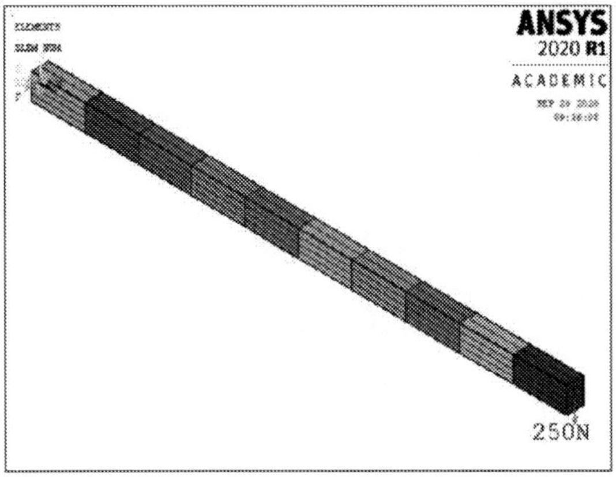

2. Solve the problem.

   Solution> Solve> Current LS

   Click 'OK' in the 'Solve Current Load Step' Box.

Postprocessing

1. Plot the deformed shape.

   General Postproc> Plot Results> Deformed Shape

   Choose 'Def + undeformed', then 'OK'.

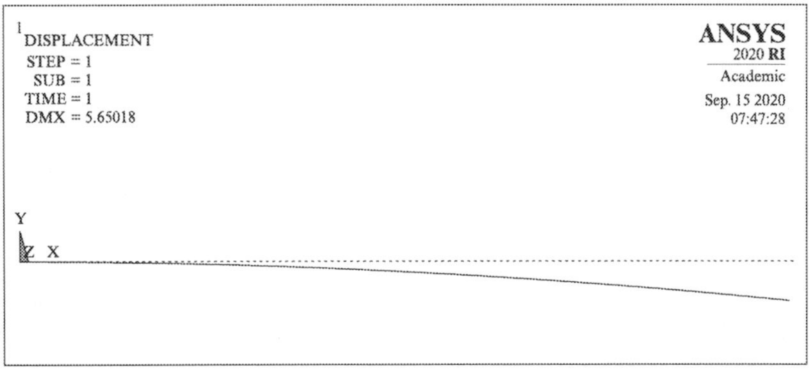

2. List reaction forces.

   General Postproc> List Results> Reaction Solution

   Use the default: 'All items'; and click on 'OK'.

3. List nodal displacements

   General Postproc> List Results> Nodal Solution> DOF Solution> Displacement vector sum

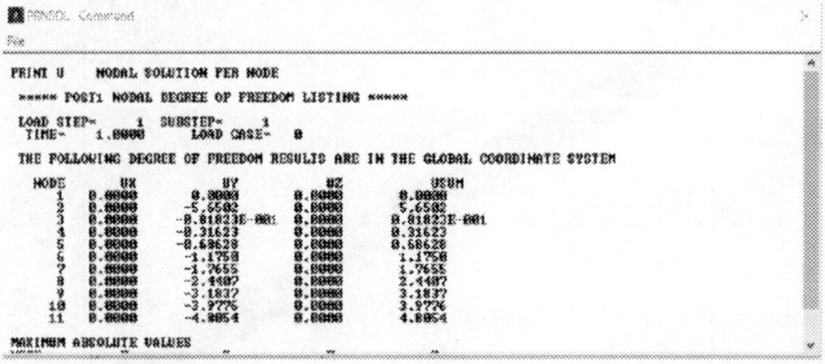

PRINT U NODAL SOLUTION PER NODE

***** POST1 NODAL DEGREE OF FREEDOM LISTING *****

LOAD STEP= 1 SUBSTEP= 1
TIME= 1.0000 LOAD CASE= 0

THE FOLLOWING DEGREE OF FREEDOM RESULTS ARE IN THE GLOBAL COORDINATE SYSTEM

NODE	UX	UY	UZ	USUM
1	0.0000	0.0000	0.0000	0.0000
2	0.0000	-5.6502	0.0000	5.6502
3	0.0000	-0.81823E-001	0.0000	0.81823E-001
4	0.0000	-0.31623	0.0000	0.31623
5	0.0000	-0.68628	0.0000	0.68628
6	0.0000	-1.1758	0.0000	1.1758
7	0.0000	-1.7655	0.0000	1.7655
8	0.0000	-2.4487	0.0000	2.4487
9	0.0000	-3.1837	0.0000	3.1837
10	0.0000	-3.9776	0.0000	3.9776
11	0.0000	-4.8054	0.0000	4.8054

MAXIMUM ABSOLUTE VALUES

4. Define element table items for subsequent plotting and listing of various stress results. For line elements, such as beam elements, to generate certain stress results, define an element table, using the ETABLE command. Note that each element is defined by two end nodes, denoted by node *i* and node *j*. Node *i* is the first of the two nodes defined when the element was created. The form of the GUI path to define a table containing the shear force, at node *i*, for each element, is:

   Main Menu> General Postproc> Element Table> Define Table> Add

   SFy for user label, Select By sequence num and SMISC, 5, click Apply

   SFy for user label, Select By sequence num and SMISC, 18, click 'OK'

   Now plot shear force diagram.

   General Postproc> Plot Results> Contour Plot> Line Ele Res

   Enter LabI = SMIS5, LabJ = SMIS18

   The entries in this command pertain specifically to the BEAM188 element.

   In the command above, 'SFyI' is simply a label chosen by the user. Practically, any other text string could be used in this field. However, it must not have more than eight characters, and it cannot be identical to any general ANSYS predefined label. 'SMISC' is the appropriate item for shear force in BEAM188 elements. Also, the '5' shown as the last entry is the correct sequence number on the ETABLE command, assuming BEAM188 elements are used, and the user wants to store the bending stresses, at node *i* for each element.

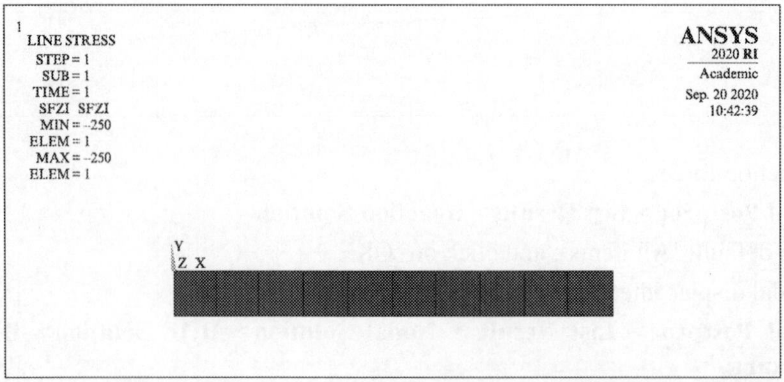

If the user also wants to define an element table in which the bending moment, at node j for each element is stored, the command that could be used is similar to that above. However, the user should choose a different identifying label 'My', and the value of the last entry for this case is '2 and 15'.

Main Menu> General Postproc> Element Table> Define Table> Add

BMzI for user label, Select By sequence num and SMISC, 2, click 'Apply'

SFyJ for user label, Select By sequence num and SMISC, 15, click 'OK'

Now plot bendind moment diagram.

General Postproc> Plot Results> Contour Plot> Line Ele Res

Enter LabI = SMIS2, LabJ = SMIS15

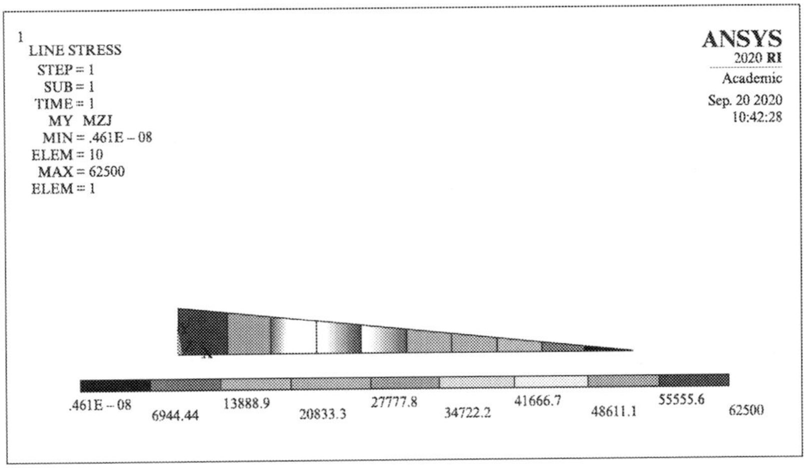

5. List element table results. Assuming all four of the above ETABLE commands were entered, then a listing of each stored stress component can be obtained with the menu path:

   General Postproc> List Results> Elem Table Data

   Then, highlight the desired labels, and click on 'OK'. The stresses corresponding to the chosen labels will be listed by element number.

6. Plot element stresses.

   General Postproc> Plot Results> Line Elem Res

   In the dialog box that opens, select the label 'MzI' for the LABI entry, and the label 'MzJ' for the LABJ entry. The optional scale factor can be left as '1' if desired, and the user can choose whether or not to include on the plot either 'Deformed Shape Only', 'Deformed + Undeformed', or 'Deformed + Undeformed Edge'.

   Of course, the other component labels could be chosen instead, if desired.

General Postproc> Plot Results> Elem Table

In the dialog box that opens, select the label 'MzI', or of course, any of the other previously defined labels could be chosen. Also, the user has the option of averaging nodal values.

7. Exit ANSYS. **Quit> Save Everything> OK**

W10: Analyze simple cantilever beam problem shown below. The beam is made of 2 different materials.

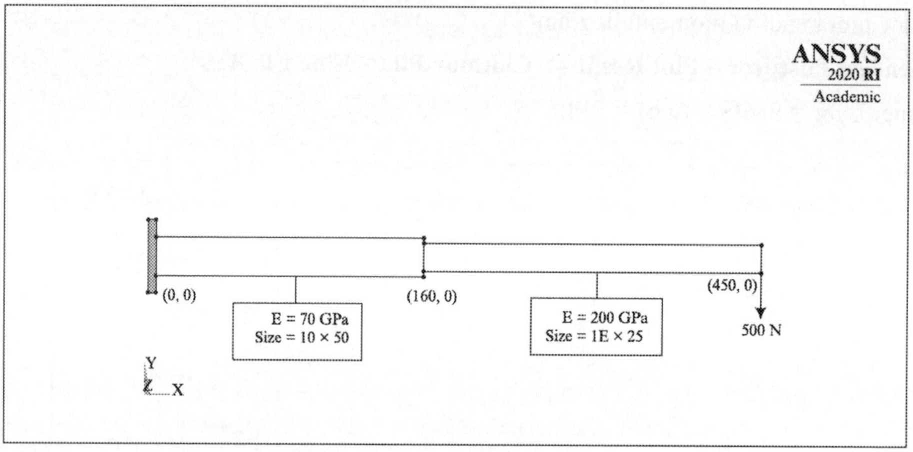

Data: Width-1 = 10 mm, Height-1 = 50 mm, Length-1 = 150, E-1 = 7e5 N/mm², μ-1 = 0.3, Width-2 = 15 mm, Height = 25 mm, Length-2 = 300, E-2 = 2e5 N/mm², μ-2 = 0.3, F = 500 N

Solution:

Preprocessing

1. Change jobname.

   File> Change Jobname

   Enter 'Stepped Cantilever Beam with point load', and click on 'OK'.

2. Define element type:

   Preprocessor> Element Type> Add/Edit/Delete

   Click on 'Add..', highlight 'Beam', then '2 node 188', click on 'OK', then 'Close'.

3. Define the section properties for the BEAM188 elements:

   Preprocessor> Sections> Beam> Common Sections

   Click 'OK' for 'Type 1 BEAM188'

   The appropriate values for the left beam are: B = 10; H = 50, Nb = 2, Nh = 5. Click 'Apply'. The values for right beam are B = 15, H = 25, Nb = 3, Nh = 5, OK

4. Define material properties.

   Preprocessor> Material Properties> Constant> Isotropic

   'OK' for material set number 1, then, enter 0.7e5 for EX, $\mu = 0.3$ then 'OK'.

   Now select 'New Model' under Material in 'Define Material Model Behavior'. Assign 2 for Define Material ID in Define Material ID window. Enter 2e5 for EX, $\mu = 0.3$ then 'OK'.

5. Create Keypoints.

   Preprocessor> Modeling> Create> Keypoints> In Active CS

   Enter 1 for node number (ANSYS would automatically number nodes if this is left blank). Enter the location as (x,y,z) = (0,0,0). Leave the entries for rotation angles blank. (For this problem, all nodes will be on the x-axis, with y = 0 and z = 0). Click on 'Apply'. Define node 2 at (x,y,z) = (150,0,0), define node 3 at (x,y,z) = (450,0,0), define node 4 at (x,y,z) = (0,30,0), define node 5 at (x,y,z) = (150,30,0), then click on 'OK' instead of 'Apply'.

6. Create beam elements between keypoints. First create left part.

   Draw first line joining the key points 1 and 2.

   Preprocessor> Modeling Create> Lines> Line> Straight Line

   Pick keypoints 1 and 2, OK

   Preprocessor> Meshing> Mesh Tool

   Select 'Lines' under 'Element Attributes', click 'Set'

   Pick the line and click 'OK' in the Line Attributes. Choose 'Yes' for Pick Orientation Keypoint(s), OK. Select keypoint 4, OK.

   In the Mesh Tool window, click 'set' for lines under Size Controls. Click on line and OK. Assign 3 for No. of element divisions under Element on Picked Lines, OK.

   Click Mesh in Mesh Tool window. Click the line, and click OK in Mesh Lines window.

   Now create right part. Draw second line joining keypoints 2 and 3.

   Preprocessor> Modeling Create> Lines> Line> Straight Line

   Pick keypoints 1 and 2, OK. Attribute second set of properties.

   Preprocessor> Meshing> Mesh Tool

   Preprocessor> Meshing> Mesh attributes> Default Attributes

   Assign 2 for Material number, 2 for section number, OK

   Select 'Lines' under 'Element Attributes', click 'Set'

   Pick the line and click 'OK' in the Line Attributes. Choose 'Yes' for Pick Orientation Keypoint(s), OK. Select keypoint 5, OK.

   In the Mesh Tool window, click 'set' for lines under Size Controls. Click on line and OK. Assign 6 for No. of element divisions under Element on Picked Lines, OK.

   Click Mesh in Mesh Tool window. Click the line, and click OK in Mesh Lines window.

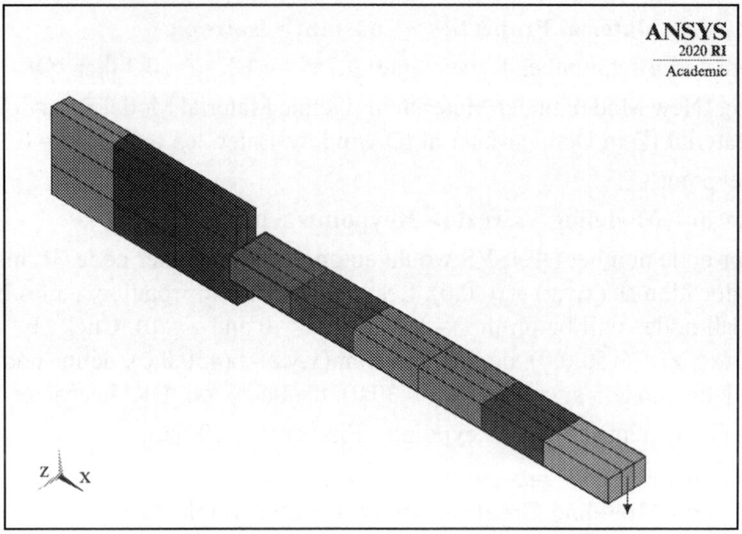

Solution:

1. Apply constraints and forces on the model.
   Solution> Loads> Apply> Structural> Displacement> On Nodes
   Pick node 1, then click 'OK' in the picking menu that has appeared. Choose ALL DOF, and use the default value of zero. Click on 'OK'.
   Solution> Loads> Apply> Structural> Force/Moment> On Nodes
   Pick node 11, then 'OK' in the picking menu, choose 'FY' as the direction of the force, and enter '500' for the force value. Click on 'OK'.

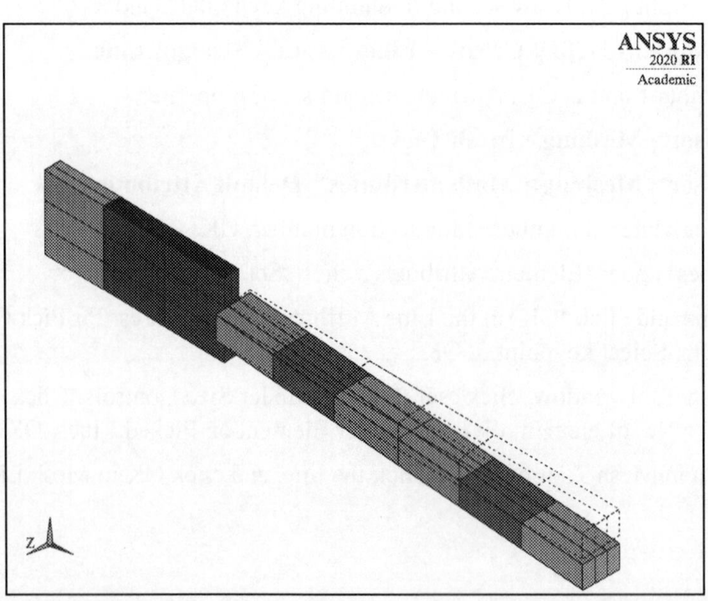

2. Solve the problem.

   Solution > Solve > Current LS

   Click 'OK' in the 'Solve Current Load Step' Box.

Postprocessing

1. Plot the deformed shape.

   General Postproc > Plot Results > Deformed Shape

   Choose 'Def + undeformed', then 'OK'.

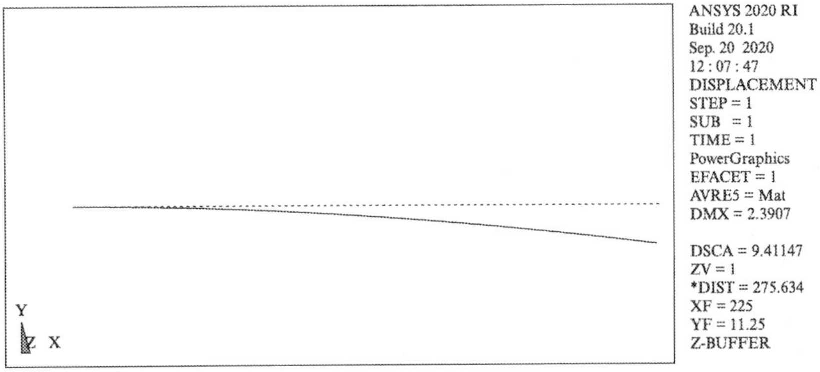

2. List reaction forces.

   General Postproc > List Results > Reaction Solution

   Use the default: 'All items'; and click on 'OK'.
3. List nodal displacements

   General Postproc > List Results > Nodal Solution > DOF Solution > ALL DOFs
4. Define element table items for subsequent plotting and listing of various stress results. For line elements, such as beam elements, to generate certain stress results, define an element table, using the ETABLE command. Note that each element is defined by two end nodes, denoted by node *i* and node *j*. Node i is the first of the two nodes defined when the element was created. The form of the GUI path to define a table containing the shear force at the 'top' of the beam, at node *i*, for each element, is

   Main Menu > General Postproc > Element Table > Define Table > Add

   SFy for user label, Select By sequence num and SMISC, 6, click Apply

   SFy for user label, Select By sequence num and SMISC, 19, click OK

   Now plot shear force diagram.

   General Postproc > Plot Results > Contour Plot > Line Ele Res

   Enter LabI = SMIS5, LabJ = SMIS18

   The entries in this command pertain specifically to the BEAM188 element.

In the command above, 'SFyI' is simply a label chosen by the user. Practically, any other text string could be used in this field. However, it must not have more than eight characters, and it cannot be identical to any general ANSYS predefined label. 'SMISC' is the appropriate item for shear force in BEAM188 elements. Also, the '5' shown as the last entry is correct sequence number on the ETABLE command, assuming BEAM188 elements are used, and the user wants to store the bending stresses at the top of the beam elements, at node *i* for each element.

If the user also wants to define an element table in which the bending stress at the top of each beam element, at node *j* for each element is stored, the command that could be used is similar to that above. However, the user should choose a different identifying label, and the value of the last entry for this case is '3'.

As additional examples, to store the bending stress at the bottom of the beam elements, at node *i* for each element, and to store the bending stress at the bottom of the beam elements, at node *j* for each element, the commands are:

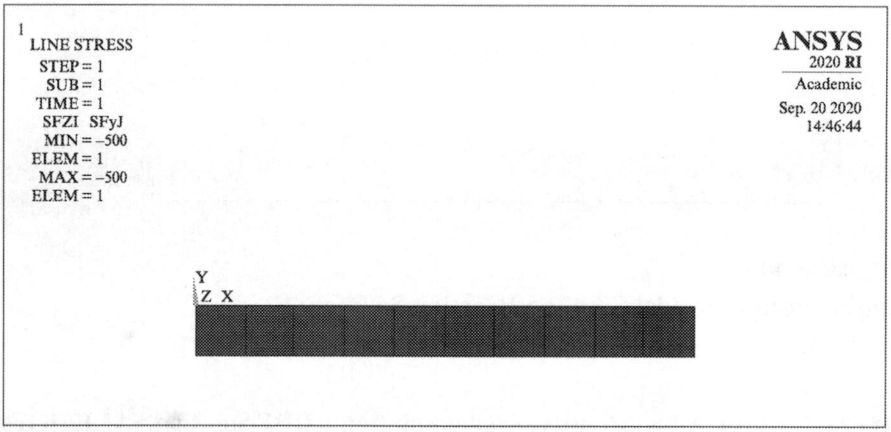

Main Menu> General Postproc> Element Table> Define Table> Add

BMz for user label, Select By sequence num and SMISC, 2, click 'Apply'

SFy for user label, Select By sequence num and SMISC, 15, click 'OK'

Now plot shear force diagram.

General Postproc> Plot Results> Contour Plot> Line Ele Res

Enter LabI = SMIS2, LabJ = SMIS15

5. List element table results. Assuming all four of the above ETABLE commands were entered, then a listing of each stored stress component can be obtained with the menu path:

   General Postproc> List Results> Elem Table Data

   Then, highlight the desired labels, and click on 'OK'. The stresses corresponding to the chosen labels will be listed by element number.

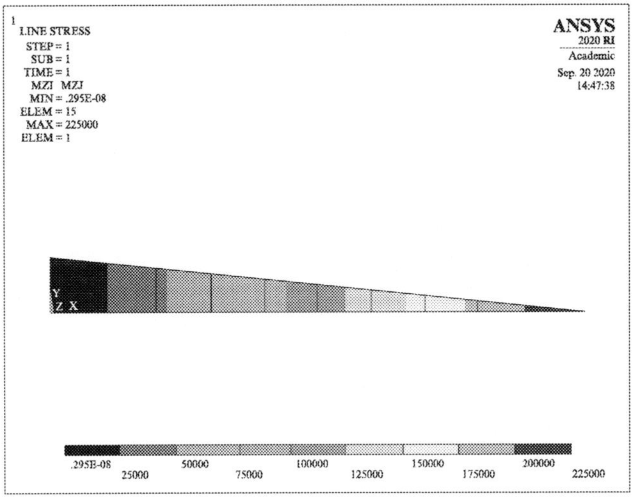

6. Plot element stresses.

   General Postproc> Plot Results> Line Elem Res

   In the dialog box that opens, select the label 'BMzI' for the LABI entry, and the label 'BMzJ' for the LABJ entry. The optional scale factor can be left as '1' if desired, and the user can choose whether or not to include on the plot either 'Deformed Shape Only', 'Deformed + Undeformed', or 'Deformed + Undeformed Edge'.

   Of course, the other component labels could be chosen instead, if desired.

   General Postproc> Plot Results> Elem Table

   In the dialog box that opens, select the label 'BMzI', or of course, any of the other previously defined labels could be chosen. Also, the user has the option of averaging nodal values.

7. Exit ANSYS. **Quit> Save Everything> OK**

W11: A uniformly distributed load of 800 N/m is applied to a solid steel beam of length 800 mm with a rectangular cross-section as shown in the figure below. The cross-section of the beam is 8 mm × 12 mm while the modulus of elasticity of the steel is 200GPa and Poisson's ratio 0.3.

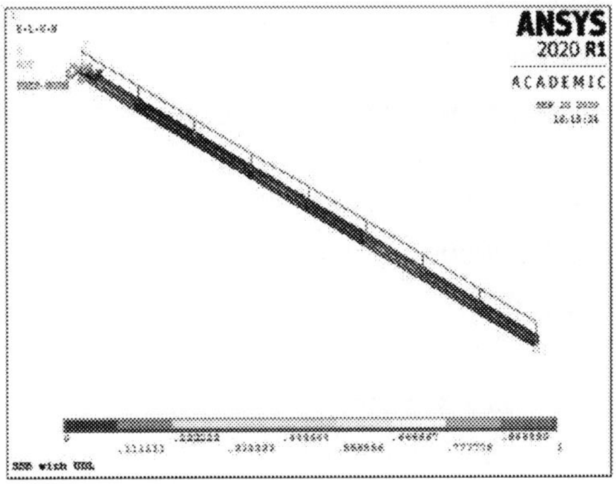

Data: Length = 800 mm, height = 12 mm, width = 8 mm, E = 2e5 N/mm^2, μ = .3, Load = 1 N/mm

Solution:

Preprocessing

1. Open preprocessor menu
   Main Menu> Preprocessor
2. Give example a Title
   Utility Menu> File> Change Title ...
   'SSB with UDL'
3. Create Keypoints
   Preprocessor> Modeling> Create> Keypoints> In Active CS
   Define 2 keypoints (the beam vertices) for this structure as given in the following table:

Keypont	Coordinates (x, y)
1	(0, 0)
2	(800, 0)

4. Define Lines
   Preprocessor> Modeling> Create> Lines> Lines> Straight Line
   Create a line between Keypoint 1 and Keypoint 2.
5. Define Element Types
   Preprocessor> Element Type> Add/Edit/Delete...
   For this problem use the BEAM188 element. This element has six or seven degrees of freedom at each node. These include translations in the x, y, and z directions and rotations about the x, y, and z directions. A seventh degree of freedom (warping magnitude) is optional.
6. Define Section Parameters
   Main Menu> Preprocessor> Section> Beam> Common Sections
   Enter B = 8, H = 12, click 'OK'
7. Define Element Material Properties
   Preprocessor> Material Props> Material Models> Structural> Linear> Elastic> Isotropic
   In the window that appears, enter the following geometric properties for steel:
   (a) Young's modulus EX: 200000
   (b) Poisson's ratio PRXY: 0.3
8. Define Mesh Size
   Preprocessor> Meshing> Size Cntrls> ManualSize> Lines> All Lines...
   For this example we will use an element length of 100 mm.

9. Mesh the frame

   Preprocessor> Meshing> Mesh> Lines> click 'Pick All'

10. Plot Elements

    Utility Menu> Plot> Elements

    Turn on element numbering and turn off keypoint numbering

    Utility Menu> PlotCtrls> Numbering

Solution:

1. Define Analysis Type

   Solution> Analysis Type> New Analysis> Static

2. Apply Constraints

   Solution> Define Loads> Apply> Structural> Displacement> On Keypoints

   Pin Keypoint 1 (i.e., UX and UY constrained) and fix Keypoint 2 in the y direction (UY constrained).

3. Apply Loads

   Apply a distributed load, of 1000 N/m or 1 N/mm, over the entire length of the beam.

   Select **Solution> Define Loads> Apply> Structural> Pressure> On Beams**

   Click 'Pick Al' in the 'Apply F/M' window.

   As shown in the following figure, enter a value of 1 in the field 'VALI Pressure value at node I' then click 'OK'.

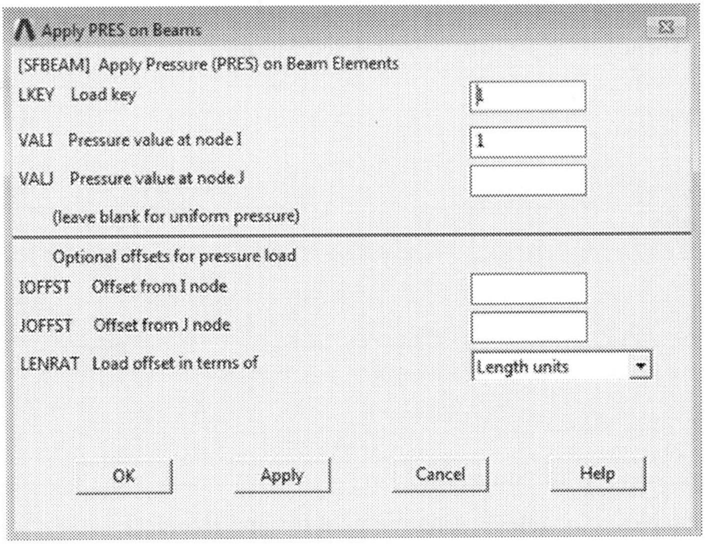

The applied loads and constraints should now appear as shown in the figure below.

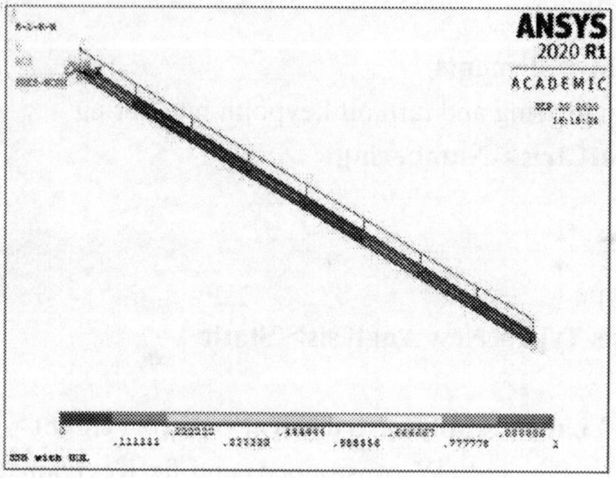

It is necessary to change some settings to have the constraints and loads appear each time the analyst selects 'Replot'. Select **Utility Menu> PlotCtrls> Symbols....** In the window that appears, select 'Pressures' in the pull down menu of the 'Surface Load Symbols' section.

4. Solve the System

   Solution> Solve> Current LS

Postprocessing

1. Plot Deformed Shape

   General Postproc> Plot Results> Deformed Shape

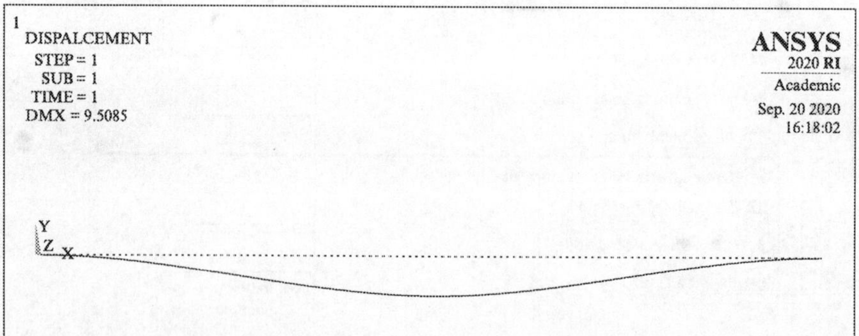

2. Plot Principle stress distribution

   Use element tables to obtain bending moment for line elements.

   1. Select **General Postproc> Element Table> Define Table**
   2. Click 'Add...'

3. In the window that appears
   (a) Enter 'BMI ' in the 'User Label for Item' section.
   (b) In the first window in the 'Results Data Item' section scroll down and select 'By sequence num'.
   (c) In the second window of the same section, select 'SMISC.'
   (d) In the third window enter '2' anywhere after the comma.
4. click 'Apply'
5. Repeat steps 2 to 4 but change 'SMAXI' to 'BMJ' in step 3a and change '2' to '15' in step 3d.
6. Click 'OK'. The 'Element Table Data' window should now have two variables in it.
7. Click 'Close' in the 'Element Table Data' window.
8. Select: **General Postproc> Plot Results> Line Elem Res...**
9. Select 'BMI' from the 'Labl' pull down menu and 'BMJ' from the 'LabJ' pull down menu

ANSYS can only calculate the stress at a single location on the element. For this example, extract the stresses from the *i* and *j* nodes of each element. These are the nodes that are at the ends of each element.

For this problem, we wanted the principal stresses for the elements. For the BEAM188 element this is categorized as SMISC, 1 for the *i* nodes and SMISC, 3 for the *j* nodes. A list of available codes for each element can be found in the ANSYS help files (i.e., type help BEAM188 in the ANSYS Input window).

As shown in the plot below, the maximum stress occurs in the middle of the beam.

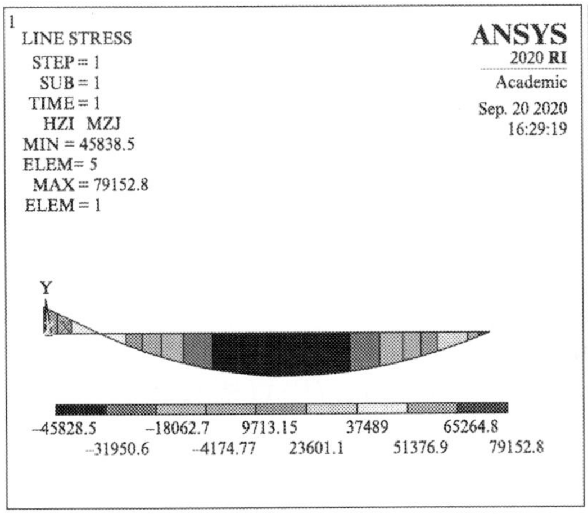

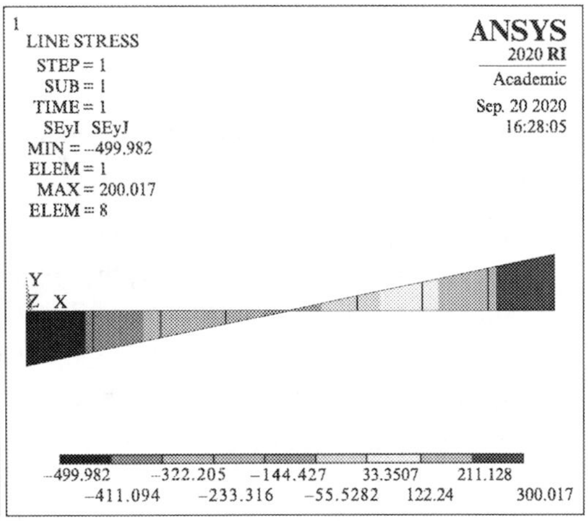

W12: Model and analyze the beam of length 6 m with a distributed load (40,000 N/mm) shown below. Width = 75 mm; Height = 150 mm; Modulus of Elasticity, E = 200 GPa; μ = 0.29.

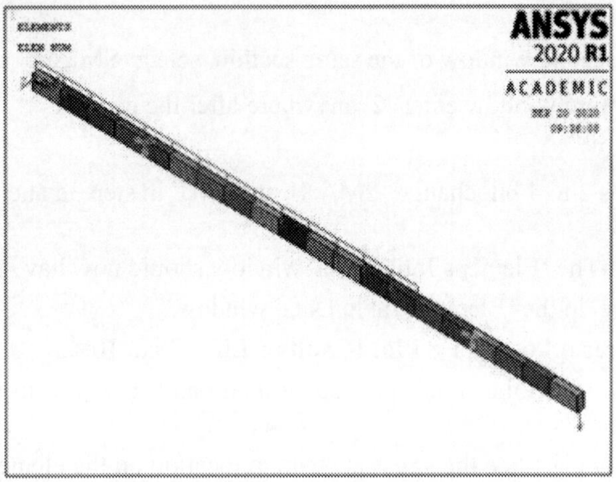

Data: B = 75 mm, H = 150 mm, F = 40,000, E = 2e5 N/mm^2, μ = 0.29

Solution:

Preprocessing

1. Change jobname
   File> Change Jobname
   Enter '*SSB with P&UDL*', and click on 'OK'.
2. Define element type
   Preprocessor> Element Type> Add/Edit/Delete
   Click on 'Add..', highlight 'Beam', then '2 node 188', click on 'OK', then 'Close'.
3. Define Material Properties
   Preprocessor> Material Properties> Constant> Isotropic
   'OK' for material set number 1, then enter the values for EX = 2e5 and NUXY = 0.3, then 'OK'.
4. Create 6 keypoints: #1 at the left end, #2 at the first pin joint from the left, #3 at the end of the distributed load, #4 at the second pin joint, #5 at the right end and #6 above 1.

   Preprocessor> Modeling> Create> Keypoints> In Active CS

   Enter 1 for keypoint number (ANSYS would automatically number keypoints if you leave this blank). Enter the location as (x, y, z) = (0,0,0). Note that keypoint 1 is located at the origin of the global x-y-z Cartesian coordinate system. Click on 'Apply'. Continue defining keypoints 2-5, using the locations based on the sketch of the beam. But, after entering the keypoint 5 location, click on 'OK' instead of 'Apply'.

Keypoint	X	Y
1	0	0
2	1200	0
3	4200	0
4	5100	0
5	6000	0
6	0	100

As a check to ensure all keypoints were entered correctly, list the keypoints:

Utility Menu> List> Keypoints

If any errors were made in defining the keypoints, redefine a keypoint by repeating the procedure of step 5. Of course, it is not necessary to redefine all keypoints simply to move one. Just repeat the keypoint creation command for the incorrectly placed one.

Turn on keypoint numbering.

Utility Menu> PlotCtrls> Numbering

Check 'keypoint numbering', then click 'OK'. The keypoint numbers may already be showing, but this will force the display of keypoint numbers on subsequent plots.

5. Create lines between keypoints.

   Preprocessor> Create> Lines> Straight Line

   A picking menu appears. Pick keypoint 1, then keypoint 2, and a line is created between the two keypoints. Continue creating lines in this way, one between keypoints 2 and 3, one between keypoints 3 and 4, and one between keypoints 4 and 5. Afterwards, click on 'CANCEL' in the picking box. The lines are already created, so this will close the box.

6. Instead of using the default mesh for each line, specify the number of element divisions per line so that all elements in the model are 300 mm long.

   Preprocessor> Meshing> Size Ctrls> Lines> All Lines

   In the box that appears, enter '300' for Element Edge length, then 'OK'.

7. Mesh the lines.

   Preprocessor> Meshing> Mesh> Lines> Pick All

   At this point, the nodes and elements are created. To see a node plot, go to the top utility menu, and choose **Plot> Nodes**. There will be a dot for each node. Now, go back and replot the lines: **Plot> Lines**.

Solution:

1. Apply constraints and loads.

   Solution> Loads> Apply> Structural> Displacement> On Keypoints

   Click on keypoints 2 and 4, then click 'OK' in the picking menu that has appeared. Choose UX and UY, and use the default value of zero. If the 'ALL DOF' label is highlighted, make sure to unselect the 'ALL DOF' label. If the 'ALL DOF' label is highlighted, unselect it by clicking on it. After confirming that only 'UX' and 'UY' are highlighted, click 'OK'. These elements have 3 dof per node: 2 translations (UX and UY) and one rotation.

Apply the force.

Solution> Loads> Apply> Structural> Force/Moment> On Keypoints

Pick keypoint 5, then 'OK' in the picking menu, choose 'FY' for 'Lab', and enter -40000 for the force value. Click on 'OK'.

Apply the distributed load.

Solution> Loads> Apply> Structural> Pressure> On Beams

DO NOT CHOOSE 'On Lines' – this does not work for beam elements!

A picking menu appears. Click on the line between keypoints 1 and 2, and the line between keypoints 2 and 3, then 'OK'. A box appears. Enter '15000' for VALI. You don't need to enter anything else. Click 'OK'.

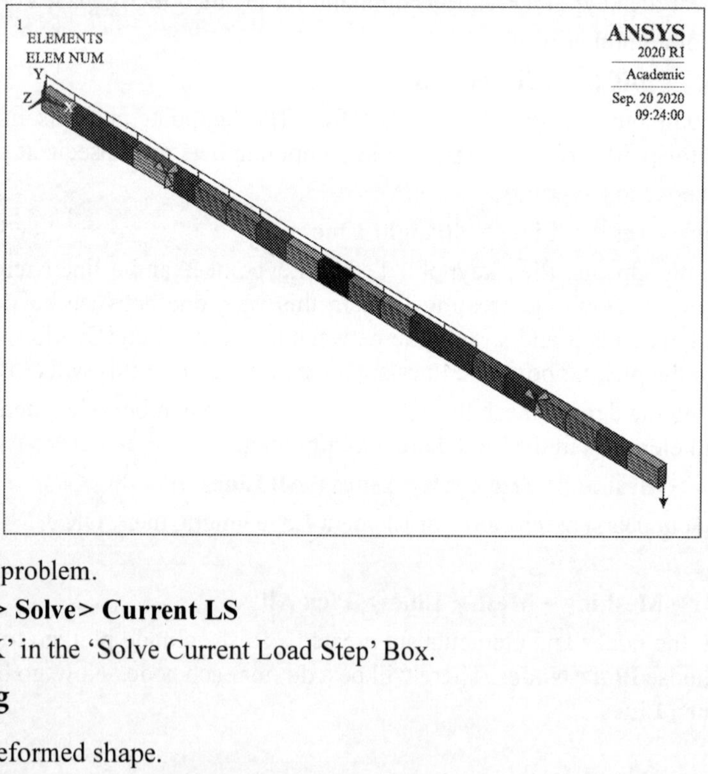

2. Solve the problem.

   Solution> Solve> Current LS

   Click 'OK' in the 'Solve Current Load Step' Box.

Postprocessing

1. Plot the deformed shape.

   General Postproc> Plot Results> Deformed Shape

   Choose 'Def + undeformed', then 'OK'.

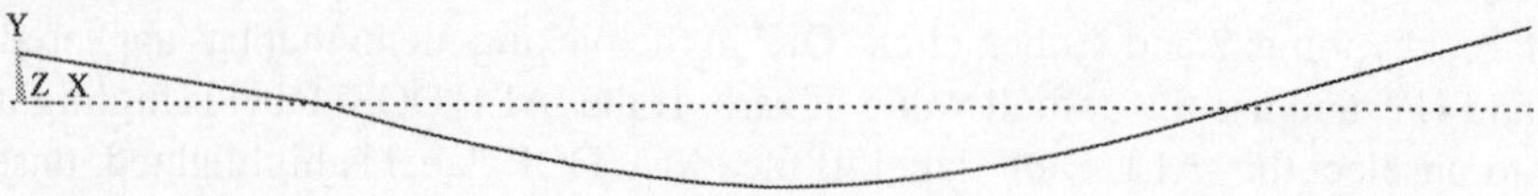

2. List reaction forces.
   General Postproc> List Results> Reaction Solution
   Use the default 'All items', and click on 'OK'.
3. List the x and y direction deflections for each node.
   General Postproc> List Results> Nodal Solution> DOF Solution> ALL DOFs

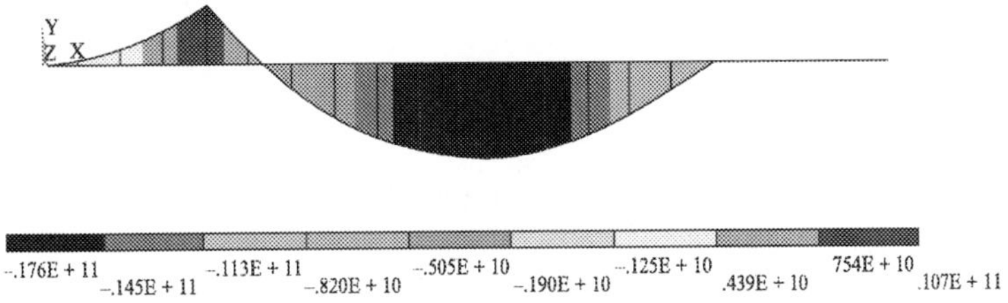

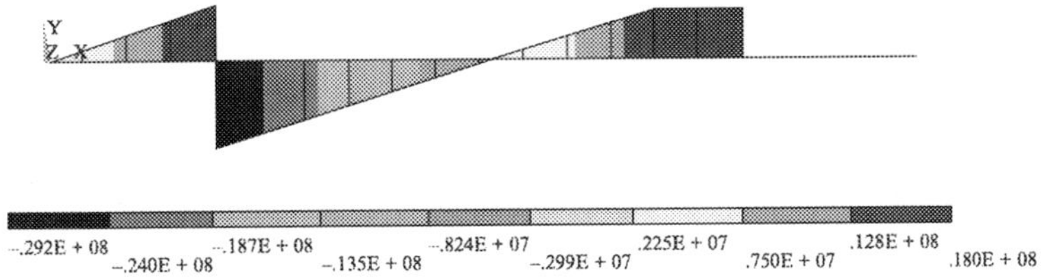

4. Exit ANSYS.
   Toolbar: **Quit> Save Everything> OK**

W13: A distributed force of 300 N/mm load acts on the cantilever beam. The right hand end is supported from the ceiling as shown below. Model and analyze the beam.

Solution: This is a simple problem that uses both beam and truss elements. The beam is represented by 10 beam elements in order to capture the maximum bending moment, which is calculated at the nodes by default. One truss element is sufficient.

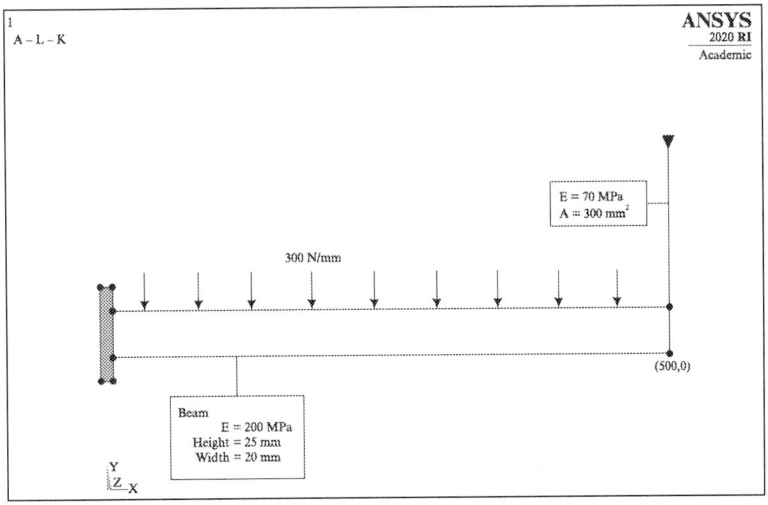

Data: Beam: Width = 20 mm, Height = 25 mm, Length = 500, E = 2e5 N/mm², $\mu = 0.3$, F = 300 N/mm

Link: E = 0.7e5 N/mm², Area = 300 mm², $\mu = 0.3$

Solution:

Preprocessing

1. Change job name.

   File > Change Jobname

   Enter '*Combined Beam and Link*', and click on 'OK'.
2. Define element types.

   Preprocessor> Element Type> Add/Edit/Delete

   Click on 'Add..', highlight 'Beam', then '2node 188', click on 'OK', click 'Options', select Quadratic Form' for Element behavior in BEAM188 element type options, then 'Close'. Now define second element type.

   Click on 'Add..', highlight 'Link', then '2node 188', click on 'OK', click 'Options', select Quadratic Form' for Element behavior in BEAM188 element type options, then 'Close'.
3. Define the section properties for the BEAM188 elements.

   Preprocessor> Sections> Beam> Common Sections>

   The appropriate values for the given geometry are: B = 13; H = 10, Nb = 6, Nh = 4 and click 'OK'.
4. Define material properties:

   Preprocessor> Material Properties> Constant-Isotropic

   'OK' for material set number 1, then, enter 2e5 for EX, then 'OK'.
5. Create keypoints.

   Preprocessor> Modeling> Create> Key Points> In Active CS

   Enter 1 for node number (ANSYS would automatically number nodes if this is left blank). Enter the location as (x,y,z) = (0,0,0). Leave the entries for rotation angles blank. (For this problem, all nodes will be on the x-axis, with y = 0 and z = 0). Click on 'Apply'. Define node 2 at (x,y,z) = (250,0,0), define node 3 at (x,y,z) = (0,30,0), define node 4 at (x,y,z) = (500,250,0), then click on 'OK' instead of 'Apply'.
6. Create beam elements between keypoints.

   Draw line joining the key points.

   Preprocessor> Modeling Create> Lines> Line> Straight Line

   Pick keypoints 1 and 2, OK

   Preprocessor> Meshing> Mesh Tool

   Select 'Lines' under 'Element Attributes', click 'Set'

   Pick the line and click 'OK' in the Line Attributes. Choose 'Yes' for Pick Orientation Key Point(s), OK. Select key point 3, OK.

In the Mesh Tool window, click 'set' for lines under Size Controls. Click on line and OK. Assign 10 for number of element divisions under Element on Picked Lines, OK.

Click Mesh in Mesh Tool window. Click the line, and click OK in Mesh Lines window.

Create Link element between key points 2 and 3.

Preprocessor> Meshing> Mesh Attributes> Default Attributes

In the Meshing Attributes window, choose 2 LINK180 for Element type number, 2 for Material number, No section for section number, OK.

Under size controls click 'Set' for Lines in the Mesh Tool window. Pick the line representing Truss. Assign I for No.of element divisions, OK.

Solution:

1. Apply constraints and forces on the model.

   Solution> Loads> Apply> Structural> Displacement> On Nodes

   Pick node 1, then click 'OK' in the picking menu that has appeared. Choose ALL DOF, and use the default value of zero. Click on 'OK'.

   Solution> Loads> Apply> Structural> Force/Moment> On Nodes

   Pick extreme right node, then 'OK' in the picking menu, choose 'FY' as the direction of the force, and enter '– 300' for the force value. Click on 'OK'.

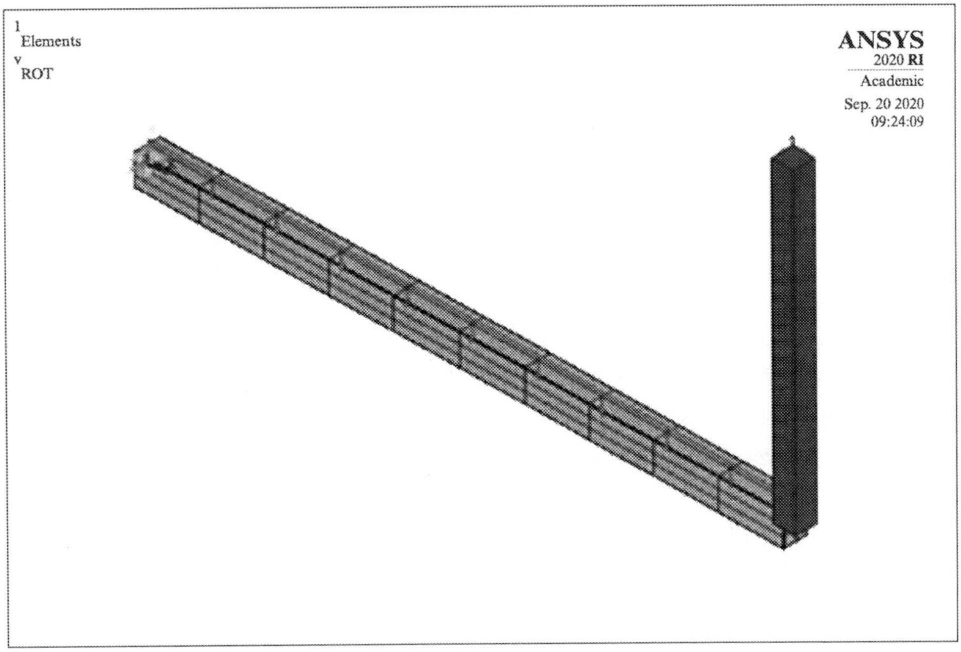

2. Solve the problem.
   Solution > Solve > Current LS
   Click 'OK' in the 'Solve Current Load Step' Box.

Postprocessing

1. Plot the deformed shape.
   General Postproc > Plot Results > Deformed Shape
   Choose 'Def + undeformed', then 'OK'.

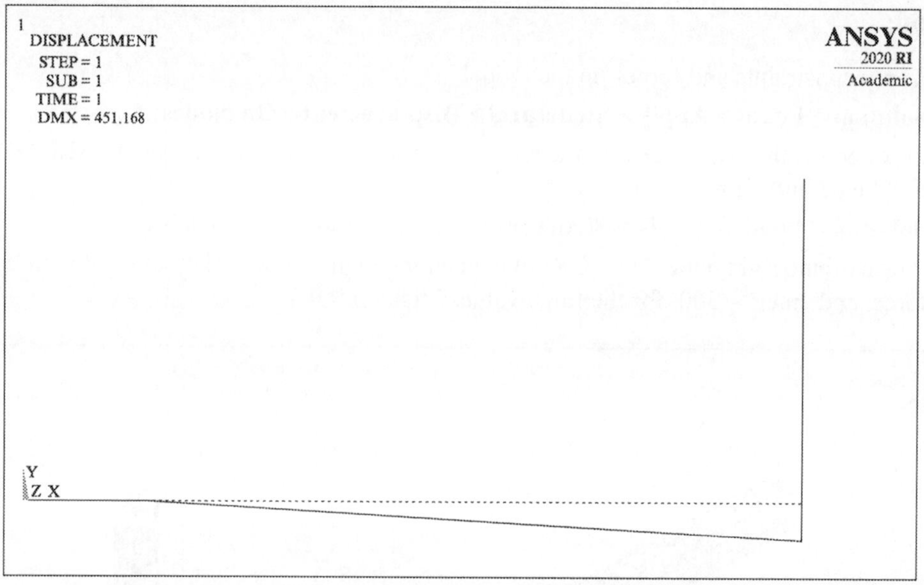

2. List reaction forces.
   General Postproc > List Results > Reaction Solution
   Use the default: 'All items'; and click on 'OK'.
3. List nodal displacements
   General Postproc > List Results > Nodal Solution > DOF Solution > Displacement vector sum
4. Define element table items for subsequent plotting and listing of various stress results. For line elements, such as beam elements, to generate certain stress results, define an element table, using the ETABLE command. Note that each element is defined by two end nodes, denoted by node *i* and node *j*. Node *i* is the first of the two nodes defined when the element was created. The form of the GUI path to define a table containing the shear force, at node *i*, for each element, is:

Main Menu> General Postproc> Element Table> Define Table> Add

SFy for user label, Select By sequence num and SMISC, 5, click Apply

SFy for user label, Select By sequence num and SMISC, 18, click OK

Now plot shear force diagram.

General Postproc> Plot Results> Contour Plot> Line Ele Res

Enter LabI = SMIS5, LabJ = SMIS18

The entries in this command pertain specifically to the BEAM188 element.

In the command above, 'SFyI' is simply a label chosen by the user. Practically, any other text string could be used in this field. However, it must not have more than eight characters, and it cannot be identical to any general ANSYS predefined label. 'SMISC' is the appropriate item for shear force in BEAM188 elements. Also, the '5' shown as the last entry is correct sequence number on the ETABLE command, assuming BEAM188 elements are used, and the user wants to store the bending stresses, at node *i* for each element.

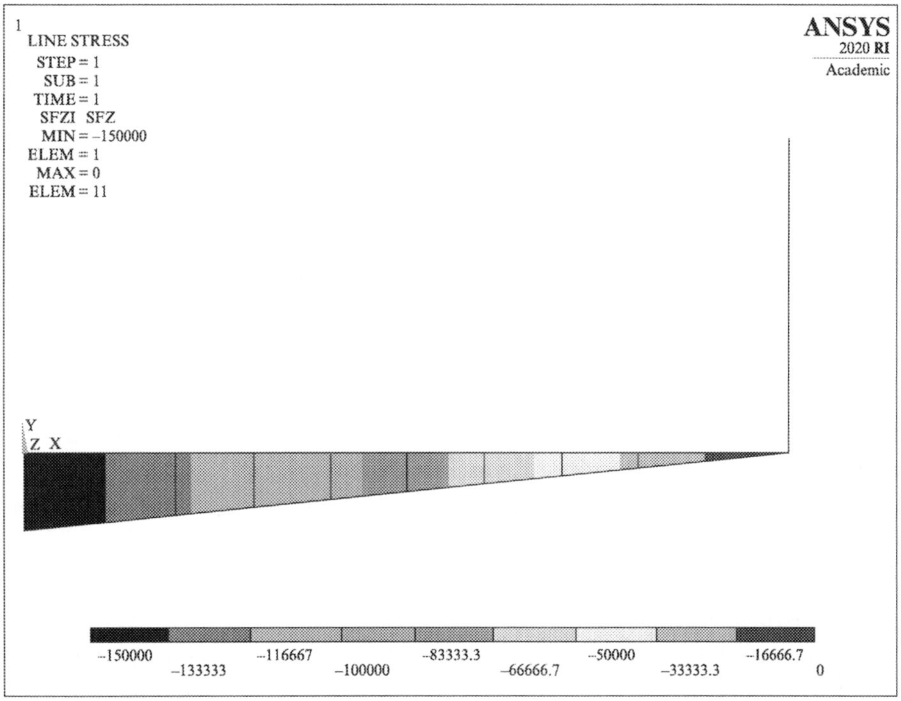

If the user also wants to define an element table in which the bending moment, at node *j* for each element is stored, the command that could be used is similar to that above. However, the user should choose a different identifying label 'My', and the value of the last entry for this case is '2 and 15'.

Main Menu> General Postproc> Element Table> Define Table> Add

BMzI for user label, Select By sequence num and SMISC, 2, click Apply

SFyJ for user label, Select By sequence num and SMISC, 15, click OK

Now plot bending moment diagram.

General Postproc> Plot Results> Contour Plot> Line Ele Res

Enter LabI = SMIS2, LabJ = SMIS15

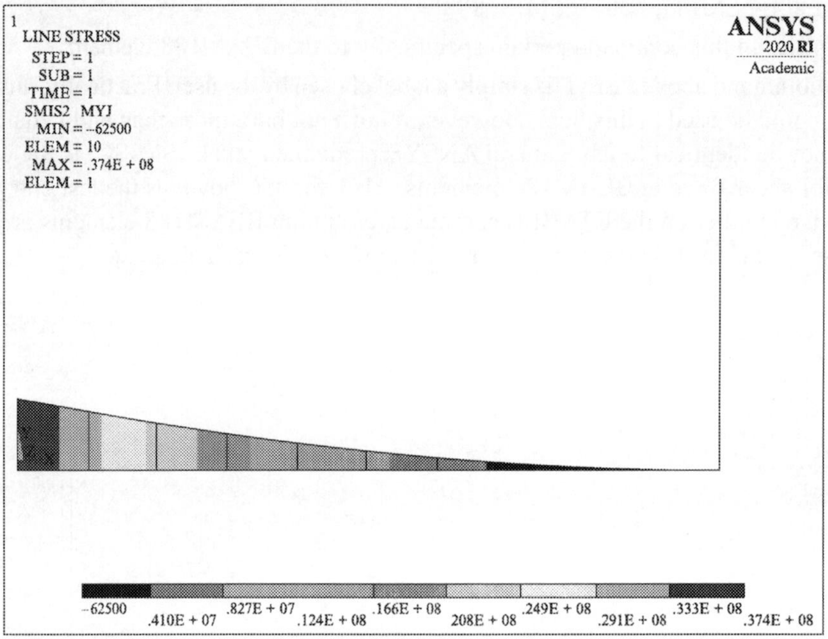

5. List element table results. Assuming all four of the above ETABLE commands were entered, then a listing of each stored stress component can be obtained with the menu path:
   General Postproc> List Results> Elem Table Data
   Then, highlight the desired labels, and click on 'OK'. The stresses corresponding to the chosen labels will be listed by element number.
6. Plot element stresses.
   General Postproc> Plot Results> Line Elem Res
   In the dialog box that opens, select the label 'MzI' for the LABI entry, and the label 'MzJ' for the LABJ entry. The optional scale factor can be left as '1' if desired, and the user can choose whether or not to include on the plot either 'Deformed Shape Only', 'Deformed + Undeformed', or 'Deformed + Undeformed Edge'.
   Of course, the other component labels could be chosen instead, if desired.
   General Postproc> Plot Results> Elem Table
   In the dialog box that opens, select the label 'MzI', or of course, any of the other previously defined labels could be chosen. Also, the user has the option of averaging nodal values.
7. Exit ANSYS. **Quit >Save Everything> OK**

4.2.3 Plates

W14: Determine accurate stresses in a plate with a hole.

Solution: The p-method is used while solving the FEM model in order to obtain the required accuracy of the computed FEM results by varying the element polynomial order.

Preprocessing

1. Start ANSYS
2. Define element type

   Preprocessor> Element Type> Add> Add> 2D triangle 146> OK> Options> Plane Stress + Thickness> Close

   This selects plane-stress element with a given thickness.
3. Define the real constants for the PLANE146, which thickness.

   Preprocessor> Real Constants> Add> Type 1 Plane 146>

   Click OK, then enter thickness.
4. Create geometry and mesh.
5. Add loads and bc to give the following model.

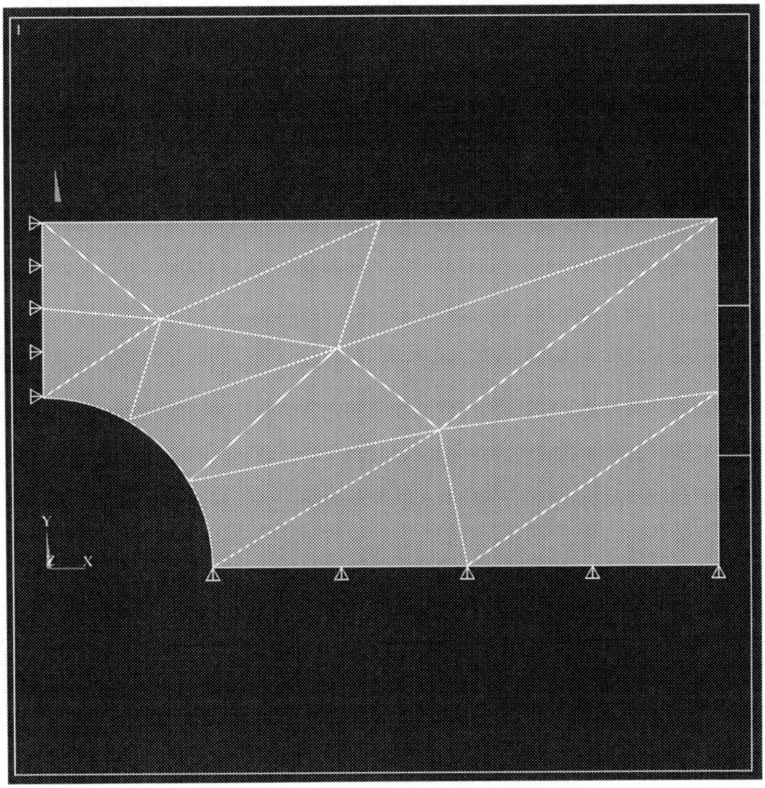

With p-method selected under preferences, the program will solve the FEM problem repeatedly, automatically adjusting the polynomial level in the elements until a user-specified solution quantity ceases to change within some specified tolerance.

The **default convergence criteria is** that the total strain energy in the model stabilize to the point that its value for a new polynomial level is within 5% of the value as found using the previous polynomial level.

Solution

1. **Solution> Current LS**

Postprocessing

1. **General Postprocessor> Read results, Last Set> Plot Results**
2. **Contour Plot> Element Solution>Stress Sx**

The initial p-level is set to 2 for each element. Select **p-levels** under plot results to see what the final p-levels are.

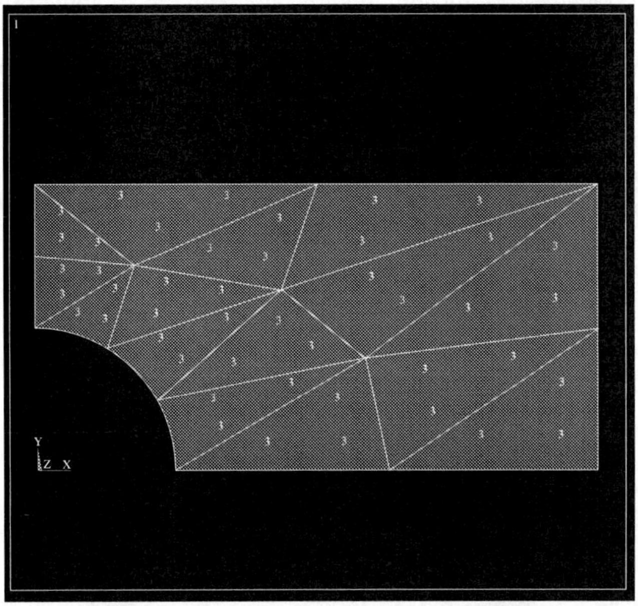

The p-level was advanced uniformly from 2 to 3 throughout the mesh. Select **p-Convergence** to evaluate the convergence.

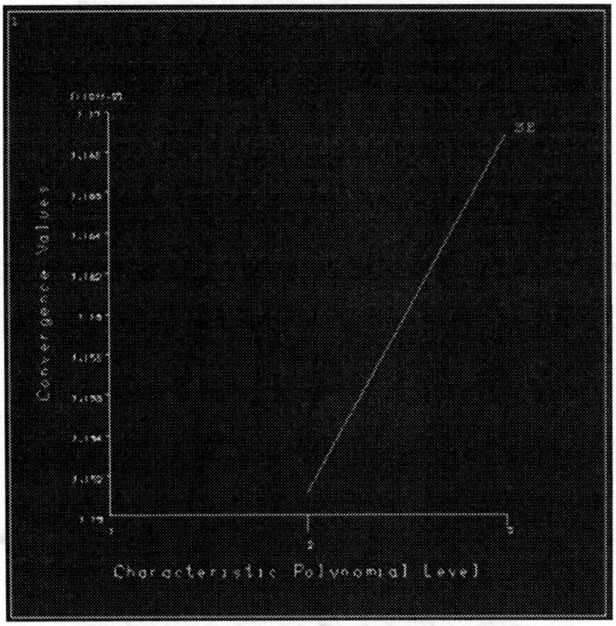

Using the default convergence criteria gives a noticeable discontinuity in stress contours across element boundaries indicating poor accuracy in the computed stresses, and the total strain energy looks as though it would continue change if the p-levels were incremented. To increase the accuracy, we will employ a different convergence criteria and solve the problem again.

The quantity of interest here is the maximum stress in the vicinity of the hole, therefore we require that the solution is repeated with increasing p-levels used for the elements until the calculated maximum stress stops changing to within some required tolerance, say 1%.

3. **Preprocessor > Loads > Load Step Opts > p-Method > Convergence Criteria > Replace > Local for solids**

   Pick the node where the maximum stress will occur.

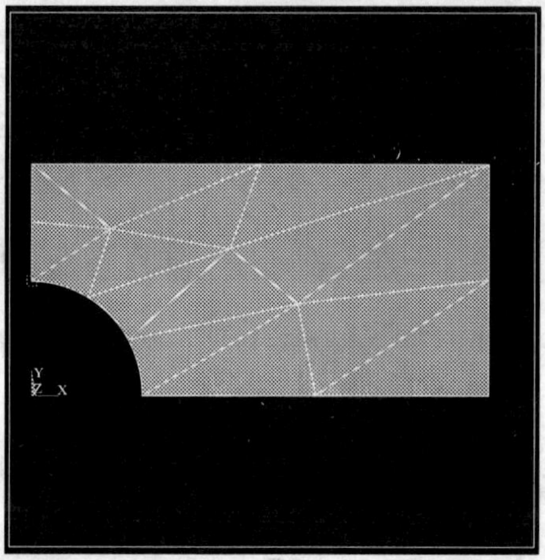

4. Select **Stress Sx** and change the convergence tolerance from 5% to 1%, OK

   Solve the problem again to obtain the stress contour plot shown next.

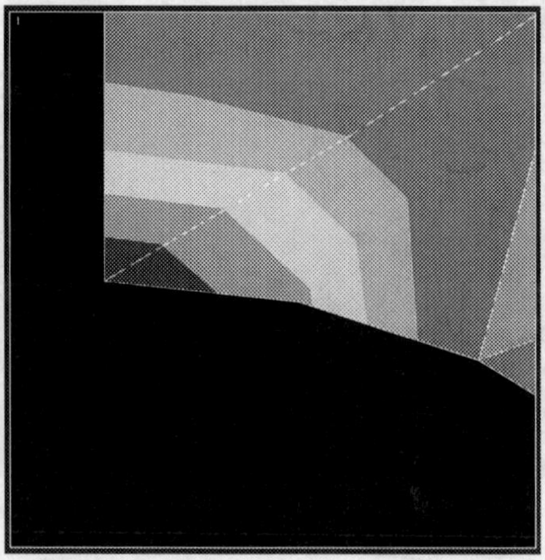

The smoothness of these stress contours gives us confidence in the solution. The p-level distribution in the mesh is shown next.

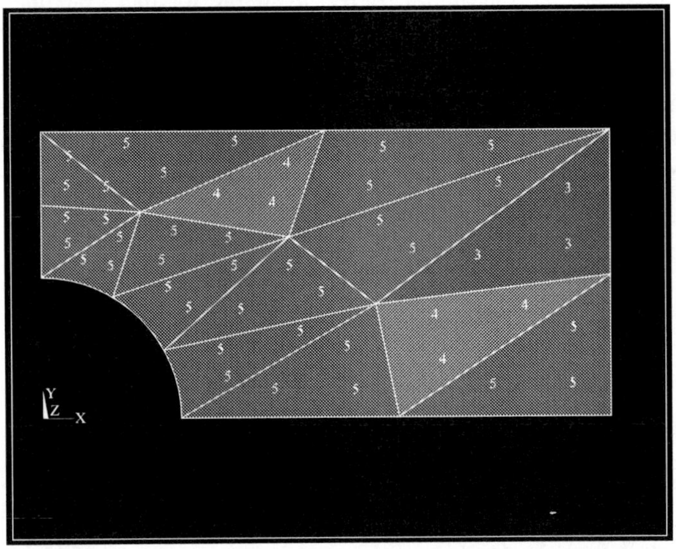

To obtain the required accuracy, it was necessary to use p-levels of 5 in most of the elements. The convergence history of Sx at the specified node is shown next.

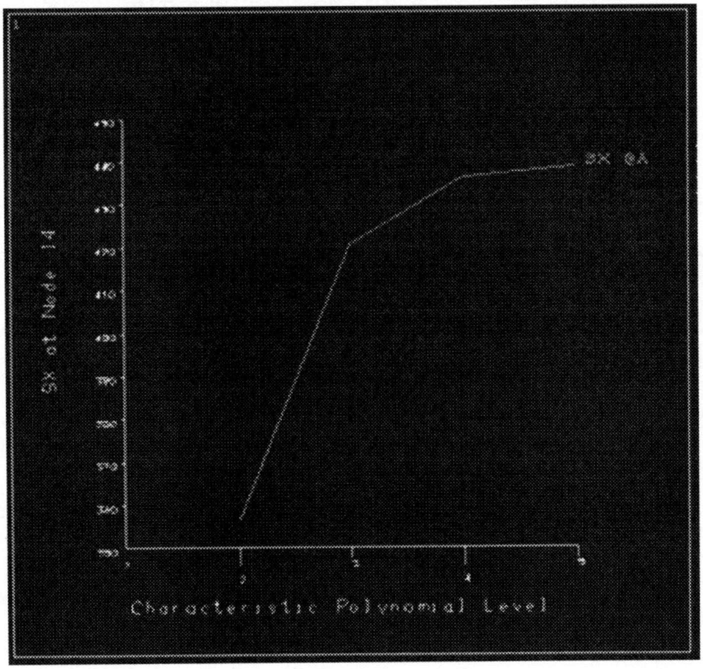

The Sx stress component at the hole is seen to converge nicely to a value in which the analyst can have confidence. Notice that the maximum p-level available for these elements in ANSYS is 8. If using p-levels of 8 does not provide the requested accuracy, a more refined mesh must first be created and the problem solved again.

W15: Mesh the quadrant of a plate to illustrate using the default settings to create a mesh shown below. Apply displacement and pressure boundary conditions and solve to obtain the Sigma-x plot shown. Refine the mesh and plot Sigma-x to study the accuracy of results.

Solution: Reducing the element size is termed 'h-refinement'.

Preprocessing

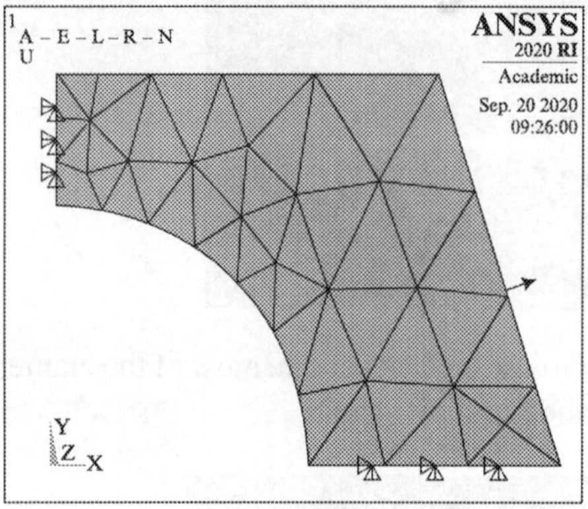

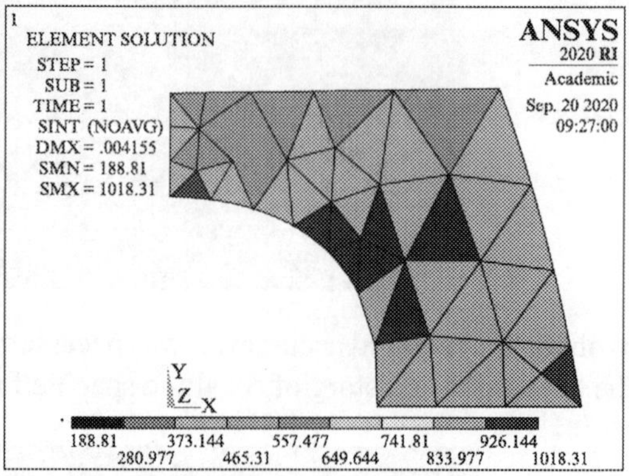

The large differences in stress contour plots across element boundaries indicate unacceptable accuracy in the results. First remove the loads (boundary restraints and applied pressure) to use the '**Refine**' option in the preprocessor.

Preprocessor> Loads> Delete> All Load Data> All Loads & Opts.

For uniform mesh refinement,

Preprocessor> Refine> All (All elements are subdivided, and the mesh shown below is created.)

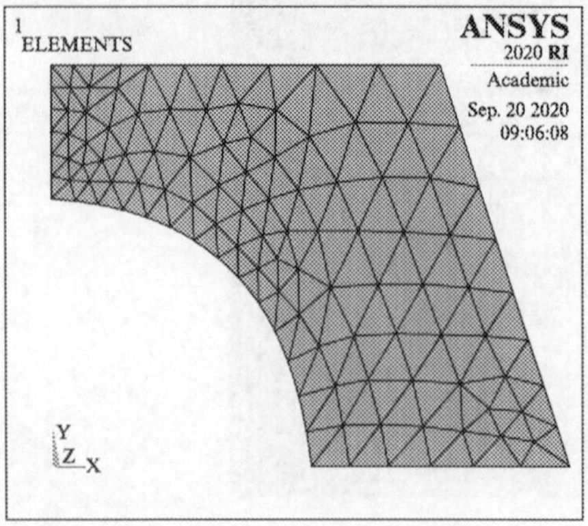

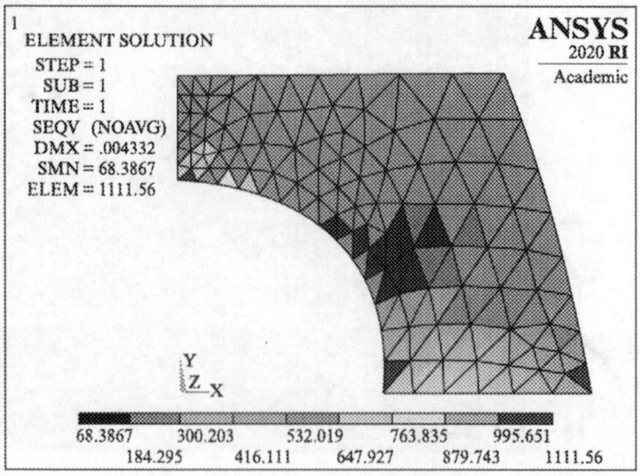

To refine the mesh selectively,

Preprocessor> Refine> Nodes (Select the three nodes near the point of highest stress.)

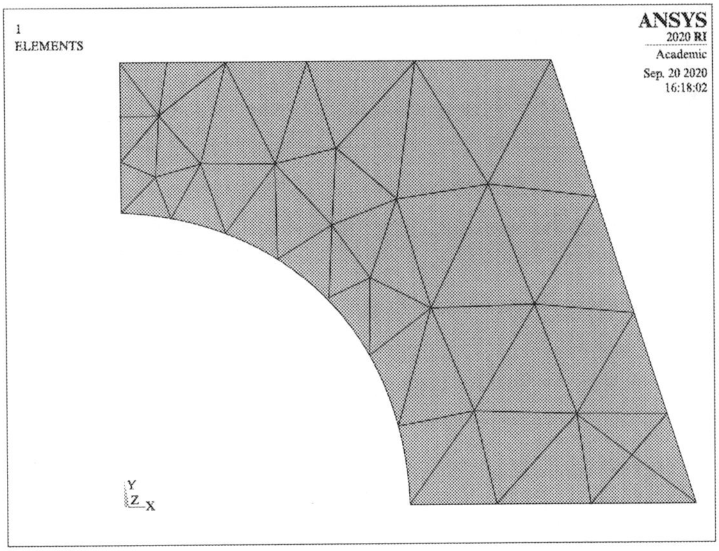

Local refinement then produces the following mesh.

Now apply displacement and pressure boundary conditions to the new mesh. Solution then gives the Sigma-x plot shown next.

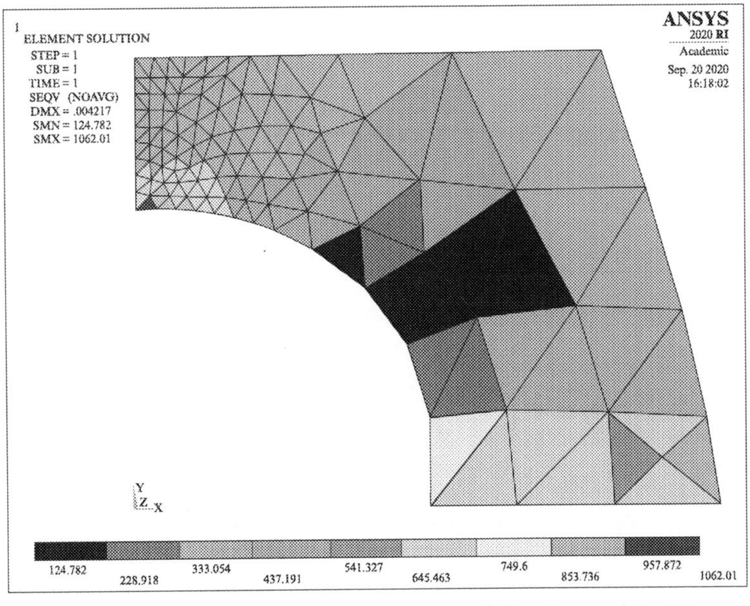

The stress contours are much smoother than those from the first mesh. The process may be repeated to obtain further accuracy.

W16: Consider a 250 mm × 250 mm steel plate that is 3 mm thick and has a 4N force applied at its center. The plate is simply supported on all sides. Determine the deflection at the load point.

Solution: The plate has two planes of symmetry, so model just one quadrant and apply appropriate boundary conditions. Use the ANSYS four-node element shell181 for the model.

Preprocessing

Start ANSYS and select the working directory for this problem.

1. Select the element.

   Preprocessor> Element Type> Add/Edit/Delete> Add> Shell> Elastic 4 Node 181

2. Specify the plate thickness

   Preprocessor> Sections> Shell> Lay-up> Add/Edit

   Thickness = 3, OK

3. Get material properties for 304 steel.

   Preprocessor> Material Props> Material Library> Import Library . . .

   C> Program Files> ANSYS Inc> v160> ansys> mat lib> Stl_AISI-304.BIN_SI

4. Create a 125 mm × 125 mm area.

   Preprocessor> Modeling> Create> Areas> Rectangle> By 2 Corners> Enter 0.0, 0.0, 125, 125

5. Mesh the area with 2 elements per edge. First set the size controls so that 2 elements per edge are created.

   Preprocessor> Meshing> Size Controls> All Lines> NDIV Number of element divisions> Enter 2

   Preprocessor> Meshing> Mesh> Areas> Free (Select the 125 × 125 square)

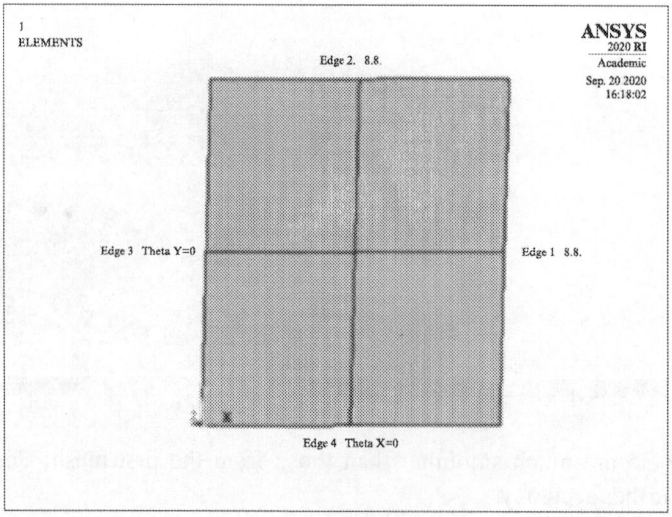

6. Apply boundary conditions and loads. Assume the global axis system is located at the center of the plate and that the simply supported edges are at x = 125 and y = 125. Edges 1 and 2 are simply supported. This means no displacement in the vertical direction and that the slope normal to the edge must be zero. Edges 3 and 4 are along lines of symmetry, and the plate must have a symmetric deformation shape across the lines of symmetry.

   Edge 1

   Preprocessor> Loads> Apply> Displacements> On Lines> Select Edge 1> Set UZ = 0> OK Then repeat and set RotX = 0

   Edge 2

   Preprocessor> Loads> Apply> Displacements> On Lines> Select Edge 2> Set UZ = 0> OK Then repeat and set RotY = 0

   Edge 3

   Preprocessor> Loads> Apply> Displacements> On Lines> Select Edge 3> Set UX = 0> OK Then repeat and set RotY = 0

   Edge 4

   Preprocessor> Loads> Apply> Displacements> On Lines

   Select Edge 4> Set UY = 0> OK. Then repeat and set RotX = 0

7. Apply one-quarter of the total load to the keypoint at the origin of coordinates.

   Preprocessor> Loads> Apply> Force/Moment> Keypoints> Select Keypoint at origin. For FZ enter – 1, click 'OK'

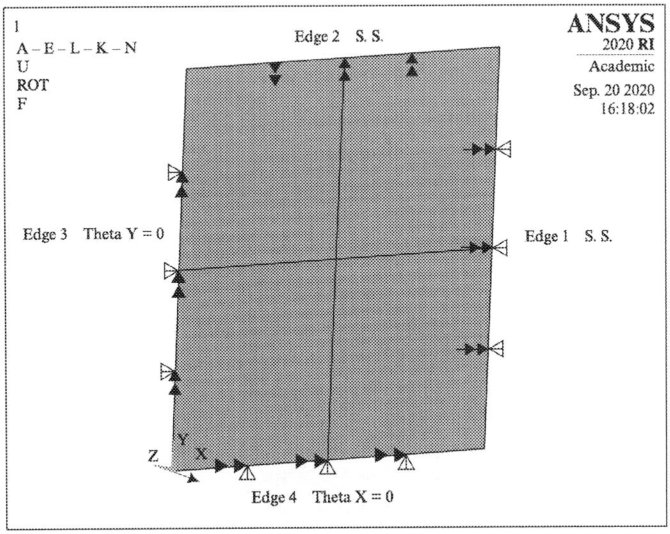

Solution:

1. Perform solution

   Solution > Current LS

Postprocessing

1. Plot deformed shape

   General Postproc > Plot Results > Deformed Shape . . . > Def + undeformed

2. Determine the deflection at the center of the plate.

   General Postproc > List > Results > Nodal Solution DOF solution > All DOFs DOF

3. As verification, plot the calculated **von Mises Stress**

   General Postproc > Plot Results > Element Solu . . . > Stress > von Mises SEQV

   The discontinuities in stress contours indicate an inaccurate solution for this mesh and element.

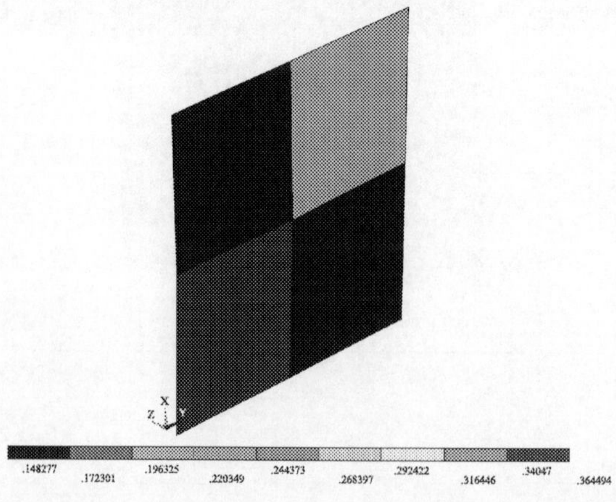

4. Clear the mesh (**Preprocessor> Meshing> Clear> Areas**) and remesh with 4 elements per edge to improve the accuracy. Then repeat with 8 elements per edge. Observe that the last mesh produces deflection with an error of only 1.1 %.The von Mises stress plot for this case is shown below.

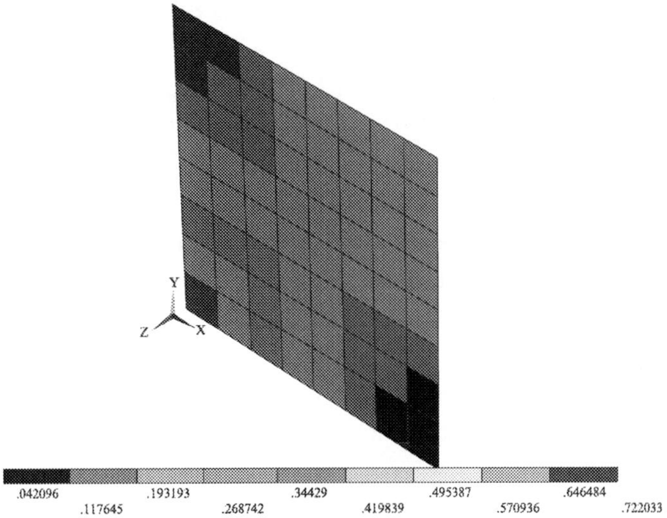

The stress contours are now much smoother in comparison with the first mesh. (Since the stresses are proportional to the spatial derivatives of the displacements, the stresses generally have greater error than the displacements.)

Increase the mesh density or switch to the eight node SHELL281 element, or do both to obtain yet greater accuracy for this problem.

W17: A steel plate measuring 150 mm × 250 mm × 10 mm has two holes drilled through the vertical centerline of the plate. The plate is constrained in the X-direction at the left and a uniform, distributed load is pulling on the right of the plate. Determine the effects of stress concentrators along the vertical centerline of the plate and plot the stress along that path.

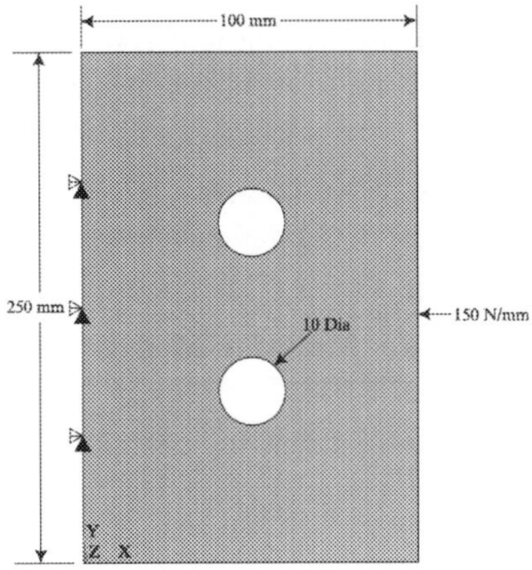

Solution:

Preprocessing

1. Give the example a Title

   Utility Menu> File> Change Title ...

   Use of Paths for Post Processing
2. Open preprocessor menu

   Main Menu> Preprocessor
3. Define Rectangular Areas

   Preprocessor> Modeling> Create> Areas> Rectangle> By 2 Corners

   Create a rectangle where the bottom left corner has the coordinates 0, 0 and the width and height are 100 and 150, respectively.
4. Create Circles

   Preprocessor> Modeling> Create> Areas> Circle> Solid Circle

   Create three circles with parameters shown below.

Circle	Parameters		
	WP X	WP Y	Radius
1	100	50	10
2	100	100	10

5. Subtract the Circles

   Preprocessor> Modeling> Operate> Booleans> Subtract> Areas

   First, select the area to remain (i.e., the rectangle) and click 'OK'. Then, select the areas to be subtracted (i.e., the circles) and click 'OK'.
6. Define the Type of Element

   Preprocessor> Element Type> Add/Edit/Delete...

   For this problem use the PLANE182 (Solid Quad 4 node 182) element. In the 'Element Types' window, click 'Options...' and set 'Element behavior' to 'Plane strs w/thk'
7. Define Real Constants

   Preprocessor> Real Constants...> Add...

   In the 'Real Constants for PLANE182' window, enter a thickness of 10.
8. Define Element Material Properties

   Preprocessor> Material Props> Material Models> Structural> Linear> Elastic> Isotropic

   In the window that appears, enter the following geometric properties for steel:

   (a) Young's modulus EX: 2e5

   (b) Poisson's Ratio PRXY: 0.3

9. Define Mesh Size
   Preprocessor> Meshing> Size Cntrls> ManualSize> Areas> All Areas...
   For this example use an element edge length of 5 mm.
10. Mesh the Area
   Preprocessor> Meshing> Mesh> Areas> Free> click 'Pick All'

Solution:

1. Define Analysis Type
   Solution> Analysis Type> New Analysis> Static
   Apply Constraints
   Solution> Define Loads> Apply> Structural> Displacement> On Lines
   Constrain the left edge of the area in the UX direction.
2. Apply Loads
   Solution> Define Loads> Apply> Structural> Pressure> On Lines
   Apply a constant, uniform pressure of – 150 on the right edge of the area.
   The model should now look like the figure below.

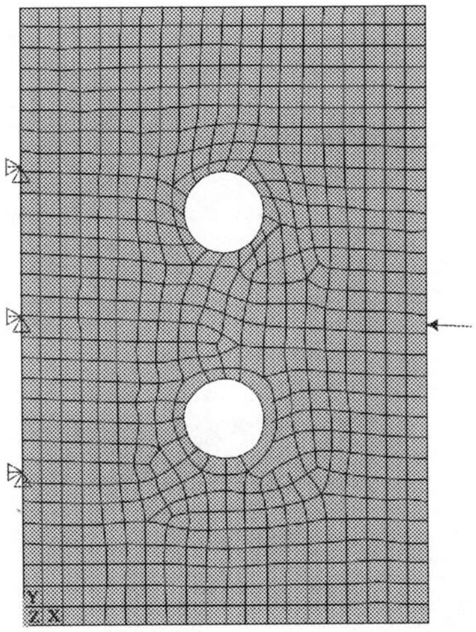

3. Solve the System
   Solution> Solve> Current LS

Postprocessing

Create a normal contour plot to see the stress distribution on the plate. It would have the distribution over the entire plate. However, if the stress near the holes is of interest, create a path through the center of the plate and plot the stress on that path. Both cases will be plotted below on a split screen.

1. Contour Plot

   Utility Menu> PlotCtrls> Window Controls> Window Layout

   Fill in the 'Window Layout' as seen below.

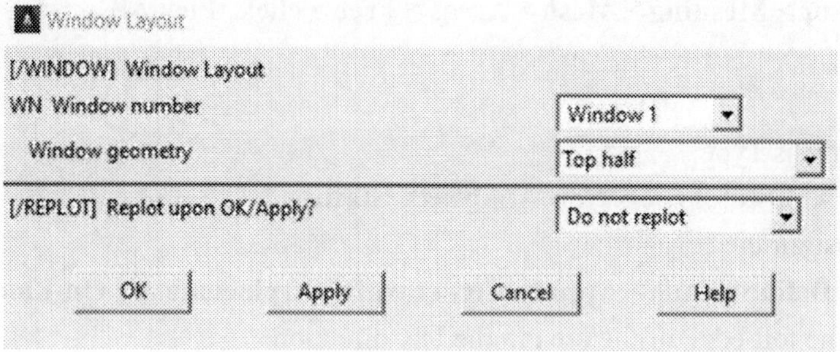

General Postproc> Plot Results> Contour Plot> Nodal Solu> Stress> von Mises

The display should now look like this.

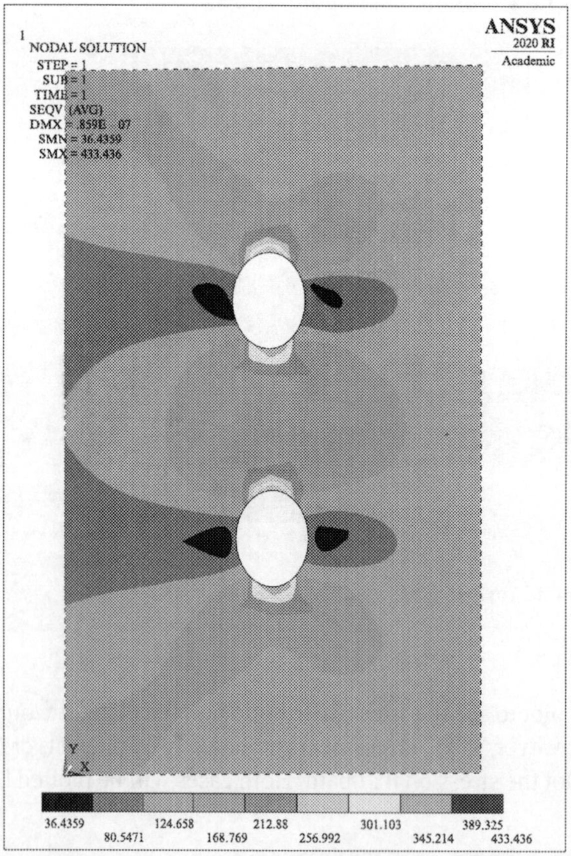

Make a couple of changes to ensure the top plot is not erased when the second plot is created.

Utility Menu> PlotCtrls> Window Controls> Window On or Off

Turn window 1 'Off'.

To keep window 1 visible during replots, select **Utility Menu> PlotCtrls> Erase Option> Erase Between Plots** and ensure there is no check-mark, meaning this function off.

To have the next graph plot in the right half of the screen, select **Utility Menu> PlotCtrls> Window Controls> Window Layout** and select **'Window 2> Right Half> Do not replot'**.

2. Create Path

   General PostProc> Path Operations> Define Path> By Location

   In the window, shown below, name the path Cutline and set the 'Number of divisions' to 800

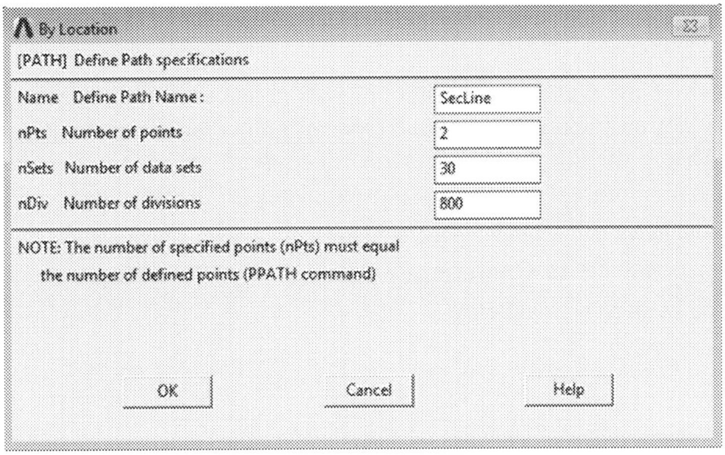

Fill the next two windows in with the following parameters.

Parameters			
Path Point Number	**X Loc**	**Y Loc**	**Z Loc**
1	50	0	0
2	50	150	0

When the third window pops up, click 'Cancel' because only two points are enabled on the path in the previous step.

3. Map the Stress onto the Path

   Now the path is defined, choose what to map to the path, or in other words, what results should be available to the path. For this example, equivalent stress is desired.

   General Postproc> Path Operations> Map onto Path

   Fill the next window in as shown below [Stress> von Mises] and click OK.

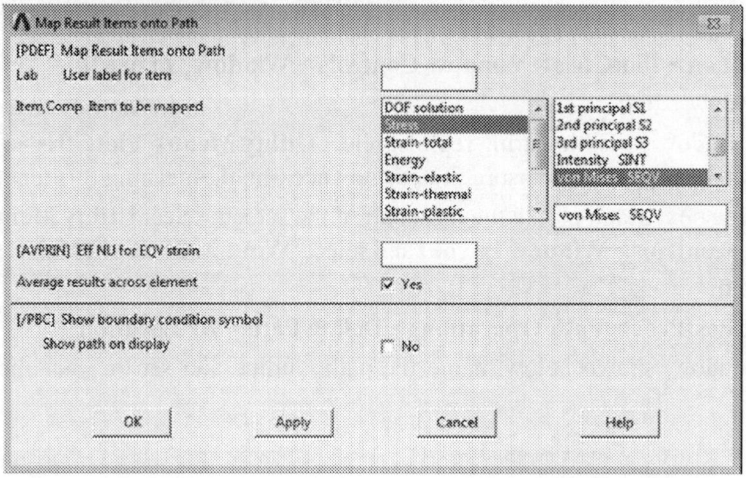

The warning shown below will probably pop up. This is just saying that some of the 800 points defined earlier are not on interpolation points (special points on the elements) therefore there is no data to map. This is of little concern though, since there are plenty of points that do lie on interpolation points to produce the necessary plot, so disregard the warning.

4. Plot the Path Data

   General Postproc> Path Operations> Plot Path Item> On Geometry

   Fill the window in as shown below.

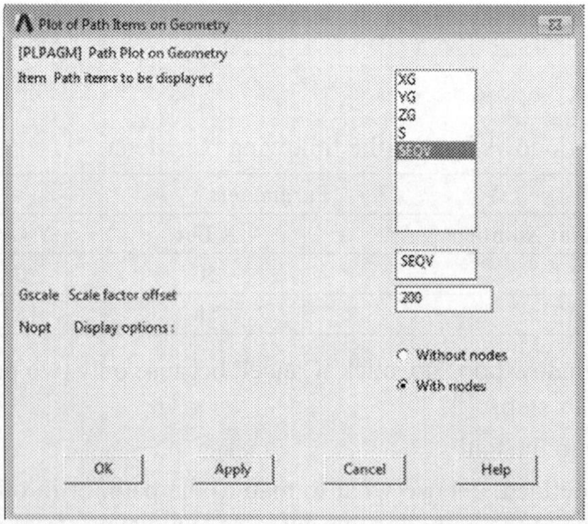

5. The display should look like the following. Note, there will be dots on the plot showing node locations. Due to resolution restrictions, these dots are not shown here.
6. This plot makes it easy to see how stress is concentrated around the holes.

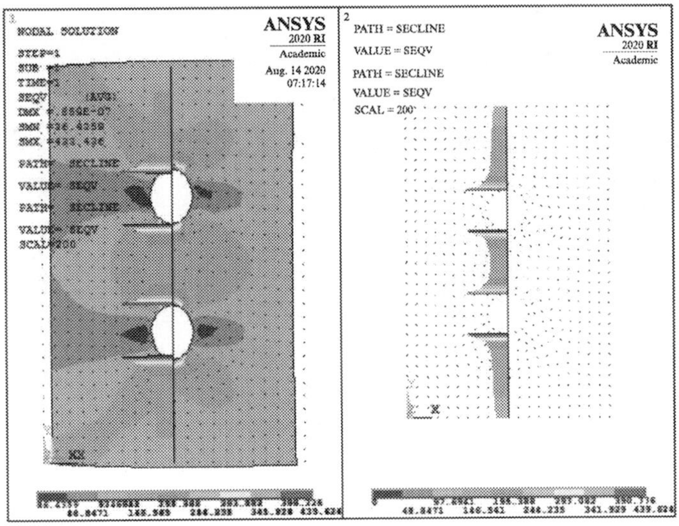

W18: Analyze the aluminum bracket shown in the figure below using a solid 8-node element. The thickness of the bracket is 5 mm. Assume the structure is made of aluminum with modulus of elasticity, E = 70GPa. This bracket will be fixed at the two small holes on the left and have a load applied to the larger hole on the right.The bracket is loaded uniformly along its top surface. The load is 2500 N/m. Plot deformed shape. Determine the principal stress and the von Mises stress.

Data: Thickness = 5 mm, E = 0.7e11. Pressure = 2500 N/m

Solution:

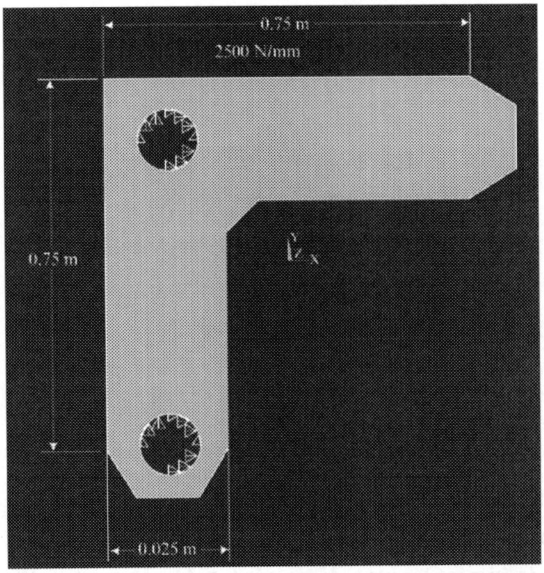

Preprocessing

1. Model the structure

   Utility Menu> Workplane> WP Settings

   The following window comes up.

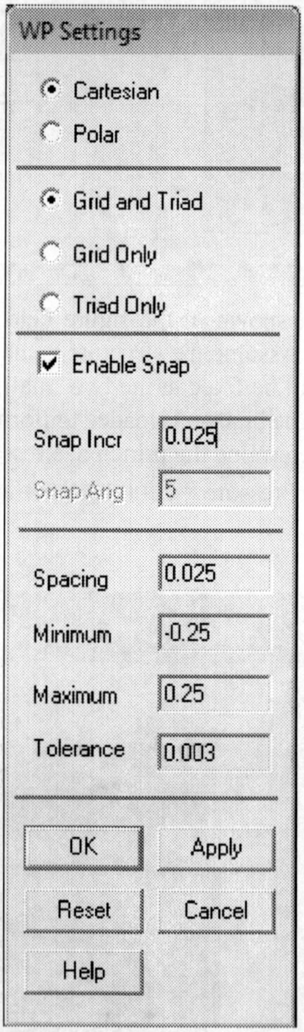

   Check the **Cartesian** and **Grid Only** buttons

   Enter the values shown in the figure above.

   Preprocessor> Modeling> Create> Keypoints> On Working Plane

   Outline a part of the bracket as shown in the figure.

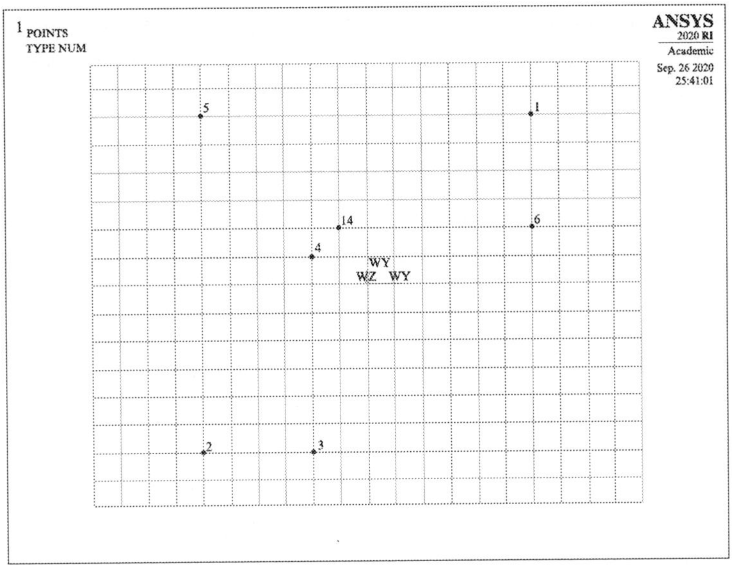

Turn on the numbering: **ANSYS Utility Menu> Plot Controls> Numbering...** Create lines between keypoints, then create area inside. Go to **Preprocessor> Modeling> Create> Areas> Arbitrary> By Lines**.

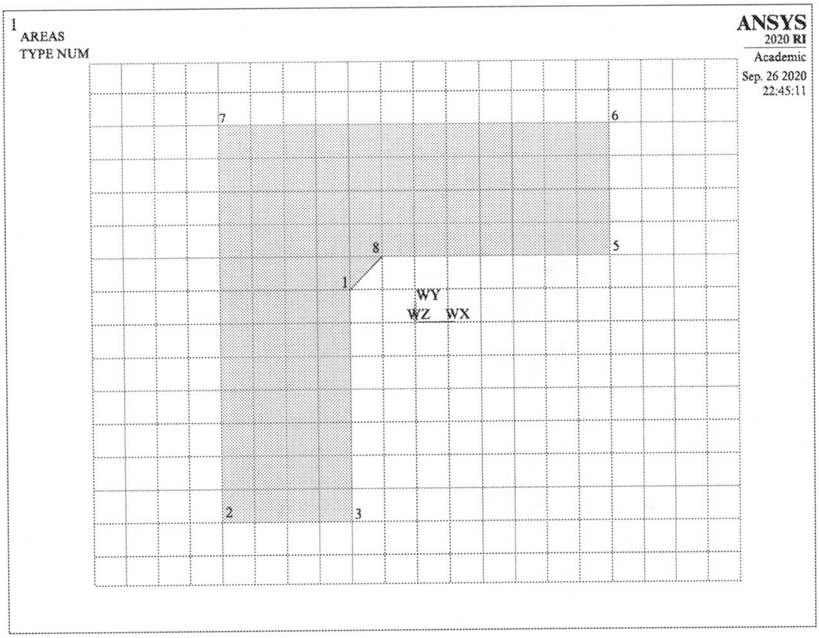

Preprocessor> Modeling> Create> Areas> Polygon> Hexagon and create the two hexagons with center at the midpoint of the right edge and the bottom edge of the bracket and the side equal to the one half of length of that edge.

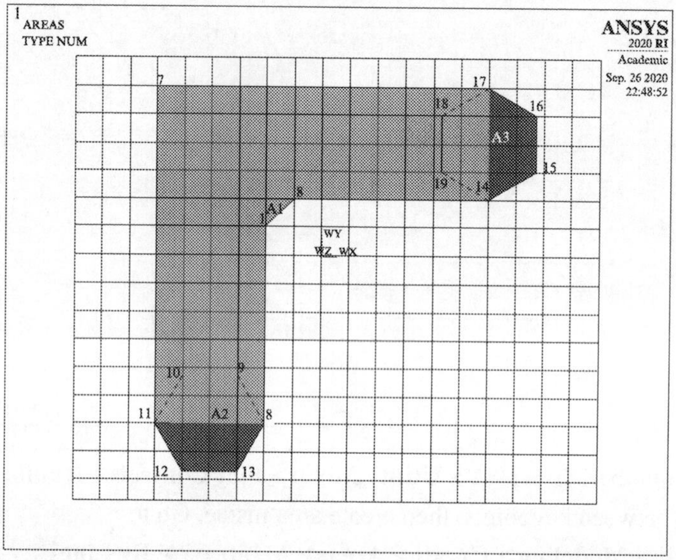

Preprocessor> Modeling> Operate> Booleans> Add> Areas and select all areas created to make a single area.

Preprocessor> Modeling> Create> Areas> Circles> Solid Circle and create a circle with center at the midpoint of the bottom edge of the vertical rectangle. Do the same thing at the top right corner.

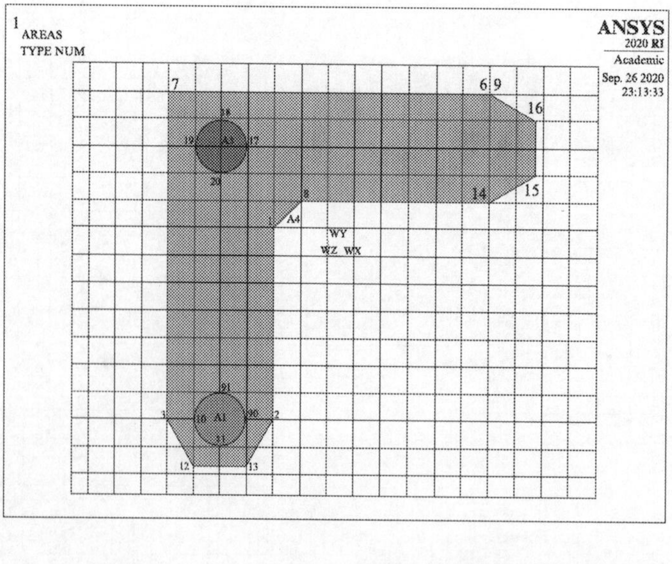

Preprocessor> Modeling> Operate> Booleans> Subtract> Areas.

First select the base area from which the smaller area will be subtracted. Say OK. Now select the smaller circles and say 'OK'. The smaller circles will now be subtracted and the figure will look like this:

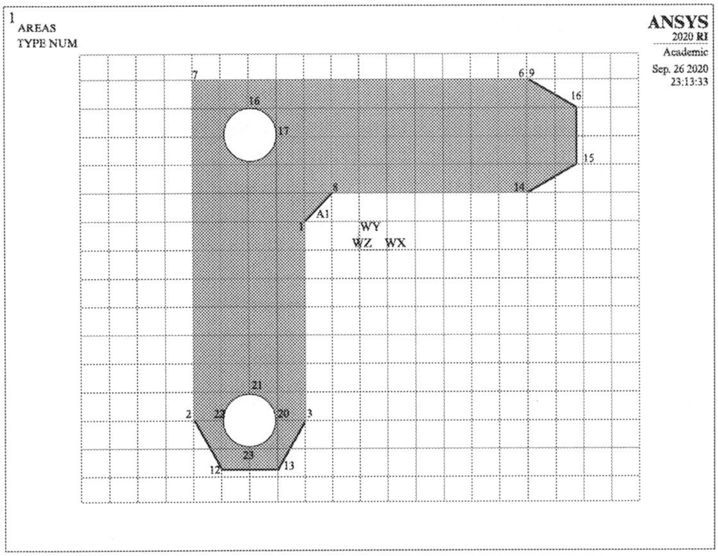

2. Define material properties

   Main Menu> Preprocessor> Material Props> Material Models. From this window, select **Structural> Linear> Elastic> Isotropic.**

   Enter 1 for the Material Property Number and click OK.

   Fill in 0.7e11 for the Young's modulus and 0.3 for minor Poisson's Ratio. Click 'OK'. Now the material '1' has the properties defined in the above table. We will use this material for the structure.

3. Define element properties

   Preprocessor> Element Type> Add/Edit/Delete... In the 'Element Types' window that opens click on Add...

   Type 1 in the Element type reference number.

   Click on **Structural Solid** and select **Quad 8 node 82**. Click 'OK'. Close the 'Element types' window.

   Click **Preprocessor> Element Type> Add/Edit/Delete...** In the 'Element Types' window that opens click on Options...

   Select **Plane strs w/thk** for K3 and click 'OK'.

   So now we have selected Element type 1 to be a Structural Solid 8 node element. The bracket will now be modeled as elements of this type.

Define the thickness for this element.
Go to **Preprocessor> Real Constants**
In the 'Real Constants' dialog box that comes up click on 'Add'
In the 'Element Type for Real Constants' that comes up click 'OK'.
Fill in the relevant values and click on 'OK'

4. Mesh the model
   Preprocessor> Meshing> Size Controls> Manual Size> Lines> Picked Lines. Pick all the lines on the outer boundary of the figure and click 'OK'.
   In the menu that comes up type **0.0125** in the field for 'Element edge length'.

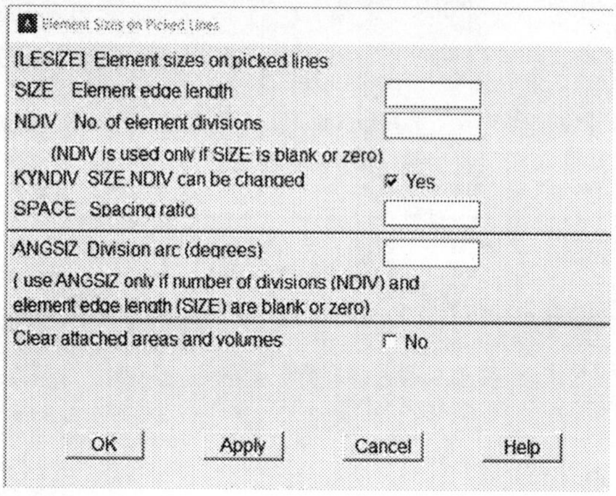

Click on 'OK'.
Preprocessor> Meshing> Mesh> Areas> Free
Select the area and click on 'OK' in the 'Mesh Areas' dialog box.
Now the bracket is divided into Solid 8 node elements.

5. Apply boundary conditions and constraints
   The bracket is fixed at the left edge.
   Preprocessor> Loads> Define Loads> Apply> Structural> Displacement> On Lines.
   Select the line on the left edge and click 'OK'. The following window comes up:

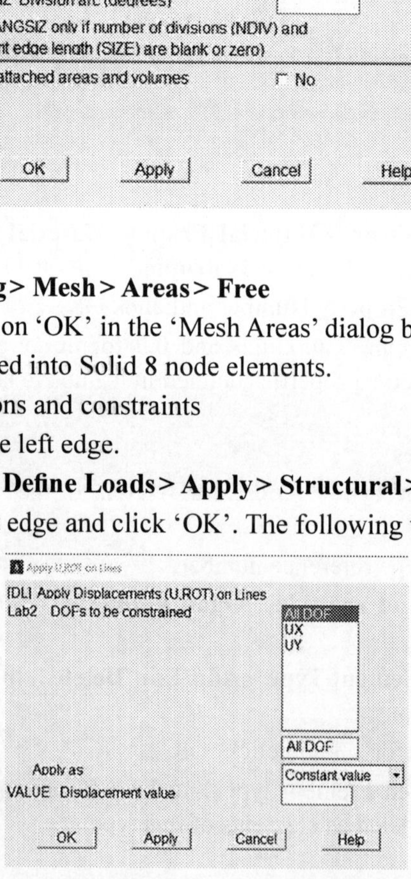

Select All DOF and click 'OK'.

Apply forces

Preprocessor> Loads> Define Loads> Apply> Structural> Pressure> On Line.

Select the top line.

Click on 'OK' in the 'Apply PRES on lines' window.

Enter the value of the pressure as shown above. Click 'OK'.

The model should look like the one below.

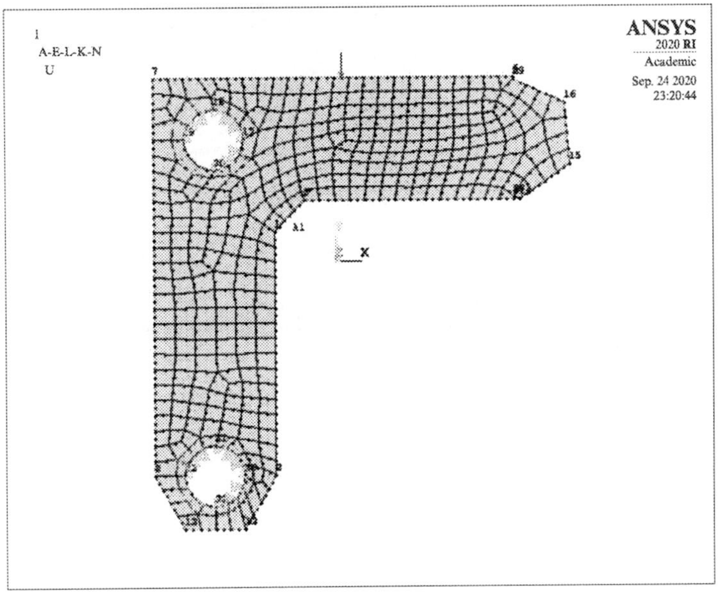

Solution:

1. **Main Menu> Solution> Analysis Type> New Analysis**

   Select static and click on 'OK'
2. **Solution> Solve> Current LS**

Postprocessing

1. List the results.

   General Postprocessing> List Results> Nodal Solution.

   Select DOF solution and All U's. Click on 'OK'.
2. Plot the displacements and stress.

   General Postprocessing> Plot Results> Deformed shape.

   Select **Def + undeformed** and click 'OK'. The output will be like the figure below.

Select a stress (**SEQV**) to be plotted and click 'OK'. The output will be like this.

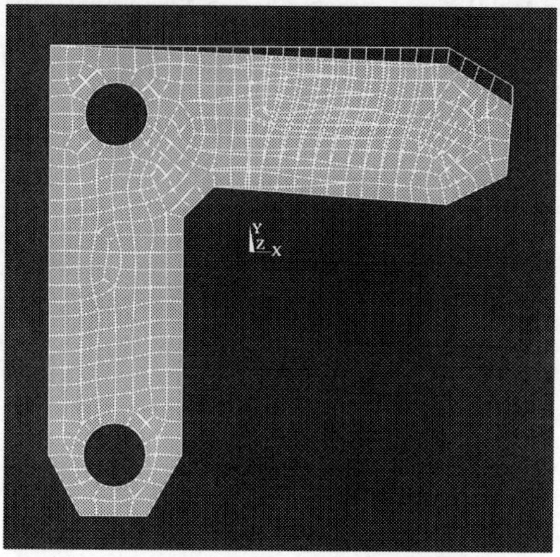

W19: Model a simple bracket shown in the following figure. This bracket is to be built from a 10 mm thick steel plate. The bracket is fixed at its left edge. Auniformly variable load of 1000N is applied to the larger hole on the right.

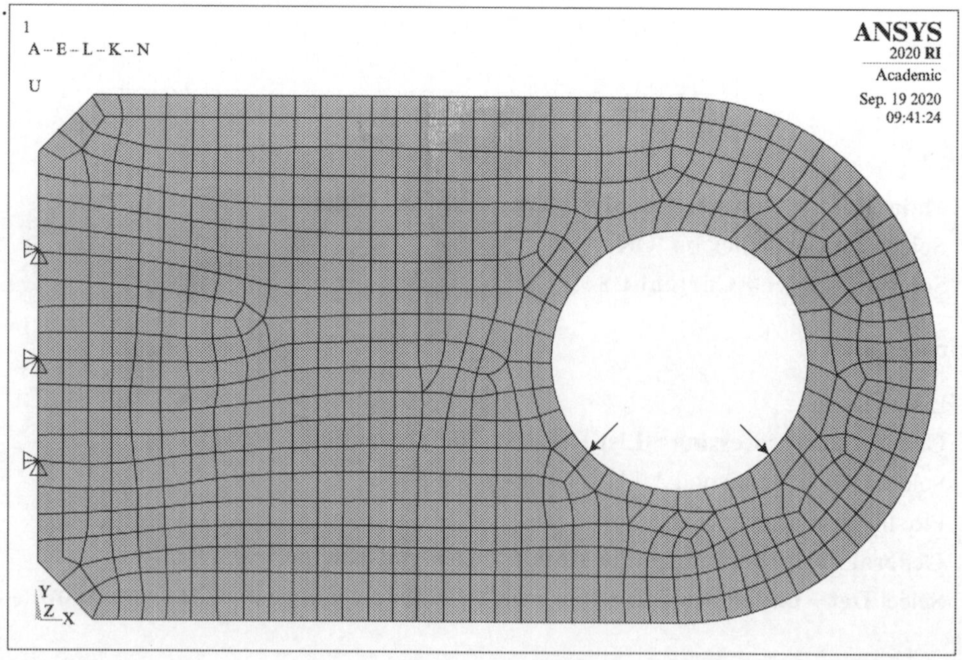

Solution:

Preprocessing

1. Give the Bracket example a Title

   Utility Menu> File> Change Title

   'Bracket-2'

2. Form Geometry

   Use Boolean operations to create the basic geometry of the bracket.

   (a) Create the main rectangular shape

   The main rectangular shape has a width of 125 mm, a height of 100 mm and the bottom left corner is located at coordinates (0, 0)

   Main Menu> Preprocessor> Modeling> Create> Areas> Rectagle> By 2 Corners

   Enter WPX = 0, WPY = 0, Width = 125, Height = 100, OK

   Now create circular portion on the right. Draw a circle of radius 50 mm with centre at (125, 50)

   Main Menu> Preprocessor> Create> Areas> Circle> Solid Circle

   The screen should now look like the following...

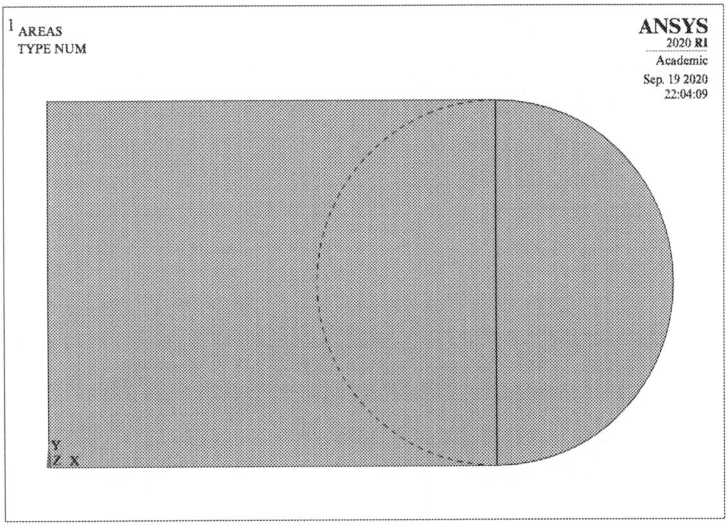

   (b) Boolean operations – Addition

   Now add these two discrete areas together to form one area.

   To perform the Boolean operation, from the Preprocessor menu select:

   Modeling> Operate> Booleans> Add> Areas

   In the 'Add Areas' window, click on 'Pick All'

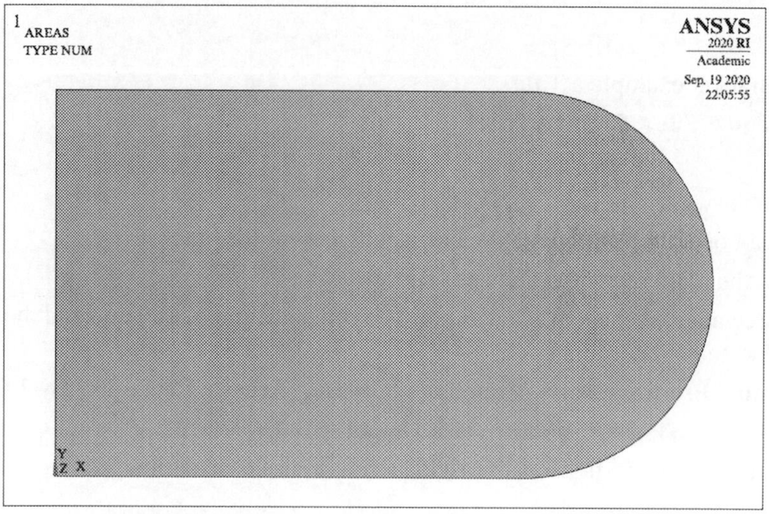

(c) Create the bolt hole

Preprocessor> Modeling> Create Areas> Circle> Solid Circle

Create the circle with the parameters: WP X = 125, WP Y = 50, Radius = 25

Now remove the bolt hole from this plate. Now select

Preprocessor> Modeling> Operate> Booleans> Subtract> Areas

Select the base area from which to subtract (the large plate that was created)

Next select the circle that is just created. Click on the circle that is just created and click 'OK'.

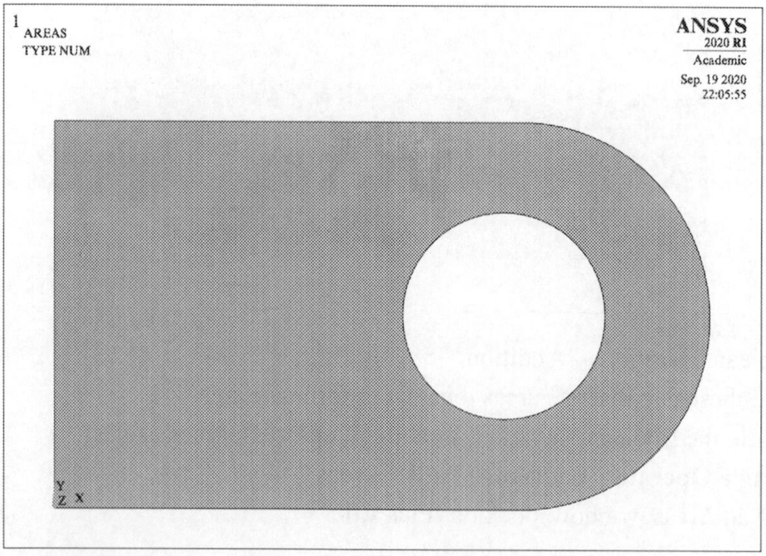

(d) Create chamfer at left side corners. Create keypoints at (10, 0), (0, 10), (90, 10) and (100, 0). Draw lines joining the corresponding keypoints.

Preprocessor> Create> Keypoints> In active CS

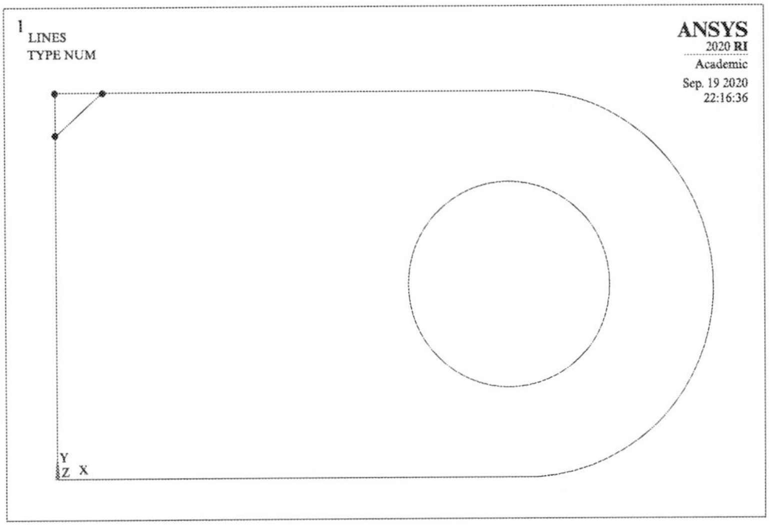

Divide the base area using the lines just created into three parts. Delete the corner triangles.

Preprocessor> Modeling> Create> Booleans> Divide> Area by Line

Select the large plate, click Apply. Click on chamfer line, then click Apply.

Preprocessor> Modeling> Delete> Areas and Below

Select two triangular areas on left side

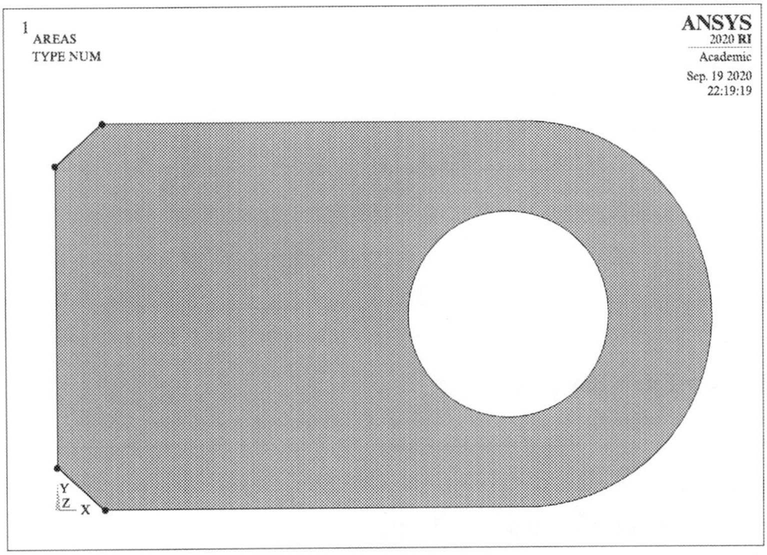

1. Define the Type of Element

   As in the verification model, PLANE182 will be used for this problem.

   Preprocessor> Element Type> Add/Edit/Delete

   Use the 'Options...' button to get a plane stress element with thickness

   Define Geometric Constants

   Preprocessor> Real Constants> Add/Edit/Delete

   Enter a thickness of 5 mm.

   Define Element Material Properties

   Preprocessor> Material Props> Material Library> Structural> Linear> Elastic> Isotropic

   Give the properties of aluminum. Enter the following when prompted: EX = 0.7e5, PRXY = 0.3

   Set Mesh Size

   Preprocessor> Meshing> Size Cntrls> Manual Size> Areas> All Areas

   Select an element edge length of 5. Again, we will need to make sure the model has converged.

   Mesh the model

   Preprocessor> Meshing> Mesh> Areas> Free and select the area when prompted

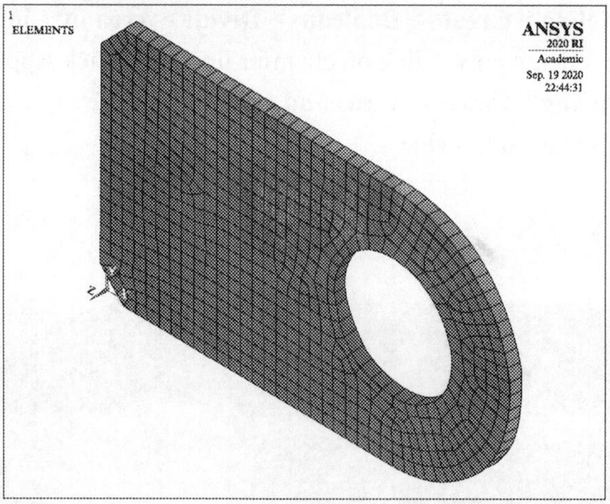

2. Save the Job

   Utility Menu> File> Save as...

Solution

1. Define Analysis Type

   Solution> New Analysis and select 'Static'.

2. Apply Constraints

   As illustrated, the plate is fixed on the left hand edge.

Solution> Define Loads> Apply> Structural> Displacement> On Lines

Select extreme left edge. OK. Highlight 'ALL DOF'

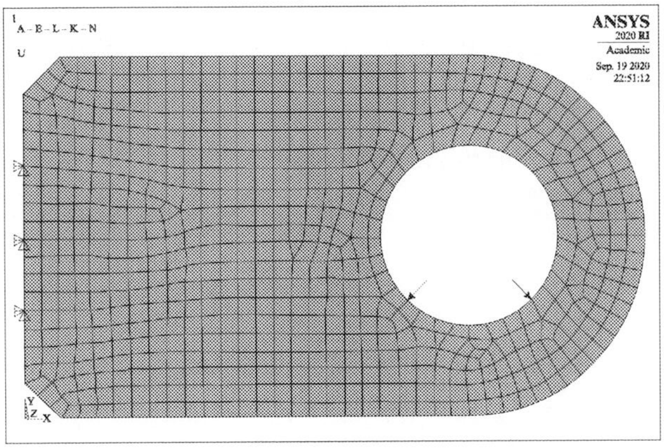

3. Apply Loads

   As shown in the diagram, there is a single vertical load of 1000N, at the bottom of the large bolt hole. Apply this force to the respective line (arc) (**Solution> Define Loads> Apply> Structural> Force/Moment> On Lines**

   Select lower left arc. Apply a force in the y direction of – 1000 to 0.

   Select lower right arc. Apply a force in the y direction of 0 to – 1000.

   The applied loads and constraints should now appear as shown below.

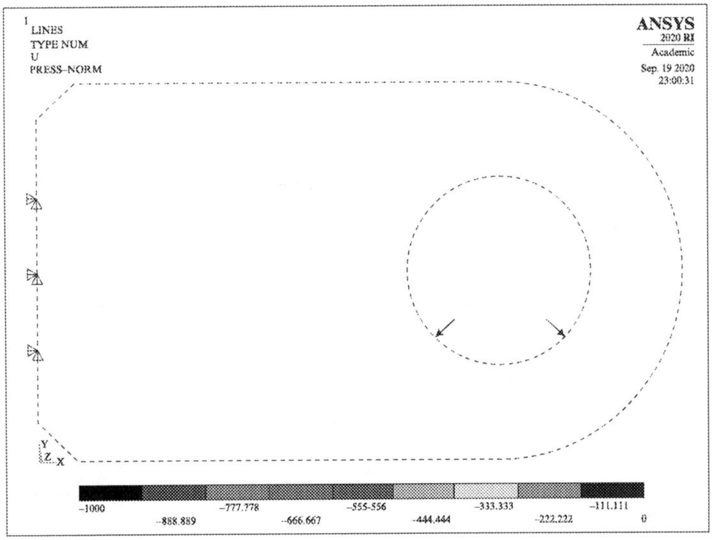

4. Solving the System

   Solution> Solve> Current LS

Postprocessing

Take a look at the deflected shape and the stress contours.

1. Deformation

   General Postproc> Plot Results> Def + undeformed to view both the deformed and the undeformed object.

   The graphic should be similar to the following.

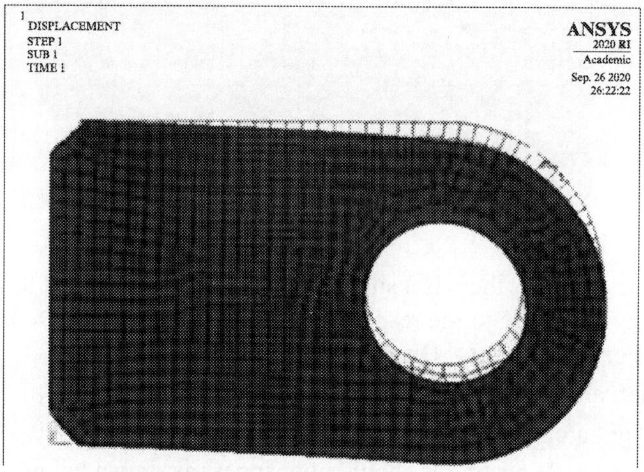

2. Deflection

   General Postproc> Plot Results> Contour Plot> Nodal Solution to plot the nodal deflections, then select DOF **Solution** – USUM in the window.

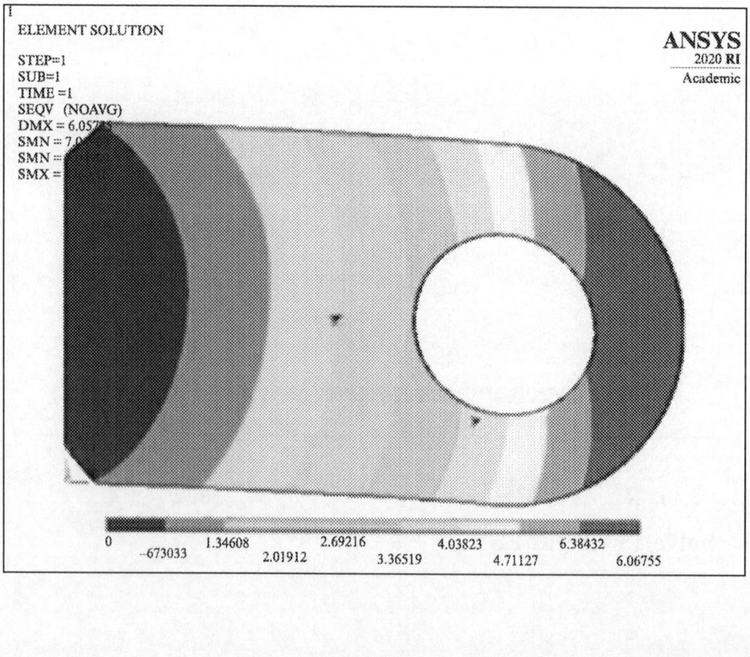

Alternatively, obtain these results as a list. (**General Postproc> List Results> Nodal Solution...**)

3. Stresses

   General Postproc> Plot Results> Nodal Solution... Then select von Mises stress in the window.

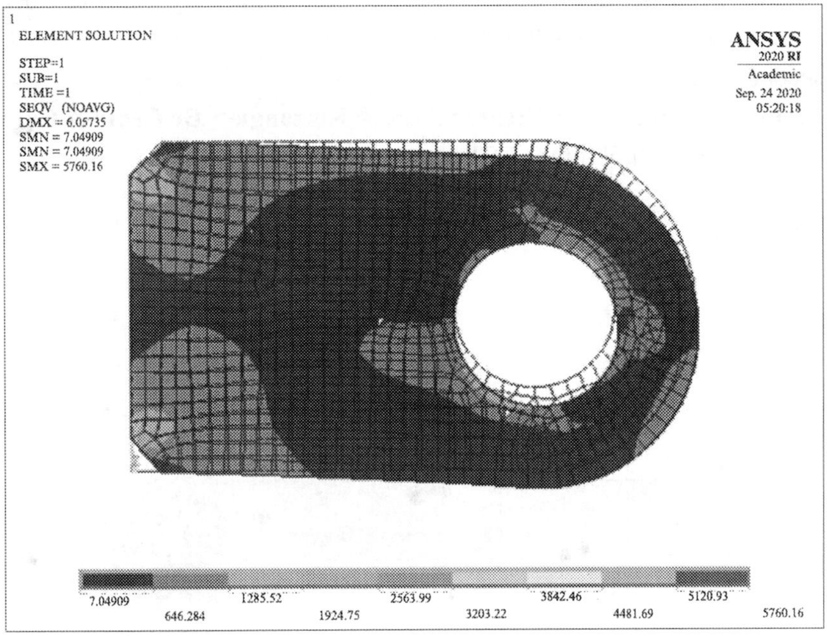

   List the von Mises stresses to verify the results at certain nodes.

   General Postproc> List Results. Select **Stress, Principals SPRIN**

4. Quit ANSYS

   Click 'QUIT' on the ANSYS Toolbar or select **Utility Menu> File> Exit...** to quit ANSYS. In the window that appears, select 'Save Everything' (and then click 'OK'.

W20: Do a finite element analysis on the wrench shown below using ANSYS.

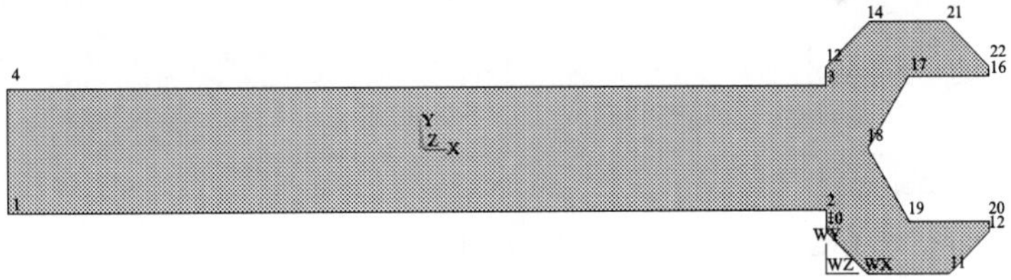

Solution:

Preprocessing

1. Draw rectangle representing handle.
   Main Menu> Preprocessor> Create> Areas> Rectangle> By Centr & Cornr
   WP X = 0, WP Y = 0, Width = 200, Height = 30, OK
2. Draw second rectangle
   Main Menu> Preprocessor> Create> Areas> Rectangle> By Centr & Cornr
   WP X = 200, WP Y = 0, Width = 40, Height = 60, OK
3. Draw a hexagon
   WPX = 230, WP Y = 0, Rad = 20, Theta = 0

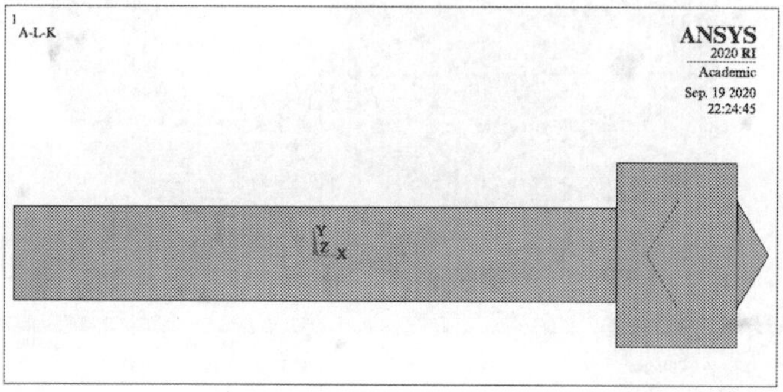

4. Chamfer the corners of the rectangle.

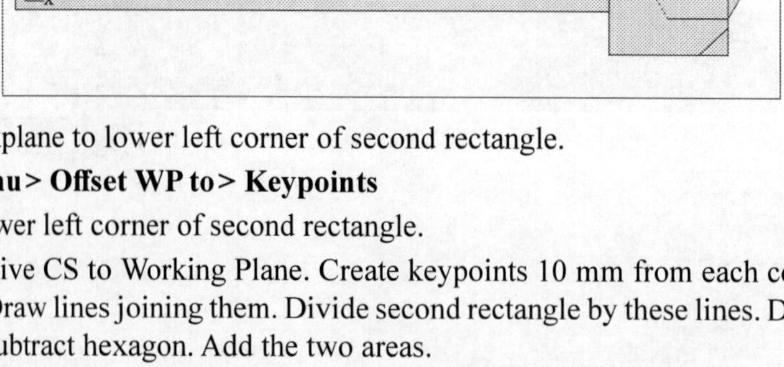

Move Workplane to lower left corner of second rectangle.
Utility Menu> Offset WP to> Keypoints
Click on lower left corner of second rectangle.
Change Active CS to Working Plane. Create keypoints 10 mm from each corner of second rectangle. Draw lines joining them. Divide second rectangle by these lines. Delete unwanted triangles. Subtract hexagon. Add the two areas.

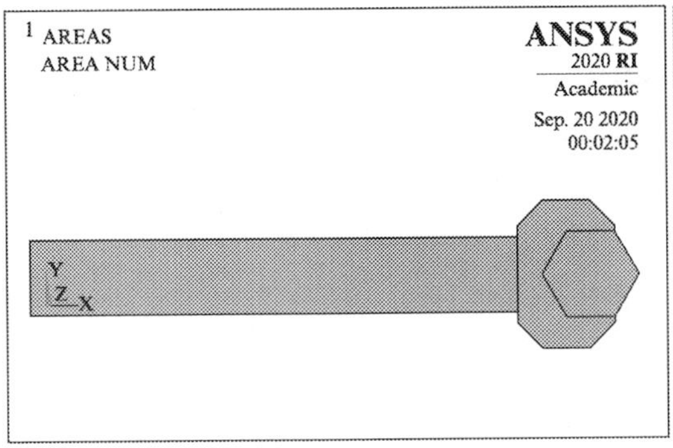

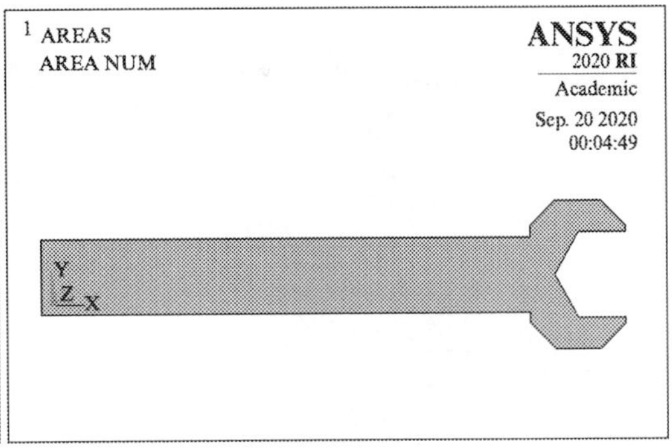

5. Mesh the Part

   Preprocessor> Add/Edit/Delete..> Add>

   Select 'Solid' and '8 node183'

   This defines element type 183 and assigns it to reference 1. Element type 183 is the plane stress quadratic displacement 8-noded element.

6. Define material properties

   Main Menu> Preprocessor> Material Props> Isotropic> OK

   In the box for 'Young's modulus', type 2e5.

   Preprocessor> Mesh> Areas> Pick All. The part will turn light blue.

   Click 'OK', Close the 'MESH' window.

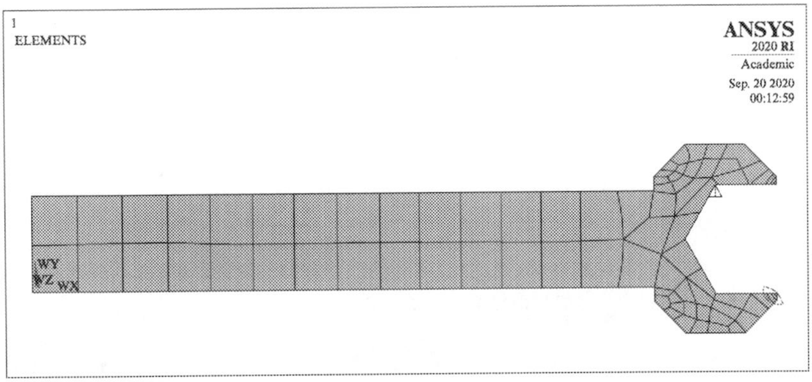

7. Apply Boundary Conditions

   Preprocessor> Loads> Apply> Displacements> On Nodes

   With the left mouse button, select two nodes in the wrench jaws where the jaws will contact the part to be twisted. Figure below shows the nodes.

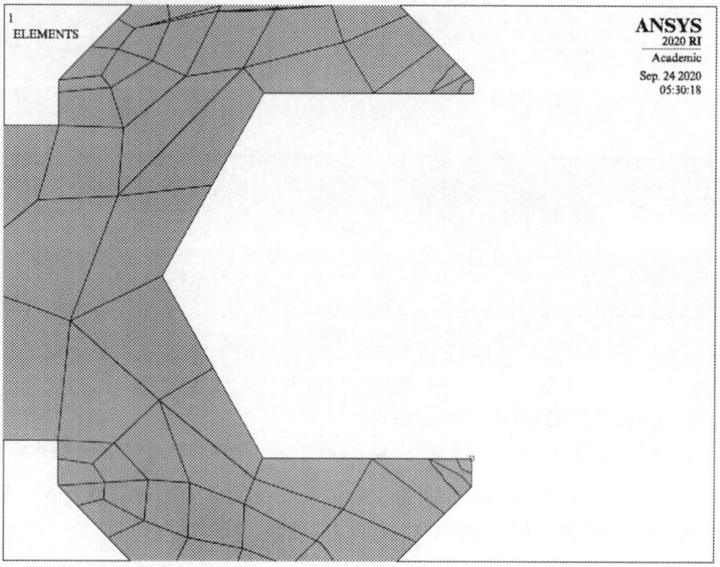

Apply U,ROT ON NODES 'DOFs to be constrained' highlight All DOF

Apply U,ROT ON NODES> 'Displacement Value' type 0

The figure should now look like figure below with arrows marking the points which are limited to zero displacement.

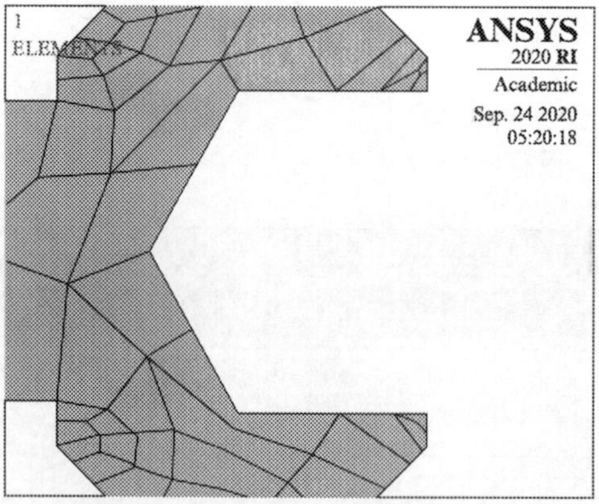

8. Apply Loads

   This section tells the program what loads are being applied to the part.

   Apply> Force/Moment> On Nodes

   With the left mouse button, select a node at the end of the handle as shown below.

   Choose the contact nodes such that the handle load must point down, then enter a negative load. An arrow should appear on the handle as shown below.

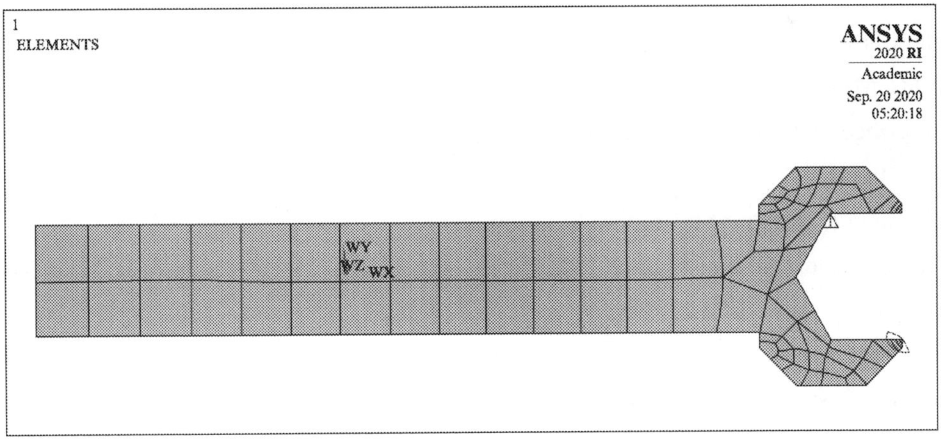

9. Print Image
   Utility Menu > PlotCtrls > Capture Image
   Click once with the left mouse button to set the window.
   File > Print
   Click next to the **Print to:** option to turn it on.
   Click in the text field just below **Print to:**
   Type: *RBC_ANSYS_Wrench Model*

Solution:

1. Solve
   Main Menu > Solution > Solve > Current LS

Postprocessing

1. Plot results
   General Postproc > Plot Results > Nodal Solu ...

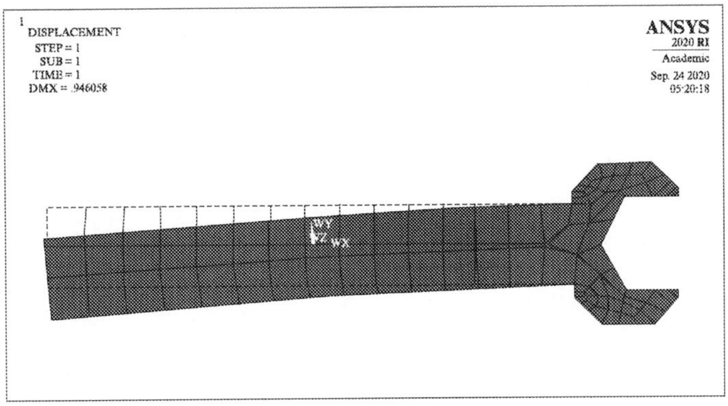

2. Print Image

   Utility Menu> PlotCtrls> Capture Image

   Click once with the left mouse button to set the window.

   File> Print

   Click next to the **Print to:** option to turn it on.

   Click in the text field just below **Print to:**

   Type: *RBC_Ansys_Wrench Results*

3. Plot stresses for a cross section of the Wrench

   General Postproc> Path Operations> Define Path> By Nodes

   Pick 2 points with the left mouse button on the figure as shown in figure below. Assign suitable name.

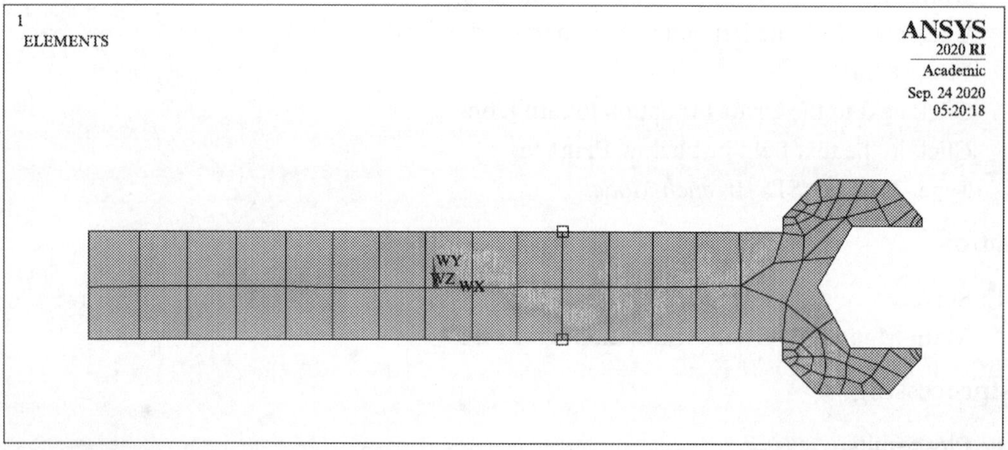

   Path Operations> Map Onto Path

   Highlight **Stress** and **X-direction SX** or whichever stress component is relevant.

   Path Operations> Plot Path Items...> On Graph

   Highlight **SX.** A graph of the stresses along the chosen path will appear as shown in figure below.

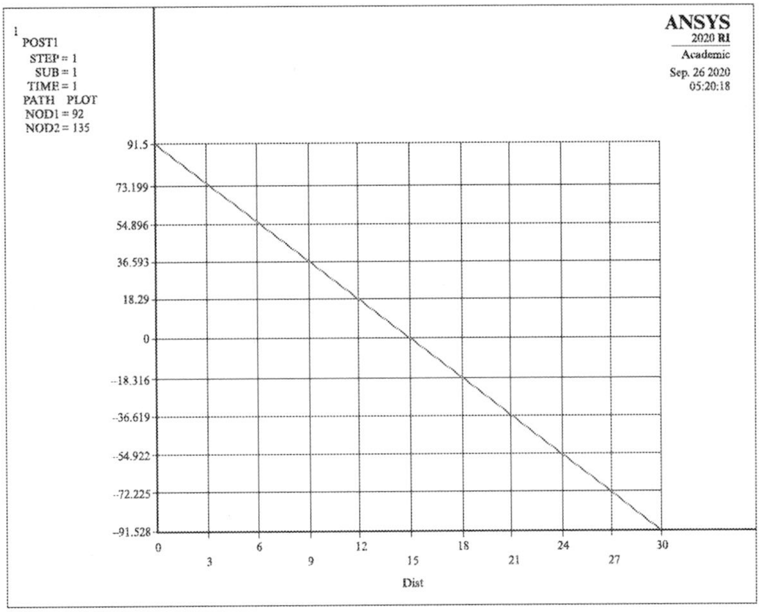

4. Exit the Program

   Utility Menu> File> Exit

   Choose the option **Quit - No Save!**

W21: A closed tube is made of steel. Point loads are applied at the center of the top and bottom plate. A 1/2 cross-sectional view of the tube is shown below. Model and analyze the tube.

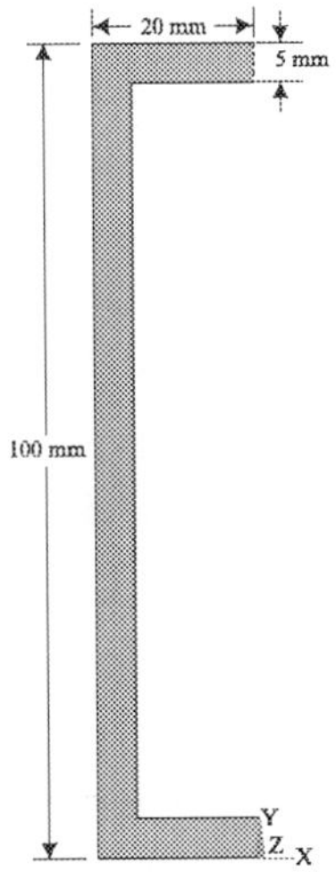

Data:

Solution:

Preprocessing

1. Give example a Title
   Utility Menu> File> Change Title ...
   '*Axisymmetric Tube*'
2. Open preprocessor menu
   ANSYS Main Menu> Preprocessor
3. Create Area
   Create keypoints
   Preprocessor> Modeling> Create> Keypoints> In Active CS>
   For an axisymmetric problem, ANSYS will rotate the area around the y-axis at x = 0. Therefore, to create the geometry mentioned above, define a C-shape.
   Define 8 corresponding to C shape as defined in the following table:

Keypoint No	X	Y
1	0	0
2	–20	0
3	–20	100
4	0	100
5	0	95
6	–15	95
7	–15	5
8	0	5

4. Draw C-shaped area
   Preprocessor> Modeling> Create> Areas> Arbitrary> By Keypoints
5. Define the Type of Element
   Preprocessor> Element Type> Add/Edit/Delete...
   For this problem we will use the PLANE183 (Structural, Solid, Triangle 6 node) element)
6. Turn on Axisymmetry
   While the Element Types window is still open, click the Options... button.
   Under Element behaviour K3 select Axisymmetric.
7. Define Element Material Properties
   Preprocessor> Material Props> Material Models> Structural> Linear> Elastic> Isotropic

In the window that appears, enter the following geometric properties for steel:

(a) Young's modulus EX: 2e5

(b) Poisson's ratio PRXY: 0.3

8. Define Mesh Size

   Preprocessor> Meshing> Size Cntrls> ManualSize> Areas> All Areas

   For this example use an element edge length of 2.5 mm.

9. Mesh the frame

   Preprocessor> Meshing> Mesh> Areas> Free> Pick All

   The model should know look like this:

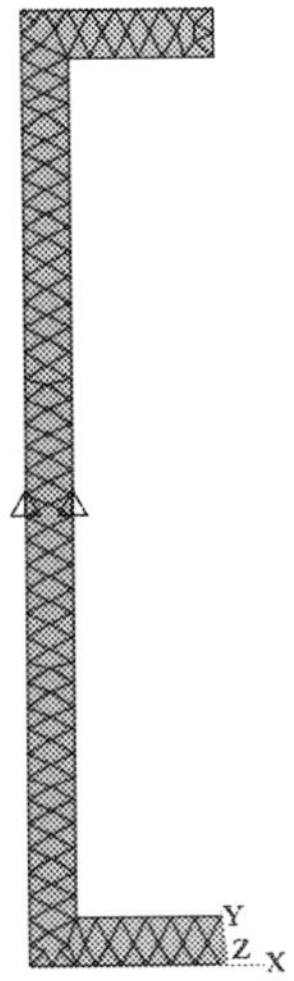

Solution:

1. Define Analysis Type

   Solution> Analysis Type> New Analysis> Static

2. Apply Constraints

   Solution> Define Loads> Apply> Structural> Displacement> Symmetry B.C.> On Lines

   Pick the two edges on the left, at x = 0, as shown below. By using the symmetry B.C. command, ANSYS automatically calculates which DOFs should be constrained for the line of symmetry. Since the element being used only has 2 DOFs per node, constrain the lines in the x-direction to create the symmetric boundary conditions.

   Utility Menu> Select> Entities

   Select Nodes and By Location from the scroll down menus. Click Y coordinates and type 50 into the input box as shown below, then click OK.

   Solution> Define Loads> Apply> Structural> Displacement> On Nodes> Pick All

Constrain the nodes in the y-direction (UY). This is required to constrain the model in space, otherwise it would be free to float up or down. The location to constrain the model in the y-direction (y = 50) was chosen because it is along a symmetry plane. Therefore, these nodes won't move in the y-direction according to theory.

3. **Utility Menu> Select> Entities**

   In the select entities window, click 'Sele All' to reselect all nodes. It is important to always reselect all entities once the task is finished to ensure future commands are applied to the whole model and not just a few entities. After clicking Sele All, click on Cancel to close the window.

4. Apply Loads

   Solution> Define Loads> Apply> Structural> Force/Moment> On Keypoints
   Pick the top left corner of the area and click OK. Apply a load of 120 N in the FY direction.

   Solution> Define Loads> Apply> Structural> Force/Moment> On Keypoints
   Pick the bottom left corner of the area and click OK. Apply a load of – 120 in the FY direction.

   The applied loads and constraints should now appear as shown in the figure below.

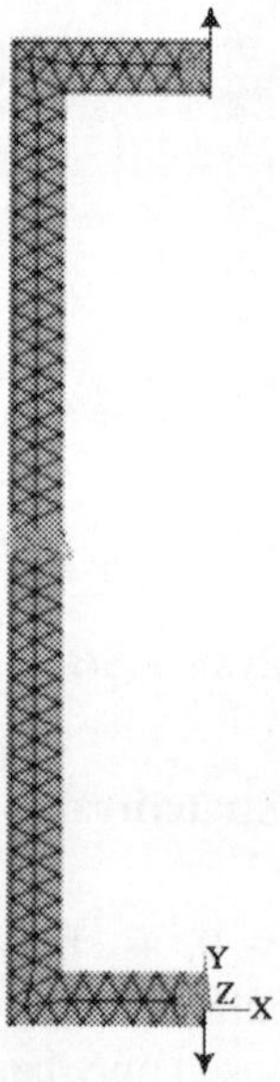

5. Solve the System

   Solution> Solve> Current LS

Postprocessing

1. Determine the stress through the thickness of the tube

   Utility Menu> Select> Entities...

Select **Nodes> By Location> Y coordinates** and type 45,55 in the Min, Max box, as shown below and click OK.

General Postproc> List Results> Nodal Solution> Stress> Components SCOMP

2. Plotting the Elements as Axisymmetric

Utility Menu> PlotCtrls> Style> Symmetry Expansion> 2–D Axisymmetric...

The following window will appear. Clicking on 1/2 expansion.

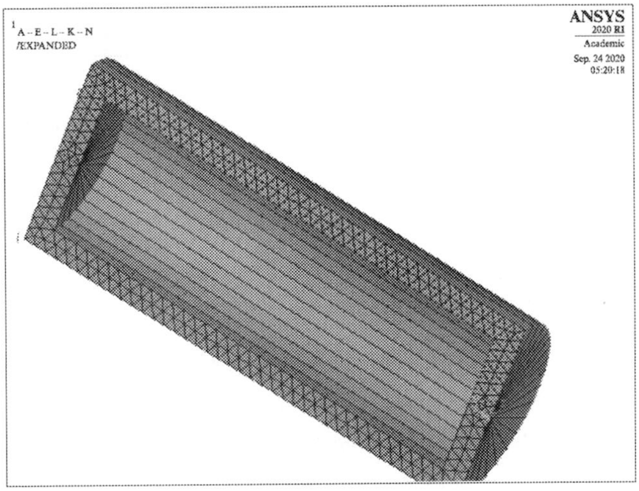

4.3 3-D PROBLEMS

4.3.1 Space

W22: A step through bicycle frame is shown in the following figure. The frame is built of hollow aluminum tubing having an outside diameter of 25 mm and a wall thickness of 2 mm. Model and analyze the bicycle.

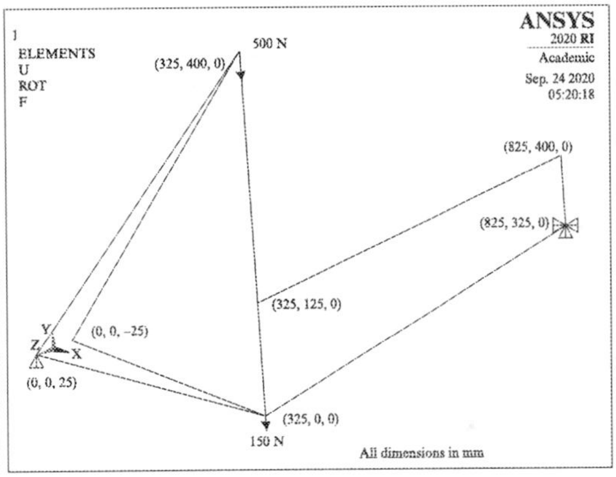

Solution:

Preprocessing

1. Start a new file
   Utility Menu> File> Clear and Start New
2. Give the Example a Title
   Utility menu> File> Change Title
   '*Step through bicycle*'
3. Define Variables

Define the vertices of the frame using variables. These variables represent the various lengths of the bicycle members. By using variables like this, it is very easy to set up a parametric description of the model. This will enable quickly redefining the frame should changes be necessary. The quickest way to enter these variables is via the 'ANSYS Input' window. Type in each of the following lines followed by Enter.

X1 = 325, X2 = 825, Y1 = 125, Y2 = 325, Y3 = 400, Z1 = 25

4. Enter Keypoints
   For this space frame example, these keypoints are the *frame vertices.*
   Define 7 keypoints for this structure as given in the following table (these keypoints are depicted by the circled numbers in the above figure):

Keypoint	Coordinates		
	x	y	z
1	0	0	–Z1
2	0	0	Z1
3	X1	0	0
4	X1	Y1	0
5	X1	Y3	0
6	X2	Y2	0
7	X2	Y3	0

Main Menu> Preprocessor> Modeling> Create> In Active CS

Now check to ensure that all of the keypoints are entered correctly

Utility Menu> List> Keypoints> Coordinates only

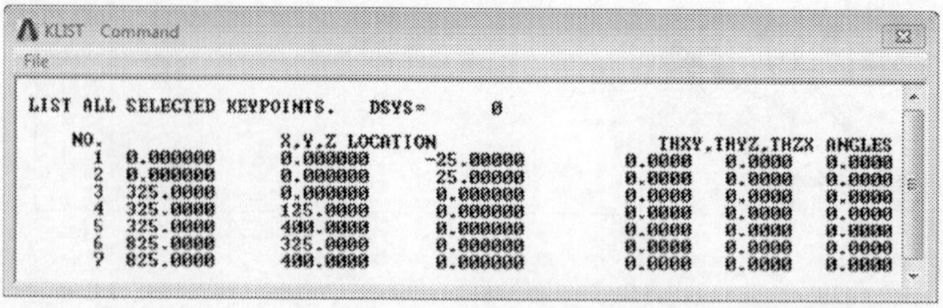

```
KLIST Command
File

LIST ALL SELECTED KEYPOINTS.   DSYS=      0

  NO.               X,Y,Z LOCATION                         THXY,THYZ,THZX ANGLES
   1  0.000000     0.000000     -25.00000            0.0000    0.0000    0.0000
   2  0.000000     0.000000      25.00000            0.0000    0.0000    0.0000
   3  325.0000     0.000000      0.000000            0.0000    0.0000    0.0000
   4  325.0000     125.0000      0.000000            0.0000    0.0000    0.0000
   5  325.0000     400.0000      0.000000            0.0000    0.0000    0.0000
   6  825.0000     325.0000      0.000000            0.0000    0.0000    0.0000
   7  825.0000     400.0000      0.000000            0.0000    0.0000    0.0000
```

5. Create Lines

   Join the following keypoints together: **Main Menu> Preprocessor> Create> Lines> Lines> Straight Line**

6. Define the Type of Element

   Preprocessor> Element Type> Add/Edit/Delete> Add

   Define the type of element (Pipe 288).

7. Define Section Properties

   Preprocessor> Sections> Pipe> Add

   Now specify geometric properties for the elements

   Outside diameter OD: 25

   Wall thickness = 2 mm

8. Define Element Material Properties

   Preprocessor> Material Props> Material Models> Structural> Linear> Elastic> Isotropic

   In the window that appears, enter the following geometric properties for steel:

   (a) Young's modulus EX: 0.7e5

   (b) Poisson's ratio PRXY: 0.33

9. Set mesh size

   As in the verification model, set the element length to 25 mm

   Preprocessor> Meshing> Size Cntrls> ManualSize> Lines> All Lines

10. Mesh the model

    Preprocessor> Mesh> Lines and click 'Pick All' in the 'Mesh Lines' Window

11. Save the Job

    Utility Menu> File> Save as...

Solution

1. Define Analysis Type

   Solution> Analysis Type> New Analysis...> Static

2. Apply Constraints

   Pin (translational DOFs will be fixed) the sixth keypoint and constrain the keypoints corresponding to the rear wheel attachment locations in both the y- and z-directions.

3. Apply Loads

   Apply vertical downward loads of 500 N at the seat post location (keypoint 5) and 150 N at the pedal crank location (keypoint 4).

The applied loads and constraints should now appear as shown below.

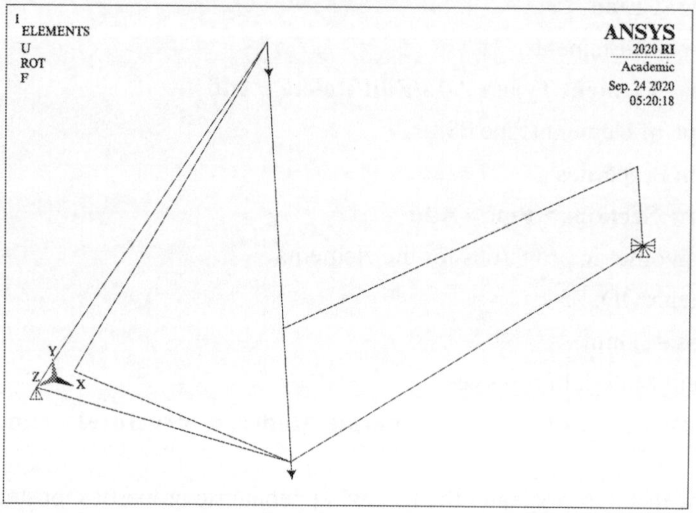

4. Solve the system
   Solution > Solve > Current LS

Postprocessing

Open the 'General Postproc' Menu to begin postprocessing.

1. Deformation
   Plot Results > Deformed Shape... 'Def + undef edge'

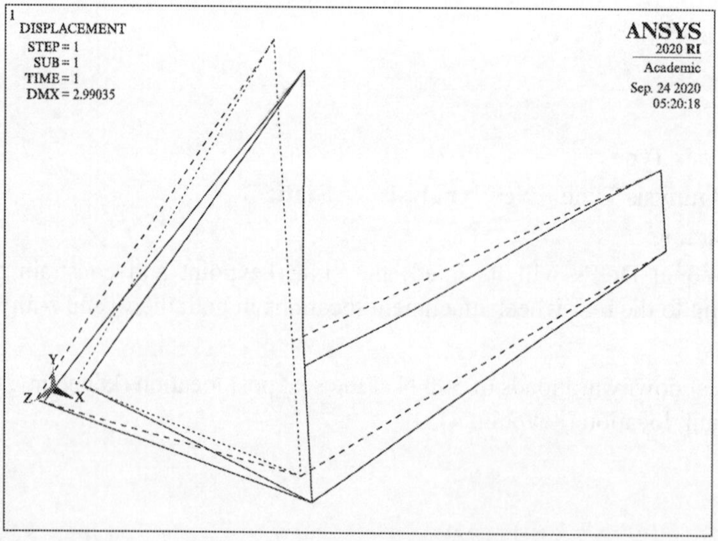

Try plotting this from different angles to get a better idea what's going on by using the 'Pan-Zoom-Rotate' menu that was earlier outlined.

Try the 'Front' view button (Note that the views of 'Front', 'Left', 'Back', etc. depend on how the object was first defined).

The screen should look like the plot below.

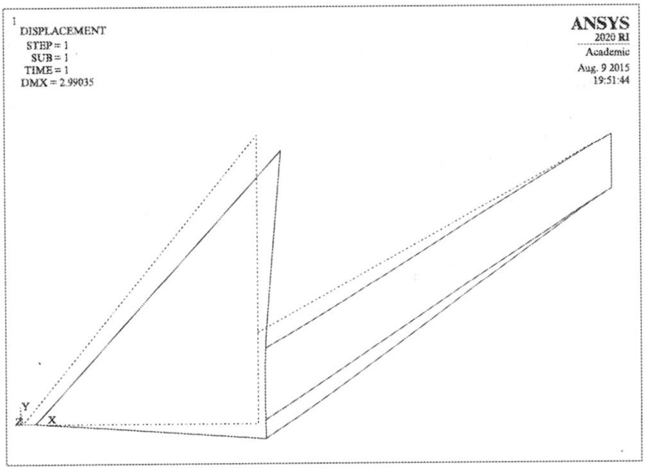

2. Deflections

   Now take a look at some actual deflections in the frame.

   List the Nodal Deflections (**Main Menu> General Postproc> List Results> Nodal Solution...**).

   Plot the deflection as well.

   General Postproc> Plot Results> Contour Plot> Nodal Solution.

   Select 'DOF solution' and 'USUM' in the window.

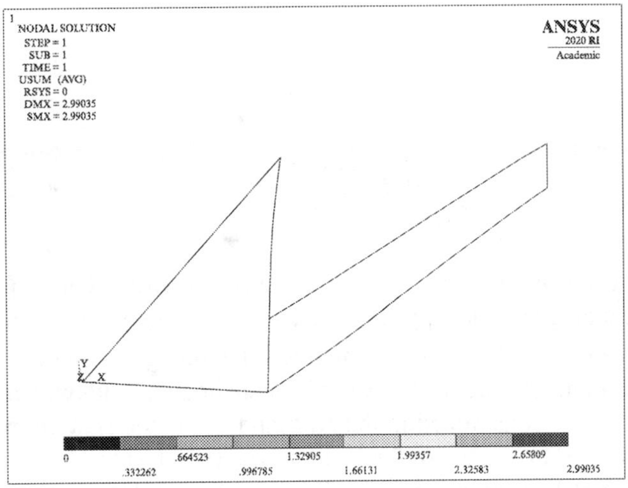

3. List Element Forces

   Also take a look at the forces in the elements in much the same way.

   Select 'Element Solution...' from the 'List Results' menu.

   Select 'Nodal force data' and 'All forces' from the lists displayed.

   Click on 'OK'.

   For each element in the model, the force/moment values at each of the two nodes per element will be displayed.

4. Produce Bending Moment Diagrams

   The bending moment diagram can be produced as follows:

   Select **Element Table> Define Table...** to define the table (remember SMISC, 6 and SMISC, 12)

   And, **Plot Results> Line Elem Res...** to plot the data from the Element Table

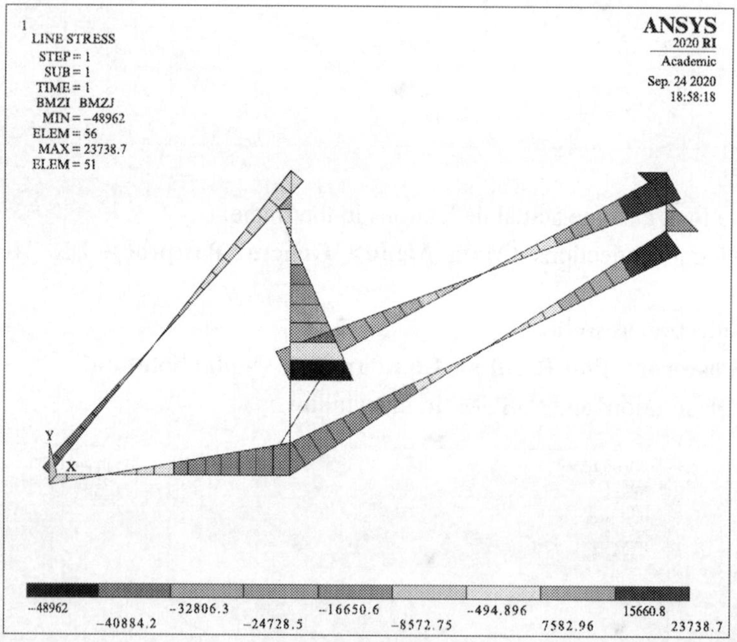

5. Quit ANSYS

   Select 'QUIT' from the ANSYS Toolbar or select **'Utility Menu'/'File'/'Exit...'** to quit ANSYS. In the dialog box that appears, click on 'Save Everything' and then click on 'OK'.

W23: A cycle frame is made up of aluminium tubing. The tubing has an outer diameter of 25 mm and wall thickness of 2 mm. Assume the structure is made of aluminum with modulus of elasticity, E = 75 GPa. The structure is constrained in the X, Y and Z directions at the bottom three corners. The cycle frame is subjected to loading of 500N at the seat (point 2) and to a loading of 150N at the pedal crank location (point 1). Determine deflection at each joint. Determine the maximum stress.

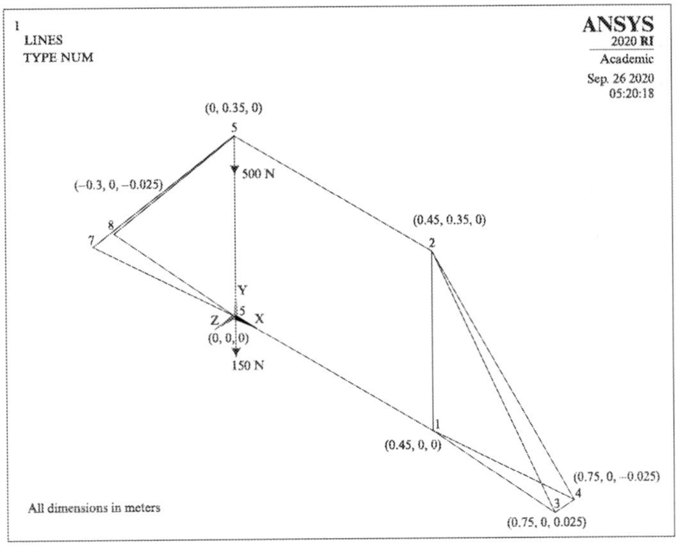

Solution: These members can be modeled as 3D beams. 3D beams experience bending in two directions perpendicular to the length of the beam.

Preprocessing

1. Model the structure

   Preprocessor> Modeling> Create> Keypoints> In active CS

   Fill in the keypoint numbers (1, 2, 3...) and the coordinates. Create all the 8 keypoints. Make sure the numbering of your keypoints matches the numbering of the joints in the figure.

   Now create lines connecting the keypoints.

   Preprocessor> Modeling> Create> Lines> Lines> In Active Coord

   Create lines by connecting keypoints. Click 'OK' when all of the lines are made

2. Define material properties

   Preprocessor> Material Props> Material Models> Structural> Linear>Elastic> Isotropic.

   Fill in 7.5e10 for the Young›s modulus and 0.3 for minor Poisson›s Ratio. Click 'OK'

3. Define element properties

   Preprocessor> Element Type> Add/Edit/Delete...> Add...

   Type 1 in the Element type reference number.

   Click on **Beam** and select **3D elastic**. Click 'OK'. Close the ‹Element types› window

   Now define the section properties for this element.

   Preprocessor> Sections> Beam> Common Sections

   B = 10, H = 10, Fill in the relevant values and click on 'OK'.

4. Mesh the model

   Preprocessor> Meshing> Size Controls> Manual Size> Lines> All Lines.

In the menu that comes up type 0.025 in the field for 'Element edge length'.

Now each element was divided to be of size 25 mm, Click on 'OK'.

Preprocessor> Meshing> Mesh> Lines

Select all the lines and click on OK in the 'Mesh Lines' dialog box.

5. Apply boundary conditions and constraints

   The cycle is constrained in all DOFs at the point 3, 4, 6, 7, 8.

   Preprocessor> Loads> Define

   Loads> Apply> Structural> Displacement> On Keypoints

   Select each keypoint on which you want to apply displacement constraints.

   Select **All DOF** and click 'OK'.

6. Apply forces

   Preprocessor> Loads> Define Loads> Apply> Structural> Forces/Moment> On Keypoints

   Select the keypoint where the force acts. Use keypoints because there are a number of nodes and therefore, it is easier to pick keypoints.

   Click on OK in the 'Apply F/M on Keypoints' window.

   Enter the value of the force in the Y direction.

   Repeat the procedure to apply the force on the other keypoint.

   The model should look like the one below:

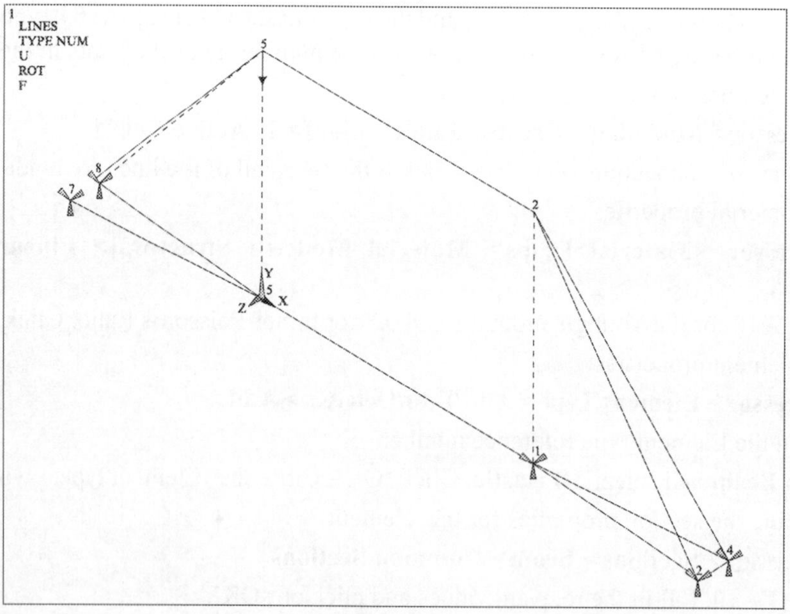

Solution

1. **Main Menu> Solution> Analysis Type> New Analysis**
   Select static and click on 'OK'.
2. **Solution> Solve> Current LS**

Postprocessing

1. List the results
   General Postprocessing> List Results> Nodal Solution.
   Select **DOF solution** and **All U›s.** Click on 'OK'.
   List the stresses for each element by clicking **General Postprocessing> List Results> Element Solution. Now select LineElem Results.**

This list can be used to determine the maximum stress but do not print this list. The easy way to determine the maximum stress is to take guesses where the maximum stress should be, Zoom to those points to get the node numbers or element numbers, then look for **SMAX** values at those nodes/elements.

2. Plot the displacements and stress.
   General Postprocessing> Plot Results> Contour Plot> Element Solution.
   Select a stress to be plotted and click 'OK'. The output will be like this.

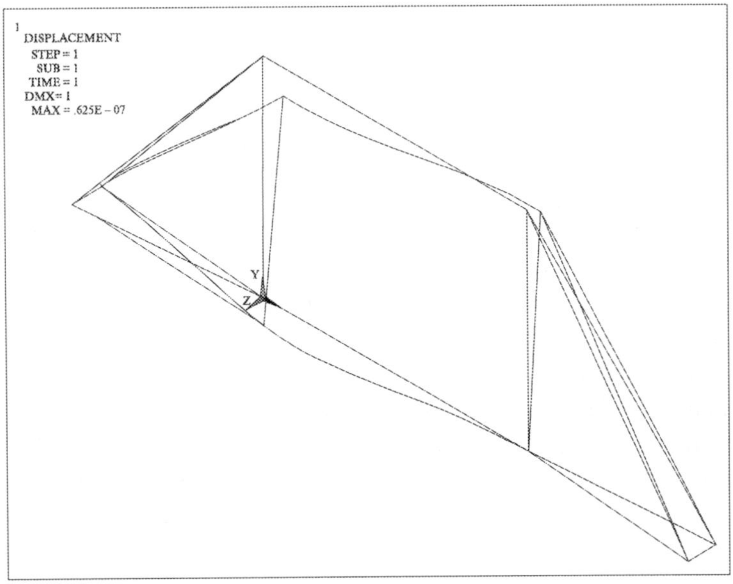

W24: Analyze of the 3D truss structure shown in the figure below. The cross section of each of the truss members is 1.56e-3 m^2. Assume the structure is made of aluminum with modulus of elasticity E=75 GPa. The structure is constrained in the X, Y and Z directions at the bottom four corners.

The tower is loaded at the top tips. The load is in the YZ plane and makes an angle of 75 with the negative Y-axis direction. The load value is 2500N. Determine deflection at each joint, stress in each member and reaction forces at the base.

Solution:

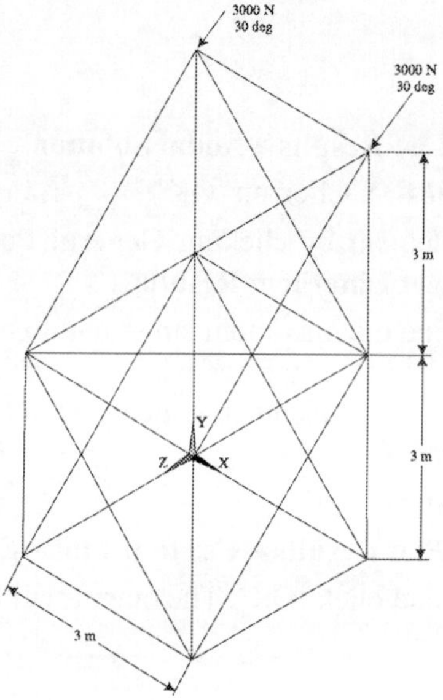

Preprocessing

1. Model the structure
   Workplane> Change active CS to..> Global cartesian
   Preprocessor> Modeling> Create> Keypoints> In active CS
   Fill in the keypoint numbers (1, 2, 3...) and the coordinates. Make sure you get the correct coordinates from the figure. Create all the 10 keypoints. Make sure the numbering of your keypoints matches the numbering of the joints in the figure.
   Now create lines connecting the keypoints
   Preprocessor> Modeling> Create> Lines> Lines> In Active Coord
   Pick the endpoints of each element to create the lines.
2. Define Material Properties
   Preprocessor> Material Props> Material Models> Structural> Linear> Elastic> Isotropic.
   Fill in 7.5e10 for the Young's modulus and 0.3 for minor Poisson's ratio. Click 'OK'
3. Define element properties
   Preprocessor> Element Type> Add/Edit/Delete...
   Type 1 in the Element type reference number.
   Click on Link and select 3D Link. Click 'OK'. Close the 'Element types' window.

4. Now define the cross-sectional area for this element.
   Preprocessor> Real Constants.
   In the 'Real Constants' dialog box that comes up click on 'Add'
   In the 'Element Type for Real Constants' that comes up click 'OK'. The following window comes up.
   Type 1.56e-3 for cross-sectional area and click on 'OK'.
5. Mesh the model
   Preprocessor> Meshing> Size Controls> Manual Size> Lines> All Lines.
   In the menu that comes up type 1 in the field for 'Number of element divisions'.
   Preprocessor> Meshing> Mesh> Lines.
   Select all the lines and click on OK in the 'Mesh Lines' dialog box.
   Now each line is a truss element (Element 1).
6. Define boundary conditions and constraints.
   The tower is constrained in the X, Y and Z directions at the four bottom corners.
   Preprocessor> Loads> Define Loads> Apply> Structural> Displacement> On Keypoints
   Select the keypoint on which you want to apply displacement constraints.
   Select UX, UY, UZ and click 'OK'.
   Apply forces
   First find the components of the force along the Y and Z directions.
   Preprocessor> Loads> Define Loads> Apply> Structural> Forces/Moment> On Nodes.
   Select the top node.
   Click on OK in the 'Apply F/M on Nodes' window.
   Enter the value of the Z-component of the force.
   Repeat the procedure to apply the Y-component of force.

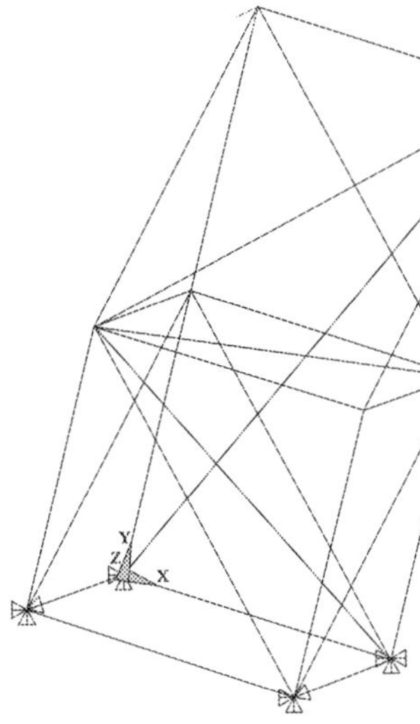

Solution:

1. **Main Menu> Solution> Analysis Type> New Analysis.**
   Select static and click on 'OK'.
2. **Solution> Solve> Current LS**

Postprocessing

1. List the results
   General Postprocessing> List Results> Nodal Solution
   Select DOF solution and All U's. Click on OK. The nodal displacements will be listed.
2. List the stresses for each element by clicking **General Postprocessing> List Results> Element Solution.** Now select LineElem Results.
3. Plot the displacements and stress.
   General Postprocessing> Plot Results> Contour Plot> Element Solution.
   Select a stress to be plotted and click 'OK'. The output will be like this.

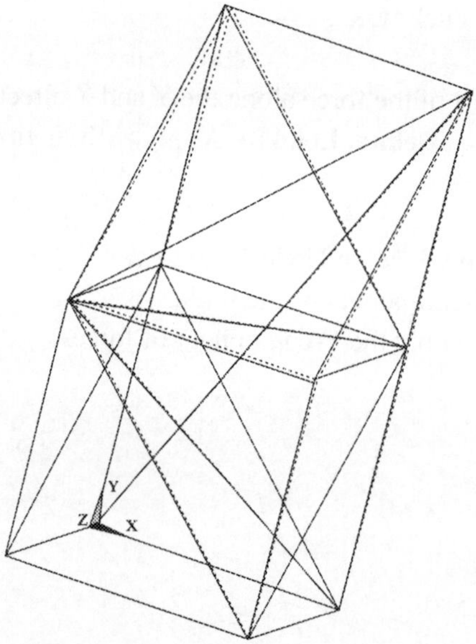

W25: Solve for the member forces and deflections of an elastic three-dimensional truss, shown below.

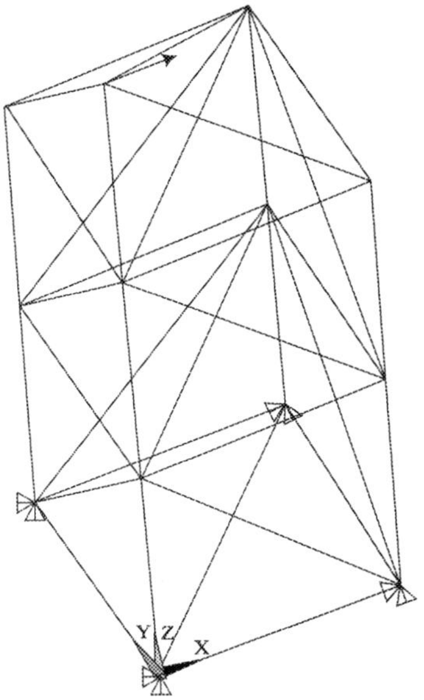

Solution:

Preprocessing

1. Select an element type.

   Preprocessor> Element Type> Add/Edit/Delete> Add>

   Choose 'Link' in the left box and '3D finit stn 180' in the right, then click 'OK'. This selects the Link 180 (two-force member) element.

2. Define real constants.

   Preprocessor> Real Constants> Add

   Enter 25 (25 mm^2) in the box marked 'Cross-sectional area'. All the members will have this cross-sectional area, so only one real constant set is needed. Now click 'OK' and close the Real Constants box.

3. Define material properties.

   Preprocessor> Material Props> Material Models> OK

   Double-click 'Structural' in the right side of the window, then 'Linear', then 'Elastic', then finally 'Isotropic'. Enter 2e5 in the box marked 'EX'. This is a ballpark value for steel. Then click 'OK'. For this problem, create nodes and elements directly rather than making an intermediate model.

4. Define nodes

   Preprocessor> Modeling> Create> Nodes> In Active CS

Enter the following nodes, clicking 'Apply' after each one:

Node Number	X	Y	Z
1	0	0	0
2	3	0	0
3	3	3	0
4	0	3	0
5	0	0	3
6	3	0	3
7	3	3	3
8	0	3	3
9	0	0	6
10	3	0	6
11	3	3	6
12	0	3	6
13	1.5	1.5	7.5

After finishing, click 'OK' and close the Nodes menu.

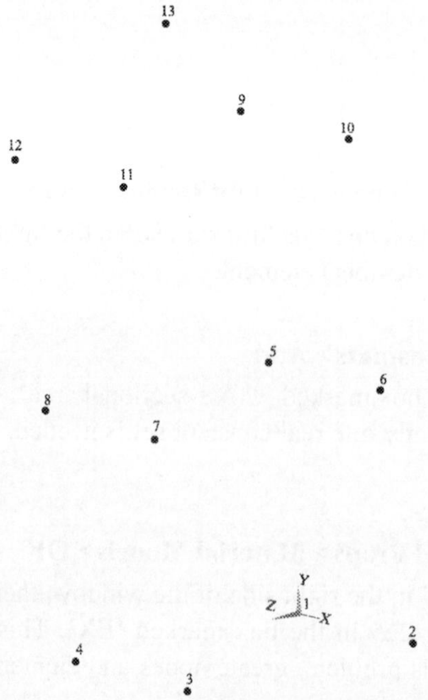

5. Define elements

   Rotate the view in the graphics window now so that all of the nodes can be seen clearly and distinctly.

Utility Menu> PlotCtrls>Pan/Zoom/Rotate

Click the ' + X', ' – X', etc., buttons until the picture on the screen is satisfactory, then click 'Close'.

Main Memu> Preprocessor> Create> Elements> Auto Numbered> Thru Nodes

First create the horizontal and vertical elements, since they will clutter the graphics window less than the diagonal elements. From this window create elements by clicking on the nodes at their ends, then clicking 'Apply'; create elements in a square pattern between nodes 1-2-3-4, 5-6-7-8, and 9-10-11-12, as shown below (note that node 1 is at the origin and is not visibly numbered).

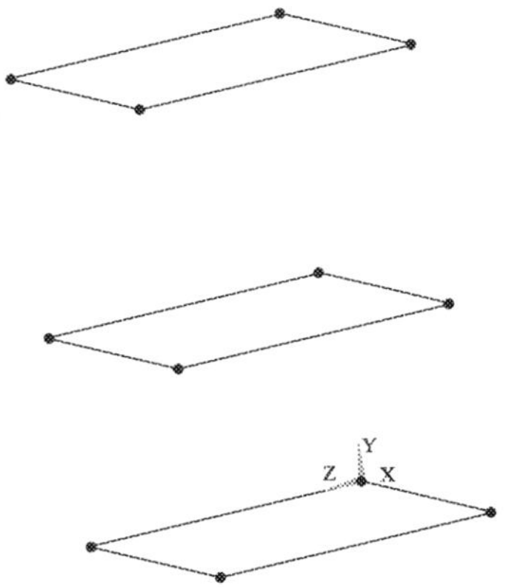

Now create vertical elements between vertically adjoining nodes, as shown below.

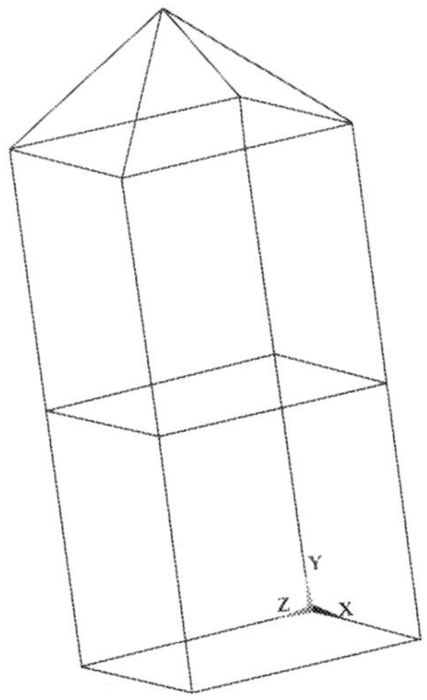

Now create diagonal elements across each horizontal square pattern of elements, going from the corner nearest the origin to the corner opposite, as shown below.

Finally, create the out-of-plane diagonal elements. These should slope down from the node in one tier closest to the origin to the adjoining nodes in the one below, and from the node in one tier farthest from the origin to those same nodes in the one below. There should also be one element from node 13 to nodes 9-12. The final configuration should match the figure given below.

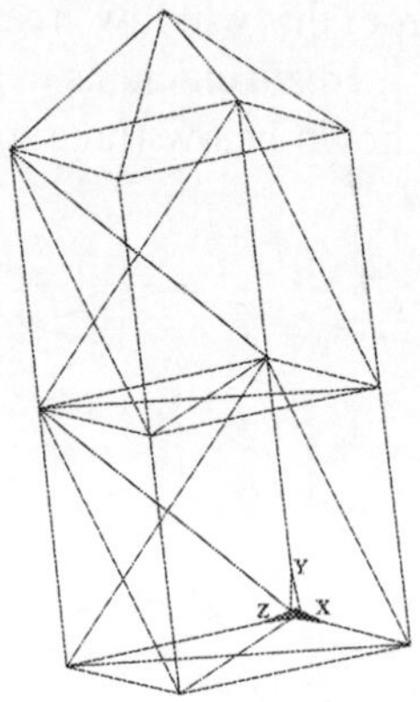

Solution:

1. Apply displacements.
   Solution >Loads> Apply> Displacement> On Nodes
   Pick nodes 1, 2, 3, and 4, then click 'OK'.
   Select 'All DOF' and enter a displacement value of 0. This fixes the nodes at the bottom of the truss.
2. Apply loads.
   Apply> Force/Moment> On Nodes
   Click node 13, then click 'OK'.
   Leave the direction of the force as 'FX', and enter a value of 1000, then click 'OK'.

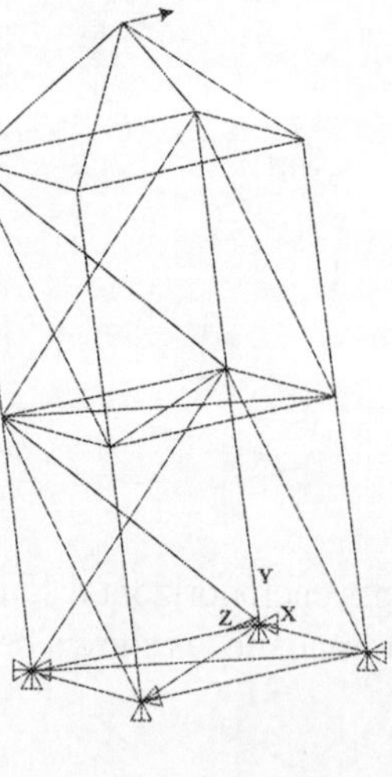

3. Solve the problem.

   Solution> Solve Current LS

   Review the information presented, and if everything seems in order, click 'OK'. ANSYS will solve the problem.

Postprocessing

1. View the deformed shape.

   Postprocessor> Plot Results> Deformed Shape> Def + undef edge> OK

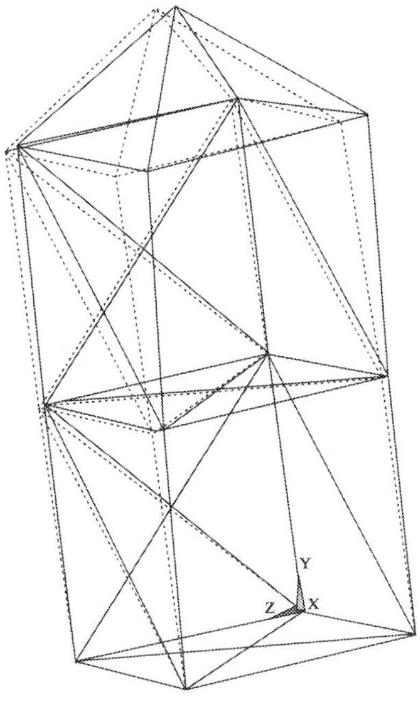

2. List member forces.

   General Postprocessor> List Results> Element Solution> Nodal force data> OK

   A listing of the forces in each element will appear.

W26: The pylon structure shown below supports a load of 116 tonnes. A block, width 2500 mm, height 3300 mm and depth 1850 mm must be able to pass underneath the structure. The maximum deflection permissible is 3 mm. The maximum stress allowed in the model is 200 MPa (localised regions of higher stress maybe allowed). The Young's modulus of steel is to be taken as 0.2e6 MPa and Poisson's ratio 0.3. Build and analyze the structure.

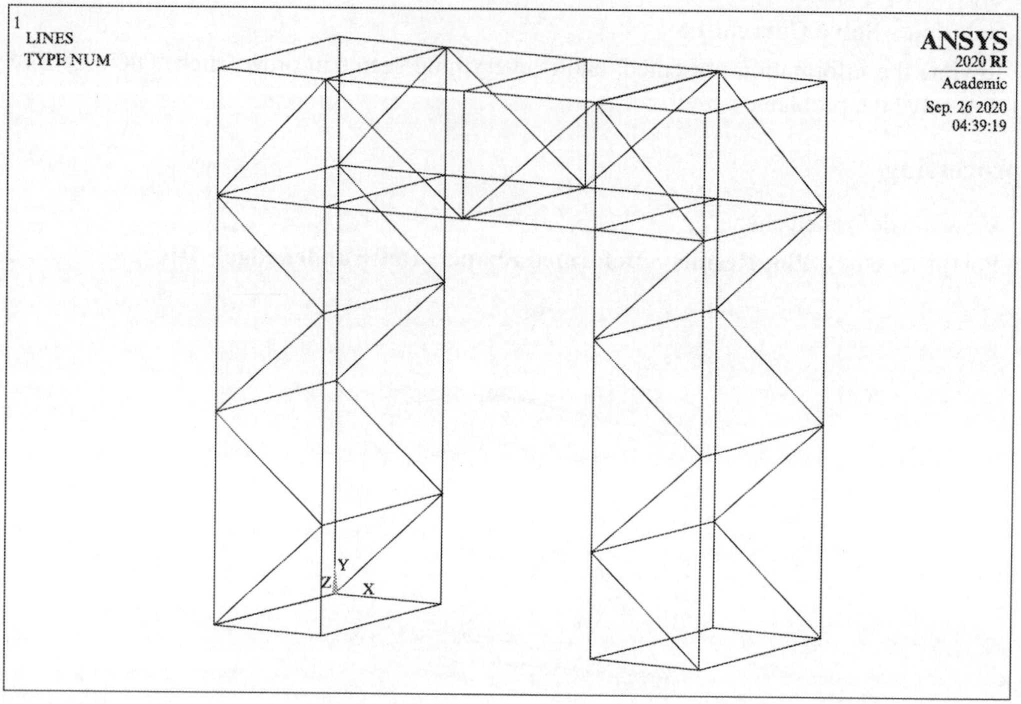

Solution: Electricity pylons are good examples of truss structures. To make a truss the most obvious way is to build a structure using beam elements. Such a structure is shown in the figure below.

Preprocessing

1. Define Element Type

   Select BEAM 188
2. Define Section Properties

   ID = 1, Name = Beam-1, B = 15, H = 20
3. Define Material Properties

   E = 2e5, Pr = 0.3
4. Create keypoints
5. Create lines
6. Mesh the model

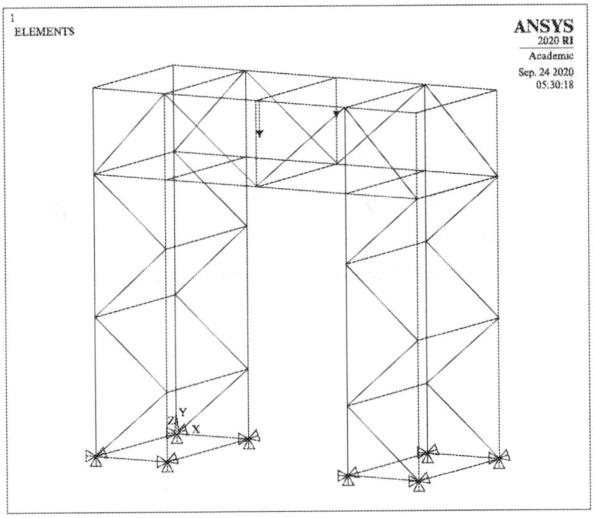

Solution

1. Solve the problem

Postprocessing

1. Plot and list results

4.3.2 Solids

W27: Apply loads to the 3-D bracket model below and solve using the PCG iterative solver. Mesh the model with SOLID 186 20 node bricks, and Young's modulus is 200 GPa.

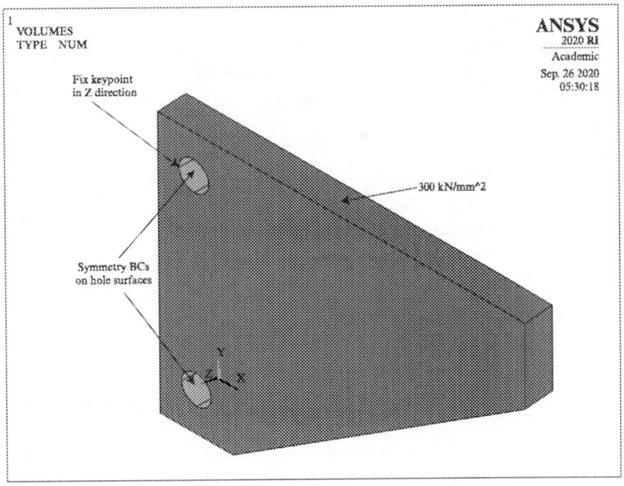

Data: E = 2e5 N/mm^2, PR = 0.3

Solution:

Preprocessing

1. Enter ANSYS. Use 'Bracket-3D' as the jobname.
2. Create the bracket model as shown in the figure.
3. Enter the solution processor and restrain translations normal to hole surfaces
   Main Menu> Solution> Loads> Apply> Structural> Displacement> Symmetry B.C.> On Areas +
   Pick the areas on the hole surfaces (area numbers 3, 4, 5, and 6), then 'OK'
4. To prevent rigid body motion along the Z axis, constrain UZ translation on left extreme keypoint as shown in figure.
   Main Menu> Solution> Loads> Apply> Structural> Displacement> On Keypoints +
5. Apply 300 kN/mm^2 pressure load to the top surface area of the 3-D bracket.
   Main Menu> Solution> Loads> Apply> Structural> Pressure> On Areas +
   Pick the top surface area (area number 11), then OK, Set VALUE = 300, then OK
6. Select the PCG iterative solver.
   Main Menu> Solution> Analysis Type> Sol'n Control ...

Pick the 'Sol'n Options' tab

Select the 'Pre-Condition CG' solver, then 'OK'

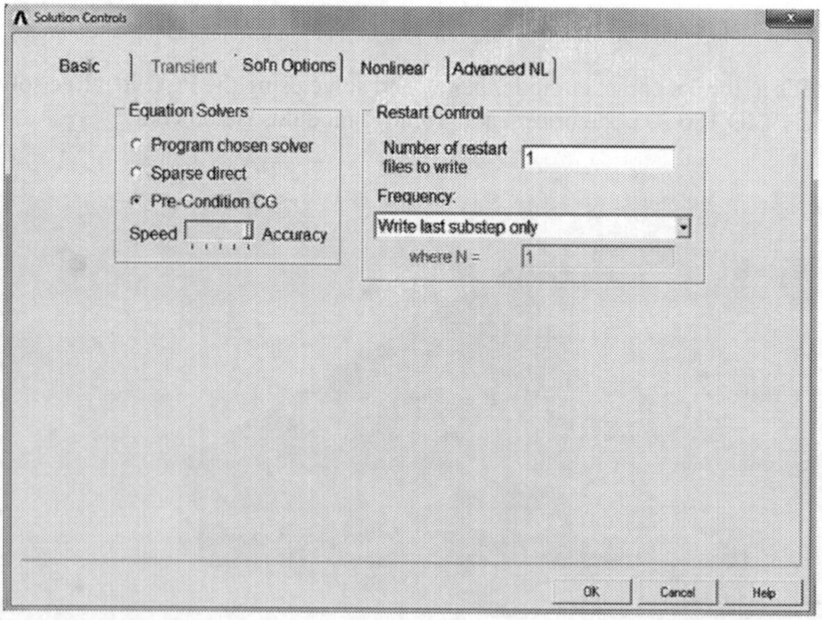

7. Save the database and obtain the solution.

   Pick the 'SAVE_DB' button in the Toolbar (or select: **Utility Menu> File> Save as Jobname.db)**

   Main Menu> Solution> Solve> Current LS

8. After the solution is completed, enter the general postprocessor and plot the von Mises stress (SEQV):

   Main Menu> General Postproc> Plot Results> Contour Plot> Nodal Solu ...

   Select 'Stress' and 'von Mises SEQV', then ' OK'

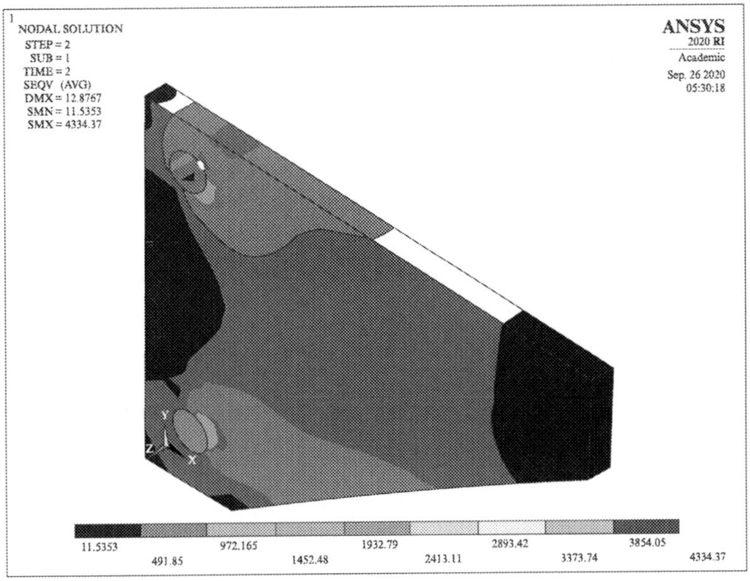

9. Save and exit ANSYS

   Pick the '**SAVE_DB**' button in the Toolbar

   Pick the 'QUIT' button in the Toolbar

W28: Plot cross-sectional results (deformation, stress, etc.) of the following example.

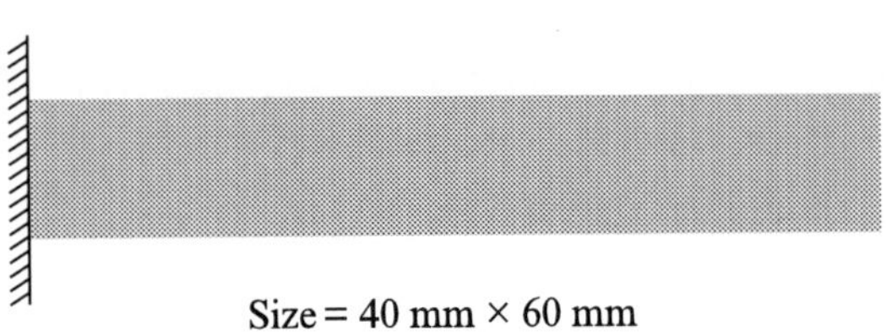

Size = 40 mm × 60 mm
Load = 7 N/mm^2
E = 200 GPs
RR = 0.3

Solution:

Preprocessing

1. Give example a Title
   Utility Menu> File> Change Title ...
   'Cross-Sectional Results of a Simple Cantilever Beam'
2. Open preprocessor menu
   Main Menu> Preprocessor
3. Create Block
   Preprocessor> Modeling> Create> Volumes> Block> By 2 Corners & Z
   Width = 40 mm, Height = 60 mm, Length = 900
4. Define the Type of Element
   Preprocessor> Element Type> Add/Edit/Delete...
   Use the **SOLID185** (3D Solid) element. This element has 8 nodes each with 3 degrees of freedom (translation along the X, Y and Z directions).
5. Define Element Material Properties
   Preprocessor> Material Props> Material Models> Structural> Linear> Elastic> Isotropic
   (a) Young's modulus EX: 200000
   (b) Poisson's ratio PRXY: 0.3
6. Define Mesh Size
   Preprocessor> Meshing> Size Cntrls> ManualSiz> Global> Size
   Element size = 20 mm
7. Mesh the volume
   Preprocessor> Meshing> Mesh> Volumes> Free> click 'Pick All'

Solution

1. Define Analysis Type
   Solution> Analysis Type> New Analysis> Static
2. Apply Constraints
   Solution> Define Loads> Apply> Structural> Displacement> On Areas
   Fix the left hand side (should be labeled Area 1)
3. Apply Loads
   Solution> Define Loads> Apply> Structural> Force/Moment> On Keypoints
   Apply a load of 10 N/mm^2 downward on the top face
4. Solve the System
   Solution> Solve Current LS

Postprocessing

- Offset the working plane for a cross-sectional view
- Select the TYPE of display for the section.

Type	Description	Visual Representation
SECT	Section display. Only the selected section is shown without any remaining faces or edges shown	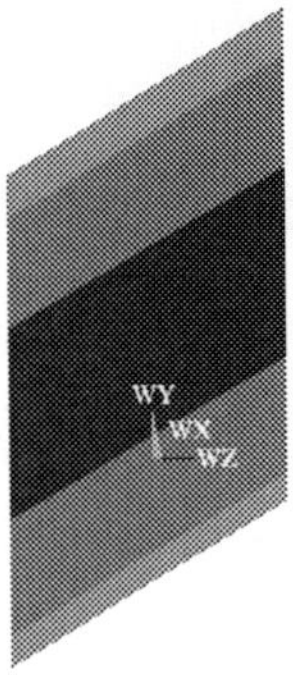
CAP	Capped hidden display. This is as though a portion of the model was cut off and the remaining model can be seen	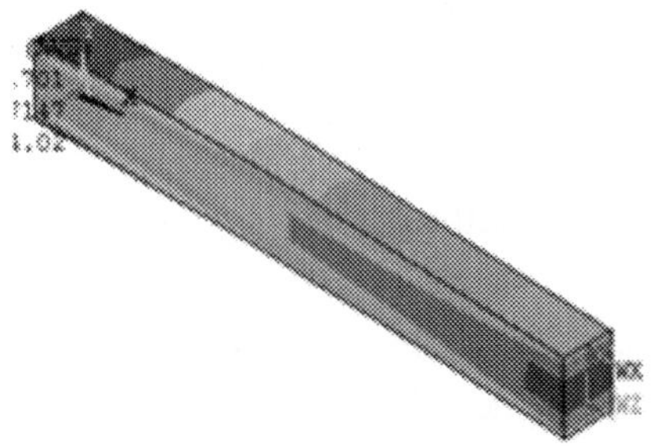
ZQSL	QSLICE Z-buffered display. This is the same as SECT but the outline of the entire model is shown.	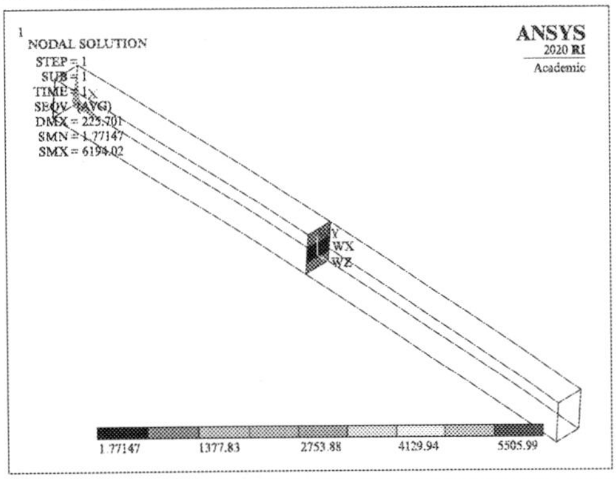

Align the cutting plane with the working plane.

1. Deflection

   Before selecting cross sections, view deflection of the entire model.

   Select: **General Postproc> Plot Results> Contour Plot> Nodal Solu**

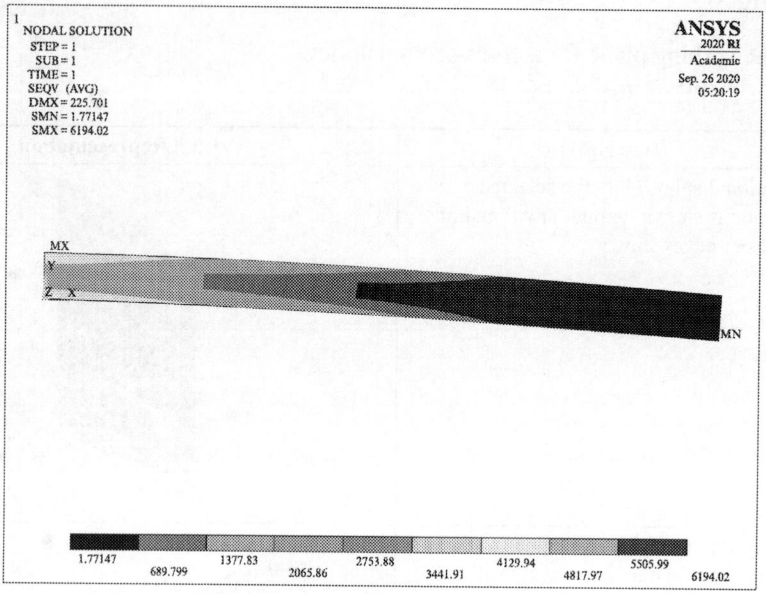

View several cross sections through the YZ plane.

Take one halfway through the beam in the YZ plane.

First, offset the working plane to the desired position, halfway through the beam

Select: **Utility Menu > WorkPlane > Offset WP by Increments**

In the window that appears, increase Global X to 20 (Width/2) and rotate Y by + 90 degrees

Select the type of plot and align the cutting plane with the working plane. (Note that in GUI, these two steps are combined.)

Select: **Utility Menu > PlotCtrls > Style > Hidden-Line Options**

Fill in the window that appears as shown below to select

/TYPE = ZQSL and **/CPLANE** = Working Plane

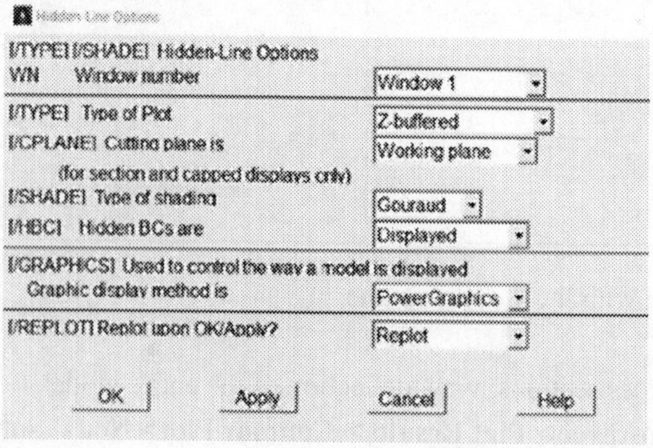
Hidden-Line Options

[/TYPE] [/SHADE] Hidden-Line Options
WN Window number — Window 1
[/TYPE] Type of Plot — Z-buffered
[/CPLANE] Cutting plane is — Working plane
(for section and capped displays only)
[/SHADE] Type of shading — Gouraud
[/HBC] Hidden BCs are — Displayed
[/GRAPHICS] Used to control the way a model is displayed
Graphic display method is — PowerGraphics
[/REPLOT] Replot upon OK/Apply? — Replot

OK Apply Cancel Help

As desired, you should now have the following:

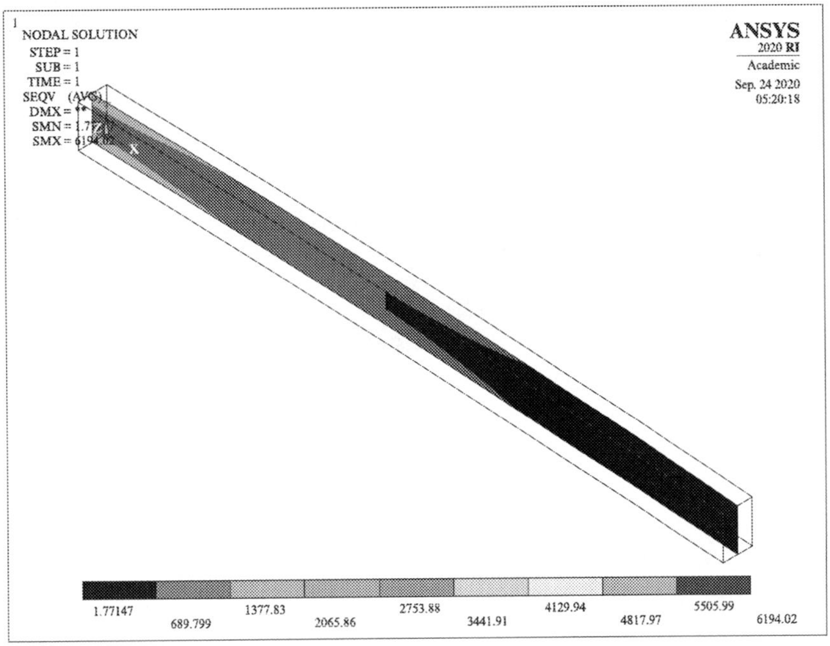

2. Equivalent Stress

   Again, view stresses within the entire model.

   First realign the working plane with the active coordinate system. Select: **Utility Menu> WorkPlane> Align WP with> Active Coord Sys** (NOTE: To check the position of the WP, select **Utility Menu> WorkPlane> Show WP Status**)

   Next change/**TYPE** to the default setting (no hidden or section operations). Select: **Utility Menu> PlotCtrls> Style> Hidden Line Options...** And change the 'Type of Plot' to 'Non-hidden'

   Select: **General Postproc> Plot Results> Contour Plot> Nodal Solu> Stress> von Mises**

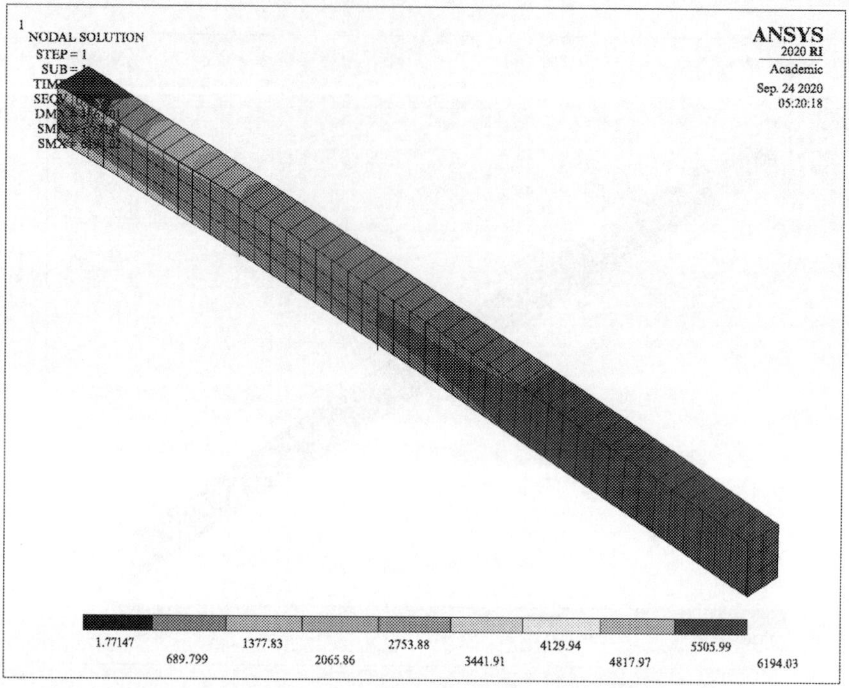

Take a closer look at the base of the beam through the XY plane.

Note that it is not necessary to rotate the WP because it is wished to look at the XY plane which is the default. Also note that the capped hidden display is used this time.

The graphics should now look similar to the following:

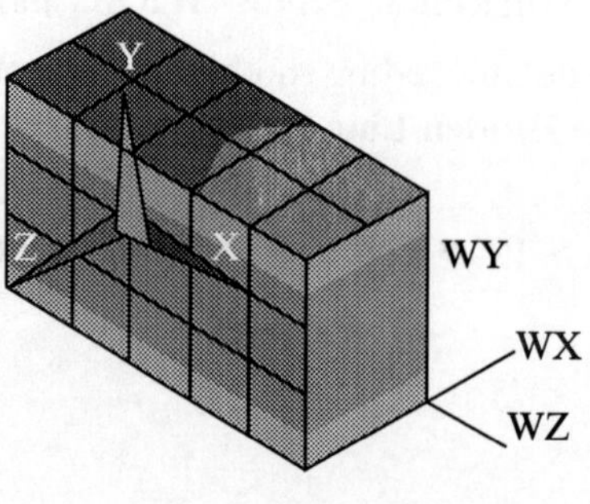

3. Animation

   See an animation of the von Mises stress through the beam.

   Select: **Utility Menu > PlotCtrls > Animate > Q-Slice Contours**

   In the window that appears, just change the item to be contoured to 'Stress' 'von Mises'

   ANSYS prompts to select 3 nodes; the origin, the sweep direction, and the Y axis. In the graphics window, select the node at the origin of the coordinate system as the origin of the sweep (the sweep will start there). Next, the sweep direction is in the Z direction, so select any node in the Z direction (parallel to the first node). Finally, select the node in the back, bottom left hand side corner as the Y axis.

An animated version of the contour slices through the beam should appear now.

W29: Model and analyze the 3-D cylinder shown below using Cyclic Symmetry.

Solution: The problem can be solved with a much smaller model because of its innate symmetry. Any slice behaves just like any other slice. This is illustrated by solving the problem working with a ten-degree slice defined in global cylindrical coordinates.

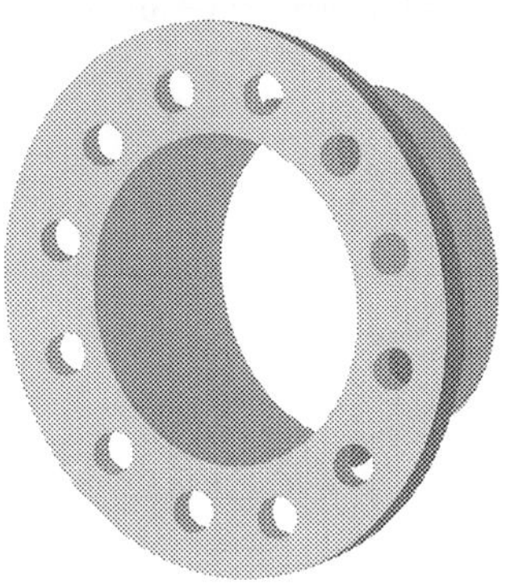

Preprocessing

1. Start ANSYS and establish jobname and working directory.
2. **Utility Menu> WorkPlane> Change Active CS to> Global Cylindrical.**

   The angular coordinate theta corresponds to a rotation about the Z axis.
3. In the Preprocessor create ten-degree wedge with 125, inner radius, 275mm outer radius, 10 theta and 25 mm height. Use **Modeling > Create > Volumes > Cylinder > Partial Cylinder** to do this.

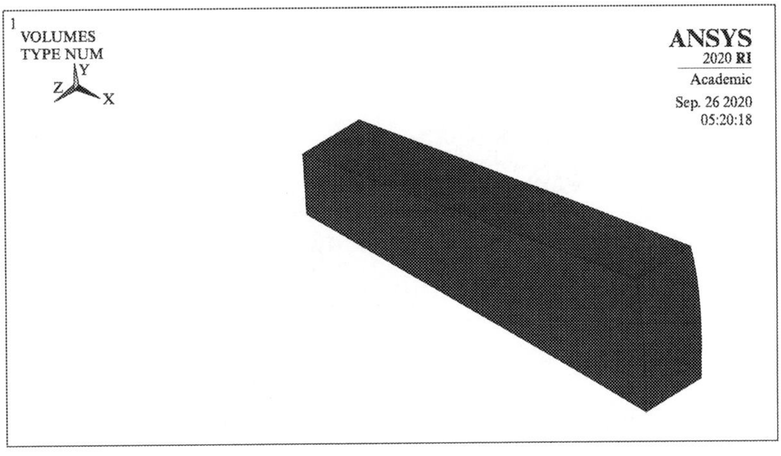

4. Displacement Boundary conditions are applied to restrain motion in the Y direction on the X-Z face of the wedge.

Preprocessor> Loads> Apply> Displacement> On Areas

Select the X-Z face and set UY = 0.

On the symmetric face of the wedge apply Symmetric Displacement conditions.

Preprocessor> Loads> Apply> Displacement> Symmetry B.C.> On Areas Select the symmetric face **OK**. The 'S' in the figure below indicates a symmetric BC.

Apply a pressure of 7 N/mm^2 to the interior surface and restrain the lower surface of the wedge from motion in the Z direction.

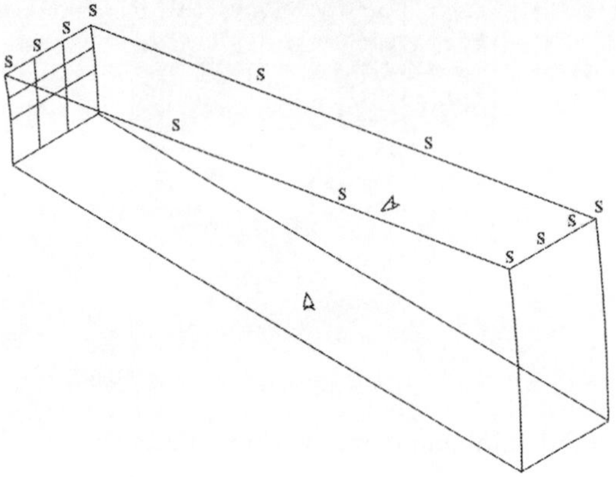

Solution:

1. Select **Solve> Current LS**

Postprocessing

Postprocessor> **Plot** the **Element Solution SEQV** to obtain the plot below. Notice that a smooth solution is obtained with much fewer elements than in the previous example.

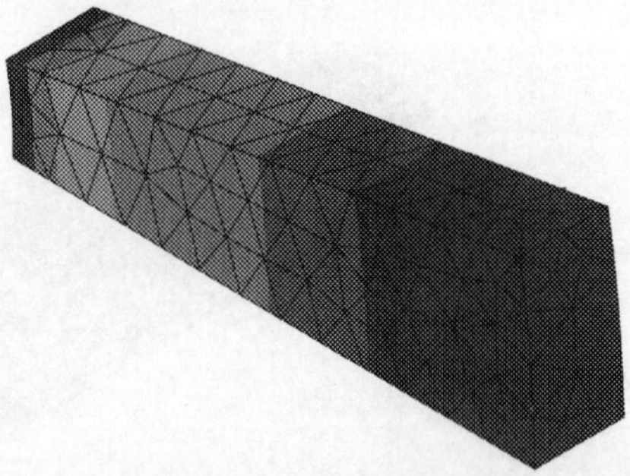

The component stresses SX, SY, SZ, SXY, etc., will be aligned with the global XYZ axis system and reflect the stress state shown on the X-Z face of the wedge.

This approach is particularly useful for analysis of items with cyclic symmetry such as the flange. The geometry, boundary conditions, loadings, and materials must be symmetric with respect to an axis of revolution. Then take a sector from the center of one hole to the center of the next. Or use a sector centered about a hole.

The following settings can be changed to give cleaner prints of the mesh plots:

Turn off min/max symbols and XYZ axis system **PlotCtrls> Window Controls> Window Options ...> Min-Max Symbols> Off; Location of Triad> Not Shown**

To get a white background:

PlotCtrls> Style> Colors> Reverse Video

PlotCnrls> Style> Background (leave all options unchecked)

Change BC symbols to darker color. **PlotCtrls> Style> Colors> BC Colors ...** Set U and Rot to darker colors.

Replot elements to cause these settings to take effect.

To obtain hard copy: **PlotCtrls> Hard Copy> To Printer**

REFERENCES

1. ANSYS Operations Guide.
2. ANSYS Basic Analysis Guide.
3. ANSYS Modeling and Meshing Guide.
4. ANSYS Structural Analysis Guide.
5. ANSYS Verification Manual.
6. ANSYS Workbook for Release 5.3.
7. Lawrence, K.L., "ANSYS Tutorial (Release 2020)", Schroff Development Corp. Publications, 2020.
8. Moaveni, S., "Finite Element Analysis: Theory and Applications with ANSYS", 2nd Edition, Prentice Hall, 2003.
9. Tadeusz Stolarski, Y. Nakasone, S. Yoshimoto, "Engineering Analysis with ANSYS Software", Butterworth-Heinemann Elsevier Butterworth-Heinemann, Amsterdam 2011.
10. Srinivas Paleti and Sambana Krishna Chaitanya, Finite Element Analysis Using Ansys 11.0, PHI Learning Pvt. Ltd., 01-Jan-2010
11. Fereydoon Dadkhah, Jack Zecher, Vibration Simulation Using MATLAB and ANSYS - Volume 10, SDC Publications - 2009

Websites

1. www.ansys.com
2. www.ansys.net/ansys
3. www3.sympatico.ca/peter_budgell/home.html
4. www.mece.ualberta.ca/tutorials/ansys/index.html
5. www.xansys.org

1. ANSYS Theory Guide [illegible]
2. ANSYS Basic Analysis Guide [illegible]
3. [illegible] Modelling and Meshing Guide [illegible]
4. ANSYS Structural Analysis Guide [illegible]
5. ANSYS [illegible] Manual
6. ANSYS [illegible] 10.0 [illegible]
7. [illegible] ANSYS Tutorial [illegible] 2020 [illegible]
8. Moaveni, S., Finite Element Analysis: Theory and Application with ANSYS, 3rd [illegible] Prentice Hall, 2011.
9. [illegible] ANSYS [illegible] Finite Element Analysis with ANSYS [illegible] 2011.
10. [illegible] Finite Element Analysis Using ANSYS [illegible] PHI Learning Pvt. Ltd. [illegible]
11. [illegible] Introduction to Finite Element Using MATLAB and ANSYS [illegible]

Websites

1. [illegible]
2. [illegible]
3. [illegible]
4. [illegible]
5. [illegible]

INDEX